Edited by
Peter M. Maitlis and
Arno de Klerk

Greener Fischer-Tropsch Processes for Fuels and Feedstocks

Related Titles

de Klerk, A.

Fischer-Tropsch Refining

2011

ISBN: 978-3-527-32605-1

Li, C.-J., Perosa, A., Selva, M., Boethling, R., Voutchkova, A. (eds.)

Handbook of Green Chemistry - Green Processes

Series: Handbook of Green Chemistry (Set III)

Series edited by Anastas, Paul T.

2012

ISBN: 978-3-527-31576-5

Stolten, D., Emonts, B. (eds.)

Fuel Cell Science and Engineering

Materials, Processes, Systems and Technology

2012

ISBN: 978-3-527-33012-6

Jansen, R. A.

Second Generation Biofuels and Biomass

Essential Guide for Investors, Scientists and Decision Makers

2013

ISBN: 978-3-527-33290-8

Stolten, D. (ed.)

Hydrogen and Fuel Cells

Fundamentals, Technologies and Applications

2010

ISBN: 978-3-527-32711-9

Edited by Peter M. Maitlis and Arno de Klerk

Greener Fischer-Tropsch Processes for Fuels and Feedstocks

WILEY-VCH

WILEY-VCH Verlag GmbH & Co. KGaA

The Editors

Prof. Peter M. Maitlis
University of Sheffield
Department of Chemistry
Sheffield S3 7HF
United Kingdom

Prof. Arno de Klerk
University of Alberta
Chemical & Materials Eng.
9107 - 116 Street
Edmonton, Alberta T6G 2V4
Canada

Library of Congress Card No.: applied for

British Library Cataloguing-in-Publication Data
A catalogue record for this book is available from the British Library.

Bibliographic information published by the Deutsche Nationalbibliothek
The Deutsche Nationalbibliothek lists this publication in the Deutsche Nationalbibliografie; detailed bibliographic data are available on the Internet at http://dnb.d-nb.de.

Print ISBN: 978-3-527-32945-8
ePDF ISBN: 978-3-527-65686-8
ePub ISBN: 978-3-527-65685-1
mobi ISBN: 978-3-527-65684-4
oBook ISBN: 978-3-527-65683-7

Cover Design Simone Benjamin, McLeese Lake, Canada

Typesetting Thomson Digital, Noida, India

Printing and Binding CPI Group (UK) Ltd, Croydon, CR0 4YY

Printed on acid-free paper

C9783527329458_171125

The manufacturer's authorized representative according to the EU General Product Safety Regulation is Wiley-VCH GmbH, Boschstr. 12, 69469 Weinheim, Germany, e-mail: Product_Safety@wiley.com.

Contents

Preface

And what is a man without energy? Nothing. Nothing at all
(Mark Twain)

Energy and persistence conquer all things
(Benjamin Franklin)

This book on Fischer-Tropsch is a study of aspects of energy: how it is produced and transformed today, with special reference to liquid fuels such as those used to drive cars, buses, planes, and other forms of transportation.

We still live in an era of relatively plentiful and cheap fuel, mostly derived from the fossilized organic materials: coal, oil, and natural gas.

New supplies are being discovered all the time and brought into use in quite surprising ways. A good example is natural gas for which it is now estimated that, because of the emergence of techniques such as fracking, the world's reserves may well be enough for around 200 years. This is close to being on a par with coal and much greater than our oil reserves. However, our assets of fossil fuels are limited and, in fairness to the next generations, we must not squander them.

We must learn to use them to buy time until a better and really sustainable source of energy becomes available.

The advantages of natural gas are considerable in comparison to those of coal or oil: it is much easier to clean and much easier to transport from where it occurs in nature to where it is required for work, warmth, and recreation. Compared to oil or coal, the main disadvantage of natural gas is that since it has a large volume for the equivalent energy content, a good pipeline infrastructure or the equivalent is needed.

For deposits that are small, in remote locations, or accumulations that are far from consumers, transportation by pipeline may not be economical. It is for these situations that the Fischer–Tropsch technology is particularly useful, since it enables the conversion into liquid products.

For coal the position is different. Although coal can be transported more simply than gas, cleaning it is a major task and ultimately it must also be converted into a refineable liquid product, before it can be turned into transportation fuels or chemicals. Fischer–Tropsch conversion is again a useful way to achieve this goal.

A considerable problem with all carbon-based fuels is that they produce carbon dioxide when burned. Atmospheric carbon dioxide is a “greenhouse gas,” which when present in large quantities is widely believed to have serious consequences for the climate of our planet.

It can be argued that one should not consider carbon-based fuels and chemicals technology for the future. Unfortunately, at present we have few viable alternatives to fossil fuels on the scale that is required to meet the energy needs of a world population that is already at around 7 billion and still increasing rapidly. Although most of our energy comes from the sun, the direct use of solar power to produce biofuels or to generate hydrogen on industrial scales is still a long way off. In the meantime, we will have to continue to rely on the power of the sun indirectly, via fossil fuels. The question then becomes: even if it is only an interim measure, how can we use our carbon-based resources in the most responsible manner?

The immediate challenge is the efficient transformation of one form of fossil fuel energy into another; in other words, how can we most efficiently transform natural gas, coal, or oil into say diesel or gasoline that we can harness to drive our machines. Even this is a vast task, but it is one that is being tackled very effectively through the Fischer-Tropsch process. That is what this book is about, an up to date review of the fundamental chemical, industrial, economic and environmental aspects of the Fischer-Tropsch process.

Acknowledgments

We have had a lot of help in producing the book and we thank our many friends and colleagues, including Paul Arwas, Norman Basco, Gian Paolo Chiusoli, Allen Hill, Brian James, Tom Lawrence, Kenichi Maruya, Peter Portious, Marco Ricci, Sally Maitlis, Julia Weinstein, and Valerio Zanotti, for reading parts and making helpful comments Above all, we give our warmest thanks to Marion, Peter Maitlis's wife, and Chèrie, Arno de Klerk's wife, for the help, patience, understanding, and love they showed while we worked on this book

University of Sheffield, UK — Peter M. Maitlis
University of Alberta, Canada — Arno de Klerk
October 2012

List of Contributors

Letizia Bua
Research Center for Non-Conventional Energy
Eni Istituto Donegani
via Fauser, 4
28100 Novara
Italy

Vincenzo Calemma
Eni S.p.A. – Refining & Marketing Division
via Felice Maritano, 26
20097 S. Donato Milanese
Milan – Italy

Lino Carnelli
Research Center for Non-Conventional Energy
Eni Istituto Donegani
via Fauser, 4
28100 Novara
Italy

Burtron H. Davis
University of Kentucky
Center for Applied Energy Research
2540 Research Park Drive
Lexington
KY 40511
USA

Arno de Klerk
University of Alberta
Chemical & Materials Engineering
9107 – 116 Street
Edmonton
Alberta T6G 2V4
Canada

Yong-Wang Li
Chinese Academy of Science
Institute of Coal Chemistry
Beijing
China

Peter M. Maitlis
University of Sheffield
Department of Chemistry
Sheffield S3 7HF
UK

Roberta Miglio
Eni SPA – Exploration & Production Division
San Donato Milanese
20097 Milan
Italy

Alessandra d'Arminio Monforte
Research Center for Non-Conventional Energy
Eni Istituto Donegani
via Fauser, 4
28100 Novara
Italy

Julius Pretorius
Alberta Innovates Technology
Futures
250 Karl Clark Road
Edmonton
Alberta T6N 1E4
Canada

Cecilia Querci
Research Center for Non-
Conventional Energy
Eni Istituto Donegani
via Fauser, 4
28100 Novara
Italy

Marco Ricci
Research Center for Non-
Conventional Energy
Eni Istituto Donegani
via Fauser, 4
28100 Novara
Italy

Roberto Zennaro
Eni S.p.A. – Exploration &
Production Division
via Emilia, 1
20097 San Donato Milanese
Milan
Italy

Part One
Introduction

Greener Fischer-Tropsch Processes for Fuels and Feedstocks, First Edition. Edited by Peter M. Maitlis and Arno de Klerk.

1
What is Fischer–Tropsch?

Peter M. Maitlis

Synopsis

Some of the fundamental and most frequently used terms are explained. Fischer–Tropsch (FT) technology involves the conversion of syngas (a mixture of CO and H_2) into liquid hydrocarbons. It is a key element in the industrial conversion processes X-To-Liquids (XTL), where X = C, coal; G, natural gas; B, biomass; or W, organic waste. For example, a gas-to-liquids (GTL) process converts natural gas into syncrude, a mixture mainly of long-chain hydrocarbons. The conversion reactions are usually catalyzed by metals (iron, cobalt, and sometimes ruthenium) often carried on oxide supports such as silica or alumina. The liquid hydrocarbons are important sources of transportation fuels and of specialty chemicals. Syngas is now mainly obtained from coal, oil, or natural gas, but will in future be increasingly made from renewable sources such as biomass or organic waste. Since the available reserves of fossil fuels are diminishing, the renewables should provide more sustainable feedstocks in the long term.

1.1
Feedstocks for Fuel and for Chemicals Manufacture

Syngas, the name given to a mixture of carbon monoxide and hydrogen, is the lifeblood of the chemicals industry and helps to provide a lot of our energy. It can be made from many sources, including coal, natural gas, organic waste, or biomass. The Fischer–Tropsch (FT) process converts syngas catalytically into organic chemicals, mainly linear alkenes and alkanes, which are used as both liquid fuels and feedstocks for making further useful chemicals. Some oxygenates can also be formed (chiefly methanol and ethanol) (see Chapters 4 and 6).

Alkene and alkane formation in the FT-Hydrocarbon Synthesis can be summarized as follows:

$$2n\mathrm{H_2} + n\mathrm{CO} \rightarrow \mathrm{C}_n\mathrm{H}_{2n} + n\mathrm{H_2O} \tag{1.1}$$

$$(2n+1)\mathrm{H_2} + n\mathrm{CO} \rightarrow \mathrm{C}_n\mathrm{H}_{(2n+2)} + n\mathrm{H_2O} \tag{1.2}$$

Greener Fischer-Tropsch Processes for Fuels and Feedstocks, First Edition. Edited by Peter M. Maitlis and Arno de Klerk.

Box 1.1 What its all about: some definitions

To avoid ambiguity, we will use the following terms with reference to the metal-catalyzed conversion of syngas into organic compounds.

Fischer–Tropsch process (FTP) will refer to the overall industrial process wherein the syngas is catalytically converted in a reactor into a mixture of primary (largely but not exclusively linear aliphatic hydrocarbons) and secondary products. Water is also a major primary product. Secondary products that are believed to be formed in the reactor from the primary products include internal alkenes, branched chain and cyclic aliphatics, some aromatics, and some oxygenates such as alcohols.

Fischer–Tropsch hydrocarbon synthesis (or FT-HS) will refer to the hydrocarbons (1-*n*-alkenes and *n*-alkanes) that are generally considered to be the primary products of the metal-catalyzed syngas conversion when the reaction is carried out under mild conditions where further secondary reactions are minimized. A subset of the FT-HS, the formation of methane, is sometimes treated separately as **methanation**.

We will use the term **Fischer–Tropsch reaction** (or FT reaction) largely in the discussions on the mode(s) by which the primary products are formed, for example, the kinetics and reaction mechanisms of the FT-HS.

We also introduce two terms. **Sustainable development** is the use of natural resources that "meet present (world) needs without compromising the ability of future generations to meet their own needs" and was coined by the Brundtland Commission. **Renewable energy** is energy that is renewed naturally. It includes traditional biomass (biofuels), hydroelectricity, wind, tidal, solar, and geothermal sources. It excludes raw materials that are depleted in use such as fossil fuels and nuclear power.

Energy has been said to be "the single most important scientific and technological challenge facing humanity in the twenty first century" [1], and we agree. There is the global requirement for more energy, especially as transportation fuels, as populations increase in number and sophistication. In addition, there is also a more specific need for new feedstocks for chemicals manufacture. As we will see, these two needs have features in common. And above all, we recognize the imperative now demanded by Society to produce both fuel and feedstocks in an environmentally acceptable and preferably sustainable manner. We also aim to correct some of the erroneous beliefs and myths present in the energy and chemicals sectors in order that our students, who will be tomorrow's academic and industrial leaders, have reliable foundations on which to build.

Mankind literally lives off energy. Most of it comes from the sun, indirectly via plants that use carbon dioxide and water to grow. Eventually they die and decay and, very slowly, over geological timescales, are turned into the fossil fuels (coal, oil, natural gas) that we extract and combust to provide heat, light, and other forms of power [2].

Box 1.2 Fossil fuel resources

In 2000, global oil reserves were estimated at about 1105 billion barrels; by the end of 2010, new discoveries had increased the proven reserves to 1383 billion or 1476 billion barrels ($\sim 200 \times 10^9$ tons) if Canadian oil sands and shale oil and gas are included. Similarly, gas reserves were estimated at 109 trillion cubic meters (Tcm) in 1990, 154.3 Tcm in 2000, and 187.5 Tcm in 2010 [3]. Based on the data for current and previous years, the US Department of Energy makes forecasts of the use and the production of energy. Currently, it projects that world consumption of marketed energy will increase from 495 QUAD (quadrillion, 10^{15} British Thermal Units or 1.055×10^{20} J) in 2007 to 590 QUAD in 2020 and then to 739 QUAD ($\sim 780 \times 10^{20}$ J) in 2035, an overall increase of 49%. Liquids (i.e., largely hydrocarbons) supply a large proportion of world energy consumption, and although their share is predicted to fall somewhat, it will still be around 32% in 2030 [4].

"Unconventional" resources (including oil sands, shale oil and shale gas, extraheavy oil, biofuels, coal-to-liquids, and gas-to-liquids) are expected to become increasingly competitive; world production, which totaled 3.4 million barrels per day in 2007, is forecast to increase to 12.9 million barrels per day and to account for 12% of total world liquids supply in 2035. The proportion of biofuels, largely ethanol and biodiesel, from the United States and Brazil, is forecast to grow slowly.

1.2 The Problems

There are two main problems with fossil fuels: the reserves are finite and slowly running out and, since all fossil fuels contain combined carbon, their combustion (oxidation) produces carbon dioxide, which accumulates in the atmosphere and which is likely to have serious consequences for the climate of our planet. Combustion also generates other materials that can harm mankind and the environment, such as CO, oxides of sulfur and nitrogen, and metallic oxide ashes, arising from incomplete oxidation and from impurities in the fuel.

For some end-uses there are many alternatives to fossil fuels, such as hydroelectric and nuclear power and others that are being developed commercially, including solar, wind, tidal, and geothermal power. The latter technologies will play their very important role mainly by providing electric power via large fixed installations. However, they will not have a direct part in providing more liquid transportation fuels or new feedstocks for the chemicals industry.

Why should Fischer–Tropsch be the approach to replace or supplement crude oil as a source of transportation fuels, gasoline (in the United States), or petrol and diesel (in the United Kingdom)? Today transportation fuels from crude oil must undergo extensive cleaning to remove materials containing heteroatoms (N, S, metals, etc.) from the raw feedstocks; if these materials are not removed, the impurities will quickly spoil and deactivate the catalyst. The amounts of hydrogen and energy needed for this cleaning have steadily increased as the crude oils have become heavier (i.e., more impure) over the years. Today, about 15–20% of the energy in the oil is required to produce environmentally acceptable transportation fuels, and the percentage can only increase as the crude becomes heavier. Thus, the energy advantage of crude oil over other fossil fuels is becoming narrower as time passes. Even today (2012), one is able to convert coal (a very "dirty" material) into transportation fuels in a Fischer–Tropsch process at a cost that is competitive with crude oil.

The environmental properties of the FT-synthesized transportation fuels meet or usually exceed those of crude oil-derived fuels. There are of course a number of other approaches that can be used for converting coal into transportation fuels. For example, the Exxon-Mobil methanol to gasoline process is able to convert coal first into syngas, then methanol, and then gasoline; however, the gasoline obtained by this process is high in aromatics and essentially no diesel range fuels are produced. Another variation converts the coal to low molecular weight alkenes and then further to gasoline and diesel range fuels; however, the diesel that is produced will be multiple branched and have a lower cetane number than the FT diesel.

Environmental concerns today cause governments to provide subsidies to allow renewable fuels to be utilized, as, for example, ethanol in the United States. Even without this subsidy, FT fuels are competitive with the subsidized renewables in some areas. In addition, improvements in gasification procedures are allowing fuels to be obtained from a mixture of renewables and coal so that the FT oil will have the environmental advantage over crude oil.

1.3 Fuels for Transportation

1.3.1 Internal Combustion Engines

The form in which the energy is available is important. Although it has been done (e.g., in wartime), it is unrealistic to try and run cars, trucks, or planes on coal, wood, or natural gas. Wikipedia has estimated that there were over 1 billion cars and light trucks on the road in 2010. As motor vehicles are now manufactured in many countries, developed as well as developing, the total must exceed 1.1 billion (10^9) quite soon. Almost all of them run on liquid hydrocarbons and it has been estimated that they burn well over 1 billion cubic meters (1 Bcm, 260 billion US gallons, or 8.5×10^8 tons) of fuel each year. The engineering has been well worked out so that the internal combustion engines are now extremely efficient for the appropriate fuel. The optimum gasoline has a high proportion of branched chain alkanes (giving a high octane number), while the best diesel has a high component of linear alkanes (with a high cetane number). It should be remembered that it will be necessary to continue to provide fuel for all the (older) vehicles at present on our roads, as well as those currently being built and planned.

1.3.2
Electric Cars

There is considerable interest in using electricity for transportation and most manufacturers are making electric cars, as they are perceived to cause less pollution in their immediate neighborhoods. However, there are some serious disadvantages. Some of the problems as well as the benefits of the electric car have been amusingly illustrated by Jeremy Clarkson, the presenter of the BBC TV's very popular car show "Top Gear," when he reviewed the projected Mini E being built by BMW [5]. This car works well but requires 5088 lithium ion batteries (weighing 260 kg) and even then has a range of only 104 miles, after which it requires charging for 4.5 h. Eventually, the batteries will need replacing, the cost of which does not bear thinking about. The wide acceptance of electric cars depends on the availability of inexpensive and high-power batteries and also on the availability of national networks of fast-charging stations, which are at present hardly on the drawing board. To get round the problems, many manufacturers add on a liquid hydrocarbon fuel motor to extend both the range and the convenience of electric cars. There are many now available or coming on to the market, for example, the hybrid (electric–gasoline) Toyota Prius or the Chevrolet Volt or Ampera.

There are several serious snags on the way to commercially viable electric cars. Not only are the batteries costly and heavy, but also the lithium they require is difficult to source. The provenance of the electricity for recharging them must also be considered. Thus, the US Energy Information Agency estimates that two-thirds of world electricity is generated from fossil fuels (coal 42%, natural gas 21%, and oil 4%), 14% from nuclear and only 19% from renewables. Furthermore, it has been estimated that the average CO_2 output for electric cars is 128 g/km compared to an average of 105 g/km for hybrids such as the Toyota Prius, when the emissions from coal- and oil-fired electricity-generating stations are included [6]. If we want to minimize CO_2 production by diminishing the use of fossil fuels, given the technology available at present (2012), the nuclear option currently seems the choice for generating sustainable electricity. But that also has serious problems as the disasters at the Chernobyl, Fukushima, and Three Mile Island nuclear plants showed.

1.3.3
Hydrogen-Powered Vehicles

Hydrogen is a very attractive source of power as the only product of combustion is water; unfortunately, large-scale commercial applications are further in the future, even though the science is well known and hydrogen is easily made by splitting water, for example, by electrolysis or solar heating. However, the cost of doing so, in terms of the energy required, makes it very expensive.

Currently, hydrogen is produced mainly by gasification/reforming; thus, hydrogen should be considered a by-product of the petrochemicals industry in the formation of carbon monoxide, for example, from hydrocarbons:

$$CH_4 + 1/2O_2 \rightarrow CO + 2H_2 \quad (1.3)$$

$$CH_4 + H_2O \rightleftarrows CO + 3H_2 \quad (1.4)$$

The water-gas shift reaction (WGSR) is then employed to increase the proportion of hydrogen, but this in turn produces carbon dioxide:

$$CO + H_2O \rightleftarrows CO_2 + H_2 \quad (1.5)$$

Thus, the conventional production of hydrogen today is always associated with the production of CO_2.

Perhaps the development of hydrogen-powered fuel cells for cars is a promising direction [7].

One requirement for viable electric or hydrogen-powered transportation systems is the availability of widespread national grids for recharging, the setting up of which will be a mammoth and vastly expensive task. And if the electricity for the grid comes from burning fossil fuels, we have not addressed the sustainability problem – merely moved it sideways to another area.

1.4
Feedstocks for the Chemical Industry

The raw materials for the organic chemicals industry are largely carbon based; in the eighteenth century, the pyrolysis of wood provided useful chemicals. In the nineteenth century, coal tar was exploited as the source of many materials, especially aromatics; while in the twentieth and twenty-first centuries, the feedstocks for many organic chemicals have been derived from oil. To that extent therefore, the supply of feedstocks for chemicals and of fuel for transportation currently run parallel and both depend on nonrenewable resources.

1.5
Sustainability and Renewables: Alternatives to Fossil Fuels

It has been estimated that more solar energy strikes the Earth in 1 h (4.3×10^{20} J) than is currently consumed by all mankind in a year (4.1×10^{20} J). That even allows a great expansion of use as there would be more than enough. Thus, there is a continuing search for usable sources of energy that are either from renewable "biofuels," and thus will not deplete our reserves, or that utilize sunlight more directly and do not involve organic intermediates, for example, some form of hydrogen generation by splitting water. The main biorenewables are fast-growing plants, trees, or algae, for example, that can be harvested and burned, directly or indirectly, with the carbon dioxide produced going back to feed more plants.

1.5.1 Biofuels

The best-known commercial example of biofuel manufacture is in Brazil where sugarcane grown on a very large scale is harvested and thereby sugar is extracted and fermented into alcohol that is distilled to be sold in filling stations (as *bioethanol*) to power motor vehicles. Brazil, with a population close to 200 million, has plentiful sunlight, cheap labor, and some government assistance. Prior to the discovery of large offshore oil and gas deposits, it also had the additional stimulus of a lack of home-produced oil fuel. It therefore turned to ethanol to power internal combustion engines, and most Brazilian cars are now able to run on either gasoline or alcohol. Currently, the home-produced ethanol takes care of some 13% of the country's motor fuel needs; the comparable figure is about 4% for the United States [8].

Large amounts of bioethanol, made from maize (corn), are produced in the United States, and ethanol commonly makes up 10% of the fuel at the pump (designated E10). However, it is now recognized that there are major problems with such agriculturally produced fuels. One is that the acreage of arable land needed to grow plants to power transport can seriously hinder the growing of food. This in turn impinges on the cost of food. The energy balance is also more complex than it may appear at first sight since, in addition to sunlight, considerable energy derived from fossil fuels is required to produce the ethanol. Much water is also required, and since water is also a scarce commodity, it must be conserved and recovered, which will also require energy.

It has been calculated that irrespective of crop, one acre of land, pond, or bioreactor can annually yield enough amount of biomass to fuel one motor vehicle or meet the calorific requirement of several people. This amount of biomass therefore makes only a very small contribution to our present road transport requirements and yet can contribute significantly to global food shortages and rising prices [9, 10]. New technology to make ethanol based on lignocellulose, and which does not depend on food crops, is being actively pursued. Thus, while biomass is used as a renewable fuel, it is not yet the cure-all the world is seeking.

Other forms of biofuels are also known, such as biodiesel made from waste fats (long-chain esters); however, this has not been promoted to the extent of bioethanol and is likely to remain a minor source of energy for transportation.

1.5.2 Other Renewable but Nonbio Fuels

The production of energy by such means that do not involve biointermediates is a very active area of science research. There are many ways to harness solar energy: using photovoltaic cells or solar furnaces, it can be turned directly into electricity. Wind and tidal power can also be similarly harnessed; however, all these sources have the disadvantage that the energy is not continuously produced and the electricity must be stored and cabled to the site where it is needed. Although the technology to mass-produce solar cells has improved and in some countries (Germany, Japan, Spain, and Israel) electricity from such devices is beginning to make a significant contribution to the national grid, the cost of solar power is currently estimated to be between 10 and 20 times that of power from burning coal. Storage on the scale needed to ensure that power is available nationally even during hours of darkness has also lagged behind. Because fossil fuels are still abundant and inexpensive, non-biorenewables are not likely to play a large role in primary power generation until technological or cost breakthroughs are achieved, or environment-driven carbon taxes are brought in.

1.6
The Way Forward

So, where do we go? If the large-scale use of electric and hydrogen-powered cars is only over the horizon and renewable biofuels will supply a small fraction of our needs for transportation, we must make the best of what we have by improving our tools to deal with our present resources. Since major discoveries of oil and gas and coal are still being made, exact numbers are imprecise, but current best estimates indicate that our planet has enough reserves of oil for about 50 years and of natural gas for perhaps 150–200 years at current consumption levels. Coal is more plentiful and some 100–200 years supply may be available. However, the important factor is how difficult (i.e., how expensive) it will become to extract these fuels: cost is very likely to determine the uses to which fossil fuels will be put in future. The other side of the argument is of course the growth in carbon dioxide. The EIA estimates that annual CO_2 emissions will rise from the 2007 level of about 29.7 billion tons to around 42.4 billion tons by 2035. This 43% increase is likely to have a significant effect on many aspects of our lives, in particular through changes in our climate.

For the twin reasons of conserving our fossil fuels and curbing the increase in CO_2 levels, our primary concern should be in using our resources better and more efficiently. One way to do that is to improve the conversions of the raw materials into conveniently usable fuels and/or chemicals. Doing that is not necessarily straightforward or obvious. Taking natural gas (which is largely methane) as an example, while direct approaches such as partial oxidation of methane to methanol or to higher alkanes may become commercially viable in the future, the best way currently is to reform the natural gas into syngas ($CO + H_2$) and then to build on that. The engineering needed for reforming is well established and there are many well worked out reactions making useful products from syngas. One of these is of course the Fischer–Tropsch hydrocarbon synthesis in which the syngas is converted into linear hydrocarbons that can be used either as fuel (diesel) or as chemical feedstocks. Our thesis therefore is that improvements to Fischer–Tropsch are desirable, possible, and necessary and should be developed as soon as practicable. Some other paths that are being followed are outlined in Section 1.7.

1.7 XTL and the Fischer–Tropsch Process (FTP)

The Fischer–Tropsch Process (FTP) is a key part of the technology that is needed to convert one type of carbon-based fuel into another. This in turn allows industry to choose which feedstock and which technique is most suitable for a given purpose. A number of composite technologies known as XTL have been developed: CTL (coal-to-liquids), GTL (gas-to-liquids), BTL (biomass-to-liquids), and WTL (waste-to-liquids). Thus, for example, GTL reforms natural gas (mainly methane) by partial oxidation into syngas:

$$CH_4 + H_2O \rightleftarrows CO + 3H_2, \quad \Delta H^\circ_{(298\,K)} + 206\,kJ/mol \tag{1.6}$$

$$CH_4 + CO_2 \rightleftarrows 2CO + 2H_2, \quad \Delta H^\circ_{(298\,K)} + 247\,kJ/mol \tag{1.7}$$

Alternatively, CTL, for example, makes syngas from coal:

$$2C + H_2O + O_2 \rightarrow CO + CO_2 + H_2 \tag{1.8}$$

The CO: H_2 ratio is adjusted by the catalytic Water-Gas Shift Reaction (WGSR):

$$H_2O + 1CO \rightleftarrows H_2 + CO_2\,(WGSR), \quad \Delta H^\circ_{(298\,K)} - 41\,kJ/mol \tag{1.5}$$

and the gases are then led to another reactor where they are contacted with a different metal catalyst (usually iron or cobalt) in the Fischer–Tropsch reaction. For cobalt and other metals, the catalytically active metal is generally deposited as nanoparticles on an oxide such as silica or alumina that was classically thought to act simply as an inert support. Iron-catalyzed reactions are generally carried out over the unsupported (massive) metal. Details of the various XTL processes are given in Chapters 2 and 5.

The product distribution of hydrocarbons formed during the FT process follows an Anderson–Schulz–Flory distribution, expressed as $W/N = (1-\alpha)^2\alpha^{n-1}$ where W is the weight fraction of hydrocarbon molecules containing N carbon atoms and α is the chain growth probability [11]. This can be visualized by plotting log (W/N) against N, and shows a monotonic decrease from lower to higher molecular mass products, indicative of a step-growth polymerization of a C_1 species (see Figure 12.3).

Methane is always the largest single product; however, by bringing α close to one, the total amount of methane formed can be minimized and the formation of long-chain hydrocarbons is increased. Very long-chain hydrocarbons are waxes, which must be cracked in order to produce liquid transportation fuels.

Although the FT process has been applied on a large scale, its universal acceptance has been hampered by high capital costs, high operation and maintenance costs, and environmental concerns. In practice, FT liquid fuels compete with natural gas that can be supplied by conventional gas pipelines and liquefied natural gas (LNG) technology. Thus, FT gas as a feedstock becomes economically viable as a supply of "stranded gas," in other words, a source of natural gas that is impractical to exploit as it is far from major conurbations.

1.7.1
Some History

The history of the FTP is a classical example of the stepwise development so characteristic of science: there was really no single "Eureka" moment. It also illustrates how closely advances in science and technology are coupled to economic and political circumstances.

In 1902, Sabatier and Senderens reported that a reaction occurred between carbon monoxide and hydrogen to give methane over a nickel catalyst; then, in 1910 Mittasch, Bosch, and Haber developed promoted iron catalysts for ammonia synthesis from hydrogen and nitrogen. That was shortly followed (in 1913) by patents issued to BASF for the production of hydrocarbons and oxygenates by the high-pressure hydrogenation of CO over oxide catalysts. In the 1920s, Fischer and Tropsch working at the Kaiser Wilhem Institute in Berlin first made *Synthol* (containing oxygenates) by hydrogenation of CO over alkalized iron, and then in 1925, they announced the synthesis of higher hydrocarbons at atmospheric pressures over Co and Ni. The interest in the process grew rapidly and workers in England, Japan, and the United States, especially at the US Bureau of Mines, devoted much time and effort to improving the methodology. Considerable engineering work and catalyst development continued in Nazi Germany, especially during the 1939–1945 World War when the process was used to make motor fuel from coal. Germany was acutely short of oil, but had copious reserves of low-quality brown coal (lignite) that could be turned into fuel for the war effort. The development of new reactor designs for the FTP continued after the War as there were fears that petroleum would be in short supply. With the discovery of large new oilfields, interest in Fischer–Tropsch waned somewhat until the 1970s brought a large increase in the price of oil and sanctions on the export of oil to South Africa. This encouraged SASOL (Suid Afrikaanse Steenkool en Olie, the South African Coal and Oil company) to expand its CTL plants in order to become more self-sufficient [12]. Although the economic and political pressures have long since changed, SASOL has actively continued to develop its processes in both CTL and GTL. They are based on FT technology using iron or cobalt catalysts, and SASOL continues to play a major role in developing new plants in other countries (including Qatar, Nigeria, Egypt, etc.). Shell has also built major FT plants (in Malaysia and Qatar) using cobalt catalysts (see Chapters 3, 5 and 9). Total world production of FT hydrocarbons has been estimated at about 10 million tons per year.

In parallel with the Fischer–Tropsch hydrocarbon synthesis, work continued on another reaction based on syngas and originally developed in Germany: the synthesis of methanol. That came to fruition in 1966 when ICI in the United Kingdom brought in the low pressure process, using a copper–zinc oxide catalyst, which still dominates the technology (see Chapter 6) and currently enables methanol production of about 30 million tons annually.

1.7.2
FT Technology: An Overview

FT processes are currently used commercially to make hydrocarbons by passing syngas over supported metal catalysts. Fe, Co, Ru, Rh, and even Ni all have FT activity, though in somewhat different ways. The most active catalyst is Ru, but it is not used commercially because of its high cost. The original and most commonly used catalyst is Fe, though Shell uses a Co catalyst to make long-chain alkanes (waxes) that are then broken down to smaller alkanes. The cobalt catalyst generally consists of very fine particles of the metal supported on an oxide surface such as silica or alumina. These nanoparticles have the advantage of high activity due to their large surface areas; however, this also makes it easy for impurities to be adsorbed that can affect the performance of the catalyst. In some cases the activity can be improved, but many substances will diminish the activity and selectivity.

Two main regimes have been used: low-temperature Fischer–Tropsch (LTFT), usually at 200–250 °C, that gives long-chain molecules, and the high-temperature Fischer–Tropsch (HTFT), at 320–375 °C, that gives shorter chain molecules. It is fairly generally agreed that the primary products of the reaction are 1-*n*-alkenes, but under harsher conditions (high pressure of hydrogen, higher temperature, or hydrogenating catalysts such as Co) *n*-alkanes result. The primary alkene products are also further hydrogenated, isomerized, dehydrogenated, cyclized, carbonylated, or even oxidized, under the reaction conditions and thus a wide spectrum of products can be formed.

The best form of the reactors to be used depends on the catalyst, the conditions, and the distribution of products that is desired. HTFT uses iron catalysts in two-phase fluidized bed reactors; LTFT uses either iron or cobalt in three-phase slurry reactors or tubular fixed bed reactors. Much of the skill in running a successful FT plant comes from the use of properly designed reactors [13].

The silica or alumina support was long believed to play little role in the basic FT reaction, though it was significant in the subsequent, secondary reactions. However, studies by surface scientists have shown that the actual FT catalysis usually takes place at the interface between the metal and the oxide, which can be either the support to or a component of the catalyst.

1.7.3
What Goes on?

In progressing from CO that has one carbon atom to an alkene or alkane, quite a complex series of reactions must be occurring. Essentially however, it is a polymerization of C_1 units. The question then arises how this occurs on a metal surface. It is only quite recently that surface scientists have had access to the tools that will allow them to begin to answer this riddle. Thus, there have been many attempts to understand the reactions that occur and many theories, the more important of which are summarized in Chapter 12.

Table 1.1 Energetics of CO hydrogenation.

$$3H_2 + 1CO = H_2O + CH_4, \quad \Delta G^\circ_{(500\text{ K})} - 94\text{ kJ/mol} \tag{1.9}$$

$$2H_2 + 1CO = H_2O + 1/3(C_2H_6), \quad \Delta G^\circ_{(500\text{ K})} - 31\text{ kJ/mol} \tag{1.10}$$

$$3H_2 + 1CO = CH_3OH, \quad \Delta G^\circ_{(500\text{ K})} + 21\text{ kJ/mol} \tag{1.11}$$

$$3H_2 + 2CO = HOCH_2CH_2OH, \quad \Delta G^\circ_{(500\text{ K})} + 66\text{ kJ/mol} \tag{1.12}$$

$$4H_2 + 2CO = CH_3CH_2OH + H_2O, \quad \Delta G^\circ_{(500\text{ K})} - 27\text{ kJ/mol} \tag{1.13}$$

$$3H_2 + 1CO_2 = CH_3OH + H_2O, \quad \Delta G^\circ_{(298\text{ K})} + 3\text{ kJ/mol} \tag{1.14}$$

$$H_2O + 1CO \rightleftarrows H_2 + CO_2\ (\text{WGSR}), \quad \Delta G^\circ_{(298\text{ K})} - 28\text{ kJ/mol} \tag{1.5}$$

1.7.4
CO Hydrogenation: Basic Thermodynamics and Kinetics

As in all chemical transformations, although the rates are governed by the kinetics of the individual steps, it is important to ensure that the thermodynamics of the steps are favorable or if one step of a sequence is unfavorable, it is coupled to a very favorable one.

As Table 1.1 indicates, the formation of hydrocarbons from CO hydrogenation is generally favored overall, but, as shown in a comparison of the free energies (ΔG°), the reaction can be thought of as driven by formation of water. Thus, making methane also involves making 1 mol of water is more favorable, but higher hydrocarbons are less favored. If free water is not formed, then the thermodynamics are much more difficult as is shown by the positive ΔG° for methanol and glycol; only when some water is also formed, as with ethanol, does the reaction become favored.

1.8
Alternatives to Fischer–Tropsch

Given that at present the best way of using fossil fuels is to reform them into syngas, some of the alternatives to FT are discussed in Chapter 6. In fact, the highest volume use of syngas is the reaction to methanol, which can be used as a fuel additive and which is also a very useful chemical and a C_1 feedstock. Examples include the Mobil process that converts methanol into gasoline over an acid zeolite catalyst (HZSM-5), and the Haldor–Topsoe A/S TIGAS process that uses dimethyl ether for the same transformation. Several processes also exist for converting methanol into olefins. These include the UOP/Norske Hydro process (with a pilot plant in Norway and a demonstration plant in Belgium) and Lurgi has a similar methanol-to-propylene (MTP) process. The Institute of Chemical Physics (in Dalian, China) commissioned the first commercial methanol-to-olefin process (DMTO) in the world in 2010. This has a production capacity of 600 000 tons of lower olefins per year (http://english.dicp.cas.cn/ns/es/201008/t20100811_57266.html.).

Other widely practiced alternatives use syngas together with an organic substrate to extend the chain lengths, as, for example, in the hydroformylation of propene to butanal and isobutanal,

$$RCH{=}CH_2 + CO + H_2 \rightarrow RCH_2CH_2CHO + RCH(CHO)CH_3 \quad (1.15)$$

A large number of related reactions of olefins with carbon monoxide, for example, giving acids and esters, are known and some of these are important industrially [14]:

$$CH_2{=}CH_2 + CO + H_2O \rightarrow CH_3CH_2CO_2H \quad (1.16)$$

$$RCH{=}CH_2 + CO + MeOH \rightarrow RCH(CO_2Me)CH_3 + RCH_2CH_2CO_2Me \quad (1.17)$$

Last but not least, the WGSR is used to greatly increase the proportion of hydrogen in the syngas, which can then be separated and used as a nonpolluting fuel or in a hydrogenation plant. Since the WGSR is an equilibrium, the trouble is that by increasing the amount of hydrogen in syngas, it also increases the amount of the very undesirable CO_2.

References

1 Chem. Eng. News (August 22, 2005), quoting Nobel Laureate Rick Smalley, in testimony to the US Senate (April 2004).
2 Lewis, N.S. (2007) *MRS Bull.*, **32**, 808–820.
3 BP Statistical Review of World Energy (June, 2011)
4 International Energy Outlook (May, 2010) Energy Information Administration, US Department of Energy, Washington, DC, www.eia.doe.gov/oiaf/ieo/index.html.
5 Sunday Times (August 1, 2010), London.
6 Daily Telegraph (September 4, 2010), London, Car Clinic, September 4.
7 Eberle, U. and von Helmolt, R. (2010) *Energy Environ. Sci.*, **3**, 689–699.
8 Ritter, S.K. (June 25, 2007) Chem. Eng. News, p. 15.
9 Walker, D.A. (2010) *Ann. Appl. Biol.*, **156**, 319–327.
10 Walker, D.A. (2010) *J. Appl. Phycol.* doi: 10.1007/s10811-009-9446-5.
11 Dry, M.E. (1981) Chapter 4, in *Catalysis Science and Technology*, vol. **1** (eds J.R. Anderson and M. Boudart), Springer, Berlin.
12 Anderson, R.B. (1984) *The Fischer–Tropsch Synthesis*, Academic Press Inc., Orlando.
13 Steynberg, A.P., Dry, M.E., Davis, B.H., and Breman, B.B. (2004) Fischer–Tropsch reactors, in *Studies in Surface Science and Catalysis*, vol. **152** (eds A.P. Steynberg and M. Dry), Elsevier BV, Amsterdam.
14 For further details of these and related reactions, see Maitlis, P.M. and Haynes, A. (2006) Syntheses based on carbon monoxide, in *Metal-Catalysis in Industrial Organic Processes* (eds G.P. Chiusoli and P. M. Maitlis), RSC Publishing, Cambridge.

Part Two
Industrial and Economics Aspects

2
Syngas: The Basis of Fischer–Tropsch

Roberto Zennaro, Marco Ricci, Letizia Bua, Cecilia Querci, Lino Carnelli, and Alessandra d'Arminio Monforte

Synopsis

Syngas (a mixture of carbon monoxide and hydrogen) is normally made industrially from natural gas or coal. The ratio of H_2 to CO can be manipulated by the water-gas shift reaction (WGSR):

$$CO + H_2O = CO_2 + H_2, \quad \Delta H^\circ = +41\ \text{kJ/mol}$$

Since the WGSR is reversible, carbon dioxide and water are also formed. The significance of *gas loops* in industrial plants is explained, as is the importance of the formation of CO_2. Potential uses are being explored.

2.1
Syngas as Feedstock

Synthesis gas (or syngas) is the name commonly given to a mixture of carbon monoxide (CO) and hydrogen (H_2); various molecular ratios are used industrially. It can be made from coal (C), natural gas (CH_4), biomass ($C_xH_yO_z$), and other organic materials such as plastic waste by a partial oxidation, often with addition of steam (H_2O) to increase the hydrogen content. Syngas has approximately half the energy density of natural gas (methane) and can be used for its heat value in steam cycles, gas engines, fuel cells, or turbines to generate power and heat. Syngas is also an intermediate feedstock for making liquid fuels and a large number of commodity chemicals, including hydrogen, synthetic natural gas (SNG), naphtha, kerosene, diesel, methanol, dimethyl ether (DME), and ammonia.

Syngas is particularly important in refineries as a source of hydrogen, which is required for hydrotreating, removal of impurities, hydrogenating olefins, and other hydroprocessing such as catalytic cracking.

Synthetic natural gas is similar to natural gas (i.e., largely methane), but is produced by gasification of different carbon sources. The gasification processes involve the catalytic reaction of carbon monoxide and/or carbon dioxide with hydrogen to

Greener Fischer-Tropsch Processes for Fuels and Feedstocks, First Edition. Edited by Peter M. Maitlis and Arno de Klerk.

give gases with high methane content, and are frequently known as *methanation*.

$$CO + 3H_2 \rightarrow CH_4 + H_2O, \quad \Delta H^\circ = -206.3\ \mathrm{kJ/mol} \tag{2.1}$$

$$CO_2 + 4H_2 \rightarrow CH_4 + 2H_2O, \quad \Delta H^\circ = -248.1\ \mathrm{kJ/mol} \tag{2.2}$$

As the methanation reactions of both carbon monoxide and carbon dioxide are highly exothermic, increases in reactor temperatures need to be avoided. This can be accomplished by recycling the reacted gas, by steam dilution, or by using isothermal reactors with indirect cooling. Catalysts with high nickel content are preferred for SNG production, similar to those in reforming catalysts.

As explained in Chapter 1 and elsewhere, in the catalytic Fischer–Tropsch (FT) synthesis, one mole of CO reacts with two moles of hydrogen to form mainly straight-chain 1-alkenes (C_nH_{2n}) together with *n*-alkanes, some internal alkenes, and minor amounts of branched hydrocarbons and primary alcohols. Side reactions are methanation, the Boudouard reaction, and coke deposition.

$$CO + 2H_2 \rightarrow -CH_2- + H_2O, \quad \Delta H^\circ = -154.1\ \mathrm{kJ/mol} \tag{2.3}$$

The product is synthetic crude oil (*syncrude*) that can be refined to produce excellent diesel fuel, lube oils, and naphtha. The most important catalysts are based on iron (Fe) or cobalt (Co). Cobalt catalysts generally have higher conversion rates, are more effective for hydrogenation, and thus produce fewer olefins and alcohols compared to iron catalysts.

Methanol is also produced by the reaction of carbon monoxide and/or carbon dioxide with hydrogen, but using catalytic systems different to those that lead to hydrocarbons (Chapter 6). Both reactions are exothermic and, somewhat surprisingly, it has been found that the methanol synthesis by hydrogenation of CO largely proceeds via carbon dioxide.

$$CO + 2H_2 \rightarrow CH_3OH, \quad \Delta H_{298} = -91\ \mathrm{kJ/mol} \tag{2.4}$$

$$CO_2 + 3H_2 \rightarrow CH_3OH + H_2O, \quad \Delta H_{298} = -49\ \mathrm{kJ/mol} \tag{2.5}$$

Side reactions can again lead to formation of by-products such as methane, higher alcohols, or dimethyl ether.

Although natural gas is the most widely used carbon source for methanol production, many other feedstocks can be used to produce syngas via steam reforming. Coal is increasingly being used as a feedstock for methanol production, particularly in China.

Dimethyl ether is now produced by methanol dehydration, requiring methanol as starting material; however, a direct production route, combining three reactions (2.6; 2.7 and 2.8) in a single reactor, is planned:

$$CO + H_2O = CO_2 + H_2, \quad \text{water gas shift} \tag{2.6}$$

$$CO + 2H_2 = CH_3OH, \quad \text{methanol synthesis} \tag{2.7}$$

$$2CH_3OH = CH_3OCH_3 + H_2O, \quad \text{methanol dehydration} \tag{2.8}$$

DME is industrially used in the production of the methylating agent dimethyl sulfate and is also used as an aerosol propellant. DME has potential as a fuel and it can be used directly in power generation or in blending with (or as substitute for) LPG or diesel, in particular because of its very high cetane rating. The boiling point of −25 °C allows fast fuel/air mixing, reduces ignition delay, and gives excellent cold starting properties. DME is an efficient alternative to other energy sources for medium-sized power plants, especially in isolated or remote locations where it can be difficult to transport natural gas and where the construction of LNG regasification terminals would not be viable.

Syngas is also the raw material for ammonia production, which is needed for the manufacture of fertilizers. chemicals, plastics, fibers, and explosives. The gas mixture is purified and the hydrogen to nitrogen ratio is adjusted to the stoichiometric 3 : 1 molar ratio needed for ammonia synthesis. The catalytic reaction is carried out at high pressure (100–250 atm) and between 350–550 °C, usually over an iron-based catalyst.

$$N_2 + 3H_2 \rightarrow 2NH_3 \tag{2.9}$$

A low (20–30%) once-through conversion is used and a part of the unconverted gas is circulated to increase the total conversion.

Other applications In addition to the mixture of alcohols that are produced as by-products in the FT synthesis, related processes that give mixed alcohols include MAS (methanol plus higher alcohols) technology, developed by Snamprogetti, EniChem, and Haldor Topsoe, which was demonstrated at an industrial level in the 1980s [1, 2]. A further application is the production of carbon monoxide, used for acetyls production (acetic acid, anhydride, etc.) or as an alternative carbon source.

Table 2.1 summarizes the different syngas specifications for the main applications.

2.2 Routes to Syngas: XTL (X = Gas, Coal, Biomass, and Waste)

When syngas is applied in the Fischer–Tropsch synthesis, the overall process is generally named **XTL, X**, depending on the carbon source, for example, **CTL**, coal-to-liquids, **GTL** (natural gas-to-liquids), **BTL** (biomass-to-liquids), or **WTL** (waste-to-liquids), as shown in Figure 2.1.

The main steps [3] that take place in a XTL complex include the syngas generation, followed by syngas cleanup and Fischer–Tropsch synthesis for synthetic fuels production. Combining coal and biomass, in a so-called CBTL process, is another possible route since the cofiring of a biomass and coal mixture is feasible in a modern gasifier.

The main technologies employed in the production of syngas starting from coal, natural gas, biomass, and some kinds of wastes (in particular, wood residues from

Table 2.1 Synthesis gas specification for different applications.

Specification	Hydrogen	Ammonia production	Methanol synthesis	FT synthesis
Hydrogen content	>99%	75%	—	60%
Carbon monoxide content	<10–50 ppm(v)	$[CO + CO_2]$ <10 ppm(v)	—	30%
Carbon dioxide content	<10–50 ppm(v)		—	
Nitrogen content	<2%	25%		
Other gases	N_2, Ar, CH_4	Ar, CH_4	N_2, Ar, CH_4	N_2, Ar, CH_4, CO_2
Balance		As low as possible	As low as possible	Low
H_2/N_2 ratio		≈3		
H_2/CO ratio				0.6–2.0
$[H_2 - CO_2]/[CO + CO_2]$ module			2	
Process temperature		350–550 °C	220–300 °C	200–350 °C
Process pressure	>20 bar	100–250 bar	50–100 bar	15–60 bar

construction or plastics) are *gasification* and *reforming* that may be grouped into two large production processes (Figure 2.2):

- Gasification from solid feedstock, such as coal, biomass, and waste.
- Reforming for gas feedstock.

The term *gasification* is applied to the conversion of any carbonaceous fuel to a gaseous product with a useable heating value. This definition excludes combustion,

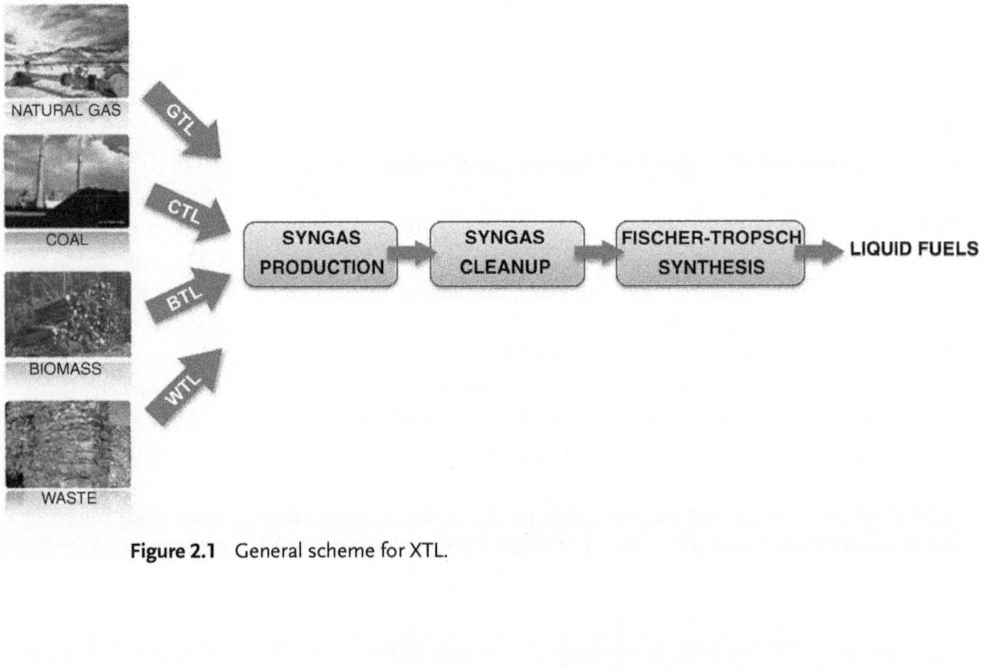

Figure 2.1 General scheme for XTL.

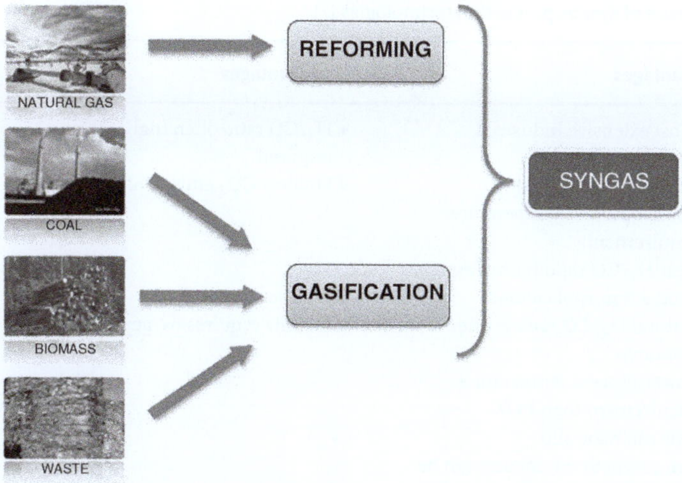

Figure 2.2 Routes to syngas.

because the product, flue gas, has no useful heating content. The dominant technology is partial oxidation to produce syngas, where the oxidant may be pure oxygen, air, and/or steam. Partial oxidation can be applied to solid, liquid, and gaseous feedstocks, including coal, biomass, residual oils, and natural gas; the last is also included despite the tautology involved in *gas gasification*. *Gasification* relates to the transformation of solid or liquid feedstock in an oxygen-poor environment (i.e., less than that needed for complete combustion), whereas *reforming* is applied to the transformation of natural gas into syngas.

2.2.1 Starting from Gas (GTL)

Syngas production represents the least efficient and most costly step of a GTL plant [4], and hence it has been an area of intense development by a number of technology providers. Six technologies for syngas production from natural gas, which have already been commercialized or are at advanced stages of development, are listed in Table 2.2 and discussed in the following sections.

(a) Steam Methane Reforming (SMR) Steam methane reforming is widely applied for hydrogen-rich syngas production, used to make ammonia, and hydrogen itself. The reactants are methane and steam and the chemical conversion, Equation 2.10, takes place over a Ni-based catalyst, with the following stoichiometry:

$$CH_4 + H_2O \rightarrow CO + 3H_2, \quad \Delta H^\circ = +206\,\text{kJ/mol} \tag{2.10}$$

$$CO + H_2O \rightarrow CO_2 + H_2, \quad \Delta H^\circ = +41\,\text{kJ/mol} \tag{2.6}$$

Operating temperatures and pressures for a conventional SMR unit are commonly 800–900 °C and 20–30 atm at the outlet. The volume percent of methane

Table 2.2 Comparison of syngas production technologies [7].

Technology	Advantages	Disadvantages
SMR	• Most extensive industrial experience • Oxygen not required • Lowest process temperature requirement • Best H_2/CO ratio for hydrogen production applications	• H_2/CO ratio often higher than required • Highest CO_2 emissions
ATR	• Natural H_2/CO ratio is often favorable • Lower process temperature requirement than POX • Low methane slip • Syngas methane content can be tailored by adjusting reformer outlet temperatures	• Usually requires oxygen
POX	• Feedstock desulfurization not required • Absence of catalyst permits carbon formation and, therefore, operation without steam, significantly lowering syngas CO_2 content • Low methane slip • Low natural H_2/CO ratio is an advantage for applications requiring ratio < 2	• Low natural H_2/CO ratio is a disadvantage for applications requiring a ratio > 2 • Very high process temperatures • Usually requires oxygen • High-temperature heat recovery and soot formation/handling adds process complexity • Syngas methane content is inherently low and not easily modified to meet downstream processing requirements
CPO	• Lower temperatures than ATR • Lower oxygen consumption	• Cost of the catalyst (usually a noble metal, in particular Rh)
HER	• Compact overall size • Application flexibility offers additional options for providing incremental capacity	• Limited commercial experience • In some configurations, must be used in tandem with another syngas generation technology
CPR	• More compact design • Very high-thermal efficiency • Less catalyst required • Suitable for offshore application	• Large number of parallel units for GTL scale

slippage is typically around 1% on a dry gas basis. The reaction (Equation 2.10) is endothermic and the need to sustain the required reaction temperatures is often a limiting factor in the design of an efficient heat transfer of the SMR reactor system. The SMR reactor is typically designed as an externally fired tubular reactor; such units have a high steam and fuel consumption and need a high capital investment. Excess steam (2.5 : 1–3.5 : 1 molar steam to carbon ratio) is required to prevent coke formation in the reactor tubes, giving a typical SMR H_2/CO ratio of 3 : 1. While this is suitable for ammonia and hydrogen production, it is much higher than the 2 : 1 ratio required for Fischer–Tropsch applications and needs to be adjusted by the WGSR (Equation 2.6).

(b) Autothermal Reforming (ATR) ATR is practiced in current commercial syngas applications by feeding a mixture of steam, methane, and oxygen over a fixed bed Ni-based catalyst. The system is adiabatic because of the presence of oxygen inside the reactor; thus, the heat required for the endothermic SMR reactions (Equations 2.6 and 2.10) is provided by the exothermic oxidation reactions (Equations 2.11 and 2.12).

In addition to the SMR reactions, a number of other reactions also occur:

$$CH_4 + 2O_2 \rightarrow CO_2 + 2H_2O, \quad \Delta H^\circ = -802\ \text{kJ/mol} \tag{2.11}$$

$$CH_4 + {}^1\!/_2O_2 \rightarrow CO + 2H_2, \quad \Delta H^\circ = -36\ \text{kJ/mol} \tag{2.12}$$

$$CO + H_2O \rightarrow CO_2 + H_2, \quad \Delta H^\circ = -42\ \text{kJ/mol} \tag{2.6}$$

The oxidation reactions take place directly through a "burner" nozzle in the vapor space above the fixed catalyst bed, where flame core temperatures may exceed 1900 °C. The SMR and CO shift reactions apparently occur sequentially in the catalyst bed, resulting in syngas exit temperatures from the ATR unit of approximately 950–1050 °C. ATR has the advantage to produce a syngas with a H_2/CO ratio close to 2 : 1 suitable for FT synthesis. Excess steam is also required to prevent soot formation, but much less than that needed for an SMR (e.g., 0.6 : 1 molar steam to carbon ratio) [5].

Although there is an additional cost needed for a cryogenic oxygen plant or an air compressor for air-blown ATR units, ATR appears to be generally more economical than conventional SMR, in particular for large-scale FT applications. Moreover, a fraction of the FT tail gas is recycled back to the ATR, in a GTL complex, to enhance the thermal efficiency of the whole process (see Section 2.6).

(c) Noncatalytic Partial Oxidation (POX) In noncatalytic partial oxidation, the oxidation reactions are predominant and require less steam than for an ATR to achieve a similar H_2/CO ratio of 2 : 1. The oxidant and the hydrocarbon feedstocks are mixed and the reactions take place in homogeneous phase. The reaction temperatures are generally higher (around 1300 °C) and more oxygen may be required in some cases. All the heat required for the syngas reaction is supplied by the partial combustion of the fuel and no external heat is required. The reactor often consists of a refractory-lined open pressure vessel. As soot formation can occur in the effluent, provision for soot removal by scrubbers is prudent.

Both Shell and Texaco have supplied technology for natural gas conversion by POX gasification for decades, and lately Lurgi is also promoting a multipurpose gasification process (MPG), which is available in a version adapted for natural gas.

(d) Catalytic Partial Oxidation (CPO) The production of syngas based on heterogeneous catalytic reactions is normally referred to as catalytic partial oxidation (CPO).

CPO represents an alternative in syngas production; the principles are similar to that of ATR, with the difference that all the reactions occur in heterogeneous phase and that the temperature is usually lower in CPO than in ATR.

The potential advantages of CPO for large-scale syngas production have been intensively investigated by Conoco-Phillips and other companies. Eni has developed the SCT-CPO (short contact time catalytic partial oxidation) for the production of hydrogen from methane or light hydrocarbons. The technology can be used for both H_2 generation in refineries and syngas production directly at the well-site for GTL applications [6]. Compared to ATR or POX, a lower oxygen consumption, essentially no use of steam, and operating temperatures under 1000 °C are among the advantages claimed in the literature for this technology. Catalysts are usually based on noble metals (Pt, Pd, Rh, and Ir). The choice of the catalyst should be driven by its cost, activity, and natural gas conversion. For example, the severe catalyst operating conditions and the importance of suppressing soot formation with minimal steam consumption appear to favor the use of rhodium (Rh)-based monolithic catalysts.

(e) Heat Exchange Reforming (HER) A two-stage process for methane reforming, heat exchange reforming (HER), is under investigation; this seems a promising approach, but has not yet been commercialized. In the first stage, SMR reactions occur inside the tube side of a heat exchanger filled with catalyst. The second stage typically consists of a conventional ATR. The heat of reaction required by the endothermic SMR reaction is provided by the hot effluent syngas from the ATR second stage. Advantages of the technology are lower capital costs and higher thermal efficiency. The companies that developed this technology (JM-Katalco, Haldor Topsøe, Uhde) initially targeted it for ammonia and methanol plant applications, but it should be suitable for GTL plants too.

A particular type of HER is the GHR (gas-heated reformer), where the reformer is heated by process gas. HERs may be classified in different ways depending on the process concept used (i.e., series or parallel arrangement). Synetix has also commercialized a second-generation HER called advanced gas-heated reformer (AGHR) with a different tube design for a methanol plant in Australia.

(f) Compact Reforming (CPR) Compact reforming is an innovative approach to SMR that has been demonstrated by Davy Process Technology and BP. The reactor design is similar to a conventional shell and tube heat exchanger: the SMR reactions occur at the tube side, containing a conventional Ni-based catalyst. Heat for the endothermic SMR reactions is provided on the shell side, where the tubes are heated by a fuel and air mixture. Compared to a conventional SMR, less catalyst is required and the design is more compact, leading to lower investment cost.

The reduced space requirements make this technology suitable for offshore application for the conversion of relatively small gas fields. A drawback of the current state of this technology is that a large number of parallel units would be required for a world-scale GTL plant. The Danish company Haldor Topsøe A/S has supplied several hydrogen units with the compact convection reformer concept

(HTCR). This reactor consists of a vertical, refractory-lined vessel, containing the tube bundle with several bayonet tubes surrounded by a flue gas guiding tube. Below the vertical section there is a horizontal combustion chamber containing the burner. Uhde has a proprietary design called "combined autothermal reforming," which combines in a single step the reforming and the noncatalytic partial oxidation.

2.2.2
Starting from Solid Feeds (CTL, BTL, and WTL)

Gasifiers can be designed to gasify almost any kind of organic feedstock, including many types of wood, agricultural residues, peat, coal, anthracite, oil residues, and municipal solid waste. The oil crisis of the 1970s provided strong motivation for using coal: the reserves were abundant and more widely spread over the world than crude oil. In this period, several coal gasifiers were developed and commercialized. In recent years, the growing interest in renewable energies has again led to similar technologies, but focused on biomass gasification.

An interesting difference between coal and biomass lies in their composition: woody biomass contains typically around 50 wt% carbon and 45 wt% oxygen, whereas coal contains 60–85 wt% carbon (depending on coal rank) and 5–20 wt% oxygen. Because of the high oxygen content, biomass gasification requires less oxygen. The coal or biomass fuels differ in many properties, such as heating values, proximate analysis (fixed carbon, volatile material, ash content, and moisture content), ultimate analysis (amounts of carbon, hydrogen, oxygen, sulfur, nitrogen, chloride, and other impurities), and sulfur analyses. The change in composition from biomass to coal is illustrated with a diagram developed by Van Krevelen (Figure 2.3) [8].

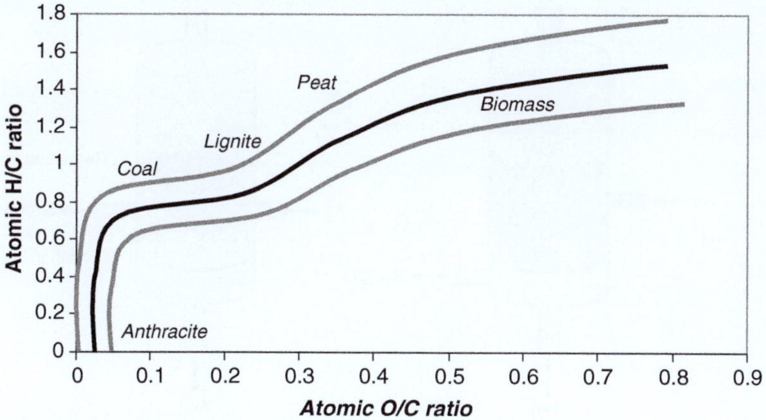

Figure 2.3 Atomic ratios for various solid fuels.

Recently, plastics gasification has been investigated in the context of converting waste to energy and many studies have been carried out on syngas production from plastics such as polyethylene.

Several gasification technologies are now available, depending on the feedstock specifications (size and chemical composition). However, the reactor type (fixed or moving) and the heating method are the main elements characterizing the families of gasifiers. These are grouped into four main technologies:

a) Fixed bed
b) Fluidized bed (FB) (bubbling or circulating)
c) Entrained flow
d) Indirect

(a) Fixed Bed The fixed bed gasifier is the oldest process for conversion of solids, both coal and biomass, into syngas. They are usually operated at temperatures around 1000 °C.

The main advantage of fixed bed gasifiers is the simple design, while the disadvantages lie in the production of a gas with low heating value (HV) and high tar content. Depending on the direction of airflow, the gasifiers are classified as updraft, downdraft, or cross-draft (Figure 2.4) [9].

(b) Fluidized Bed Fluidized bed gasifiers have been extensively used for coal gasification over many years and are now used for biomass: their advantage over fixed bed gasifiers is the uniform temperature distribution achieved in the gasification zone. The reacting bed is formed by fine-grained material into which some gas (air, oxygen, steam, or a mixture) is introduced. The gas is the fluidizing agent and it ensures good mixing of the hot bed material, the hot combustion gas, and the

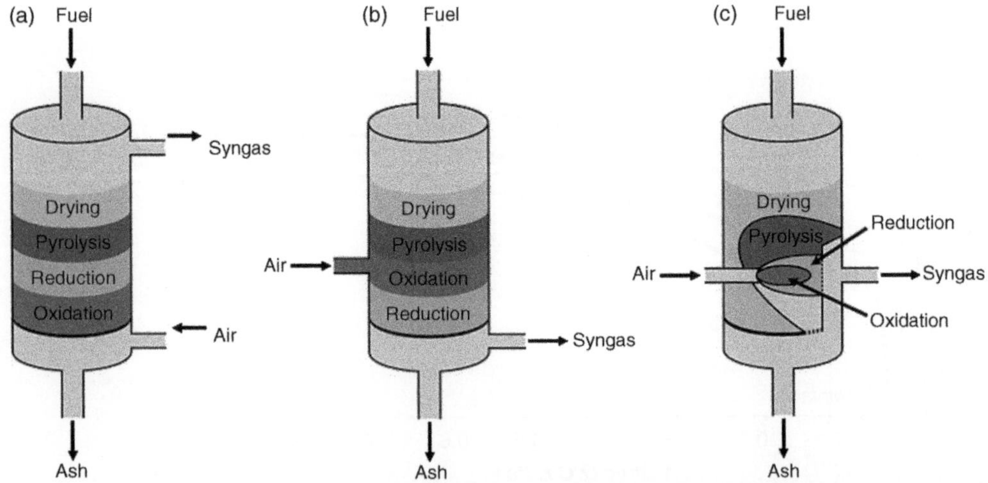

Figure 2.4 Fixed bed gasifiers: updraft (a), downdraft (b), and cross-draft (c) [10].

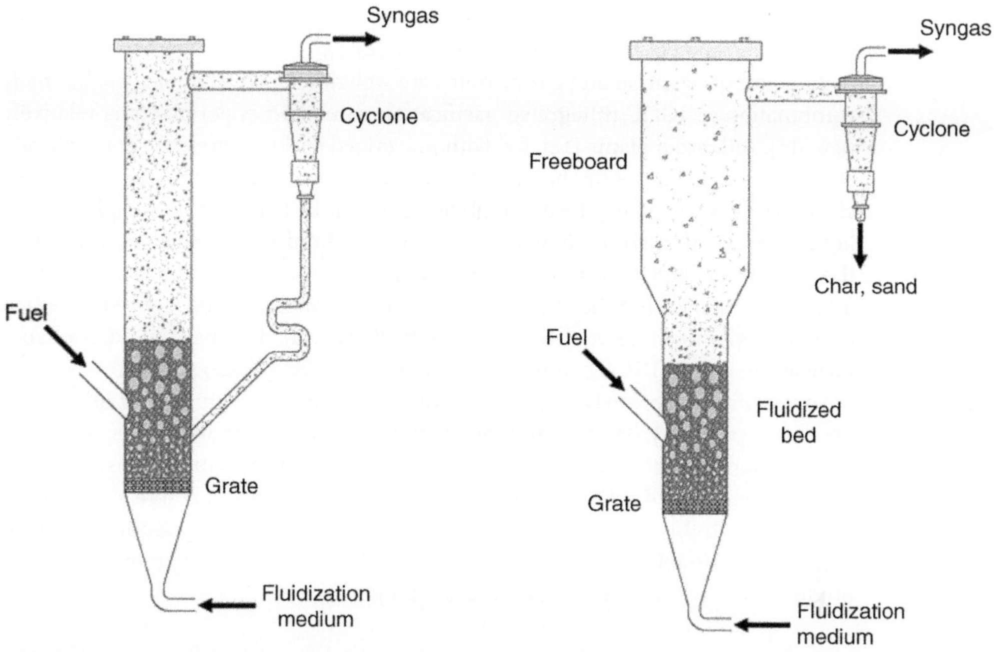

Figure 2.5 Fluidized bed, circulating and bubbling [10].

biomass feed, and consequently achieves a uniformity of temperature. The two main types of FB gasifier in use are the circulating fluidized bed (CFB) and the bubbling fluidized bed (BFB) (Figure 2.5). A new type of FB technology is the fast, internally circulating gasifier (FICFB), which is still under development.

In the CFB, the solid is dispersed between the reaction vessel and a cyclone separator, where the ash is removed, while the bed material and char are cycled back to the reaction vessel. This type of gasifier is suitable for high-capacity throughputs and can be operated at elevated pressures, clearly an advantage when the applications are at high pressure, as in the Fischer–Tropsch synthesis loop.

The BFB gasifier is based on a vessel with a grate at the bottom that, this confines, in the upper part of the reactor, the fine-grained material fluidized with an airflow that is fed at the bottom. The temperature is maintained around 700–900 °C by adjusting the air/biomass ratio. The gas produced has a low tar content, typically < 1–$3\ g/Nm^3$.

The major disadvantage of the FB gasifiers is the risk of slagging of the bed material by the ash content of the biomass. The alkali metal content can be used as an indicator of this risk, for example, biomass derived from herbaceous plants has a higher risk of bed agglomeration.

(c) Entrained Flow Gasifier Entrained flow (EF) gasifiers generally use gas, powder, or slurry as fuel, mixed with a steam/oxygen stream at high temperatures (>1200 °C), which makes this technology less suitable for biomass feedstocks. The biomass, in fact, must first be ground to powder or in some cases pyrolyzed to gas, pyrolysis oil, and coke, before conversion to a slurry.

Both coal gasification and gas turbines are well-developed technologies, but their combination in IGCC (integrated gasification combined cycle) plants is relatively new [11]. Industrial plants [12], fed with pulverized coal or petroleum coke, already exist (e.g., the Shell gasifier-based IGCC at Buggenum in the Netherlands). Almost all can work with a slurry feed at high ash content. In Italy, four IGCC plants with liquid feedstock entrained flow gasifiers are in operation: the Shell tar gasifier in the Eni refinery at Sannazzaro de Burgundi (Pavia), two with Texaco tar gasifiers in SARLUX, Sarroch (Cagliari) and API Falconara (Ancona), and one with Texaco asphalt gasifiers in ISAB ENERGY Priolo (Siracusa). In Spain at Puertollano, Elcogas operate an IGCC complex based on coal Prenflo (Uhde) gasifier.

The gasification process usually takes place at high temperatures (>1200°C) and pressures (>20 bar). These conditions result in a leach-resistant molten slag and, most important, an almost tar-free syngas, which simplifies the downstream gas cleaning steps and does not require intermediate compression before synthesis. A drawback with this type of gasifier is that it produces appreciable amounts of heat that is less desirable and must be utilized for electric power generation and steam production to achieve reasonable process efficiencies.

(d) Indirect Gasifier The char indirect system consists of two different reactors: the gasifier and the combustor (Figure 2.6). Both are CFB gasifiers and the bed material, usually sand, is removed from the cyclones and exchanged between the two

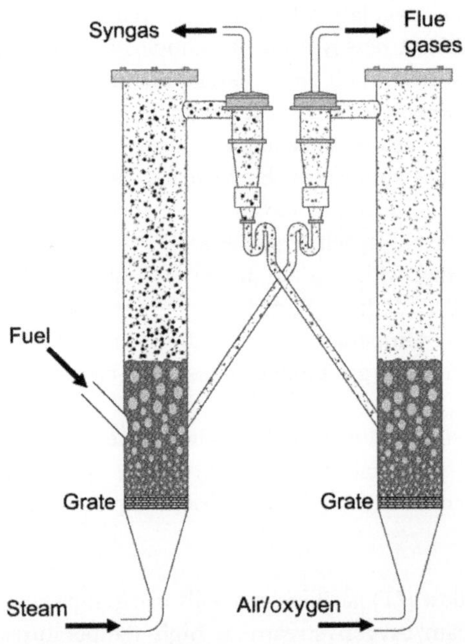

Figure 2.6 Char Indirect, two stages of gasifier [10].

Table 2.3 Comparison of solid gasification technologies.

Technology	Advantages	Disadvantages
Fixed bed	• Simple design	• Production of a low HV gas with high tar content • Scale-up limited
Fluidized bed	• Temperature uniformity • Particle size (<50 mm), suitable for biomass too	• Risk of slagging • Low temperatures limit reaction kinetics: more suitable for biomass and low-rank coal than hard coal
Entrained flow	• High temperatures and short residence time • Compact design	• Particle size very small (<100 μm), not totally suitable for biomass

reactors. In the combustor, the residual char received from the gasifier is burned and the sand is heated. When the hot sand enters the gasifier, it provides the heat necessary for gasification.

Some of the characteristics of different gasification technologies are summarized in Table 2.3.

2.3 Water-Gas Shift Reaction (WGSR)

The water-gas shift reaction was discovered by the Italian physicist Felice Fontana in 1780, but it became industrially important when it was used to produce hydrogen for the Haber–Bosch ammonia synthesis.

In the reversible WGSR, carbon monoxide reacts with water over a catalyst to produce hydrogen and carbon dioxide [13].

$$CO + H_2O = H_2 + CO_2, \quad \Delta G_{298} = -28.64\,\text{kJ}, \quad \Delta H = -41.1\,\text{kJ/mol} \tag{2.6}$$

The reaction is used industrially together with steam reforming of methane or higher hydrocarbons to produce hydrogen. The reaction is slightly exothermic (Equation 2.6) and can be considered in equilibrium under most conditions.

$$\log k_p = \log \frac{pCO_2 \cdot pH_2}{pCO \cdot pH_2O} = \left(\frac{2073}{T} - 2.029\right) \tag{2.13}$$

The temperature dependence of the equilibrium constant K_p is described by Equation 2.13 [14, 15].

The equilibrium constant decreases with the increase of temperature (Figure 2.7) by a factor of about 80 when the temperature is increased from 200 to 600 °C, thus favoring higher CO conversions at lower temperatures. Industrially a two-stage process – a high-temperature shift (HTS) and a low-temperature shift (LTS) – is

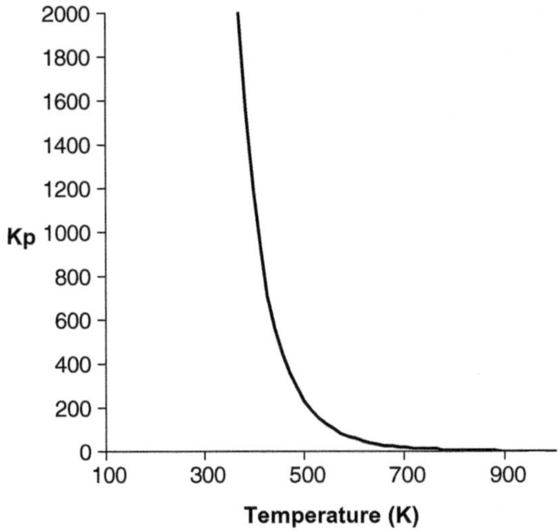

Figure 2.7 Change of WGSR equilibrium constant K_p with temperature.

used to take advantage of the kinetics and thermodynamics. In stage one, the HTS reactor operates at 320–500 °C using a long-lived iron–chromium catalyst. The reaction is rapid due to the fast kinetics at high temperatures, but conversion is limited by the thermodynamic equilibrium to a CO concentration of 4–5 wt% in the gas produced. In the second stage, a copper-based catalyst is usually employed in the LTS reactor at about 200 °C; at the reactor outlet, the CO concentration is less than 0.2 wt%.

The WGSR Fe_2O_3–Cr_2O_3 catalyst that has been used for more than 60 years works very well at high temperatures, but it has poor performance at low temperatures. These catalysts are used for their poison resistance and good selectivity and usually contain 80–90% of Fe_2O_3, 8–10% of Cr_2O_3, and small quantities of other promoters such as Al_2O_3, copper, or MgO. The Cr_2O_3 is very important to prevent the iron oxide sintering (especially during the activation phase) for good catalyst activity and to increase catalyst life from months to years. Chromium ions occupy tetrahedral sites in a magnetite spinel lattice; this strains the spinel structure, leading to iron oxide particle size decreases and surface area increases, thereby enhancing catalytic activity.

Due to the low activity of Fe–Cr catalysts below 350 °C, Cu–Zn oxides are the catalysts used for LT WGSR. However, they are extremely susceptible to sulfur poisoning, and thus sulfur species in the feed must be lower than 0.1 μg/g. Sulfur-tolerant shift catalysts, based on a cobalt–molybdenum formulation, are now available and can be operated at both high and low temperatures. As they are used in the sulfide form, a minimum level of sulfur must be present in the syngas feed.

When the FT reaction is run as part of a GTL system, the syngas has a H_2/CO ratio very close to 2 and methane reforming is enough to adjust it. In the CTL and BTL processes, the ratio depends on the gasifier type, the operating temperature, and the amount of steam added to the feedstock, but it is in the range of 0.6–1.7. In this case, the H_2/CO ratio is adjusted by reducing the carbon monoxide and increasing the hydrogen level, using the WGSR (Equation 2.6).

An important factor determining the choice of cobalt or iron catalyst for a GTL conversion is the technology of the syngas production process as that determines the H_2/CO ratio. Commercially available technologies produce syngas from methane with a H_2/CO ratio of 1.7–3.8; however, Fe catalysts, because of their WGSR activity, perform best at H_2/CO ratios of 0.6–1.5. In contrast, as Co-based catalysts do not have WGSR activity, they have optimum performance at higher H_2/CO ratios (1.9–2.2).

When a WGSR unit is present in an XTL plant, the syngas stream leaving the gasifier is cooled and scrubbed to remove tars. When a sulfur-intolerant WGSR catalyst is used, the syngas is fed to the guard column to remove residual sulfur. The syngas is then split into two streams, and steam is added to give the right steam/carbon monoxide ratio before passing to the shift reactors. The amount of steam added is more than stoichiometric, because it also controls any dangerous temperature rise during the exothermic shift reaction. The large steam addition to the syngas entering the shift unit tends to decrease the plant efficiency and increase the size of the unit. For this reason, a fraction, sometimes one-half, of the syngas bypasses the HT shift and recombines with its effluent to be cooled before entering the LT shift reactor [16].

Figure 2.8 shows the effect of the reaction temperature of the WGSR reaction on the equilibrium conversion of CO, H_2O, CO_2, and H_2 in a typical biomass-derived

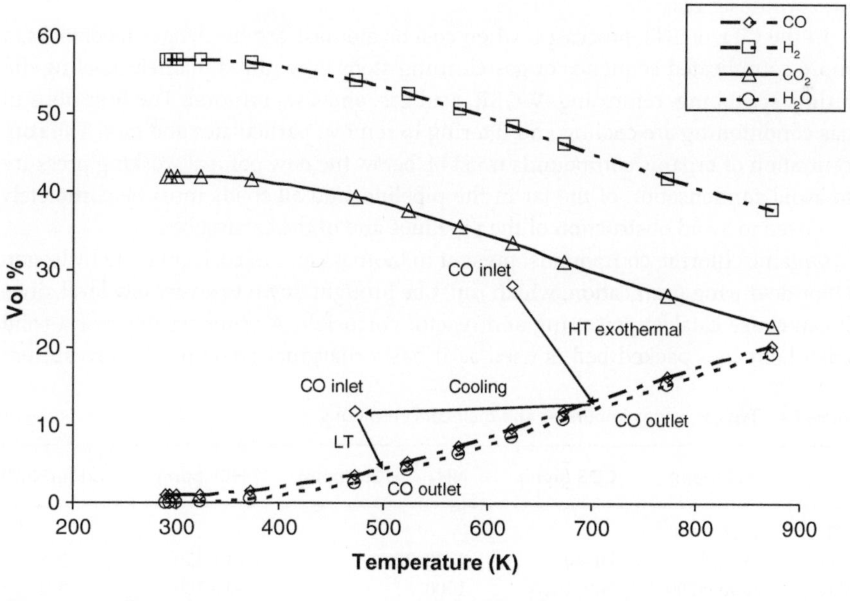

Figure 2.8 Effects of temperature on H_2, CO, CO_2, and H_2O conversion in the WGSR reaction.

syngas. The arrows show the variation of CO concentration if a HTS and LTS combination is used. The temperature rises because of the exothermic shift reaction in the adiabatic reactor bed.

2.4 Synthesis Gas Cleanup

The impurities present in syngas depend on the carbon source used for its production as well as on the reforming or gasification process and include particulates, tars, nitrogen compounds, HCl, H_2S, and COS. Typical impurities and their concentrations in the syngas produced for XTL processes are listed in Table 2.4.

H_2S and COS are strong poisons for the reformer, WGSR, and Fischer–Tropsch catalysts as they deactivate Fe, Co, and Ni catalysts by forming metal sulfides. All sulfur in the natural gas feedstock for a GTL plant must therefore be eliminated before syngas production. The typical process steps are, sequentially, desulfurization, saturation, prereforming, autothermal reforming, and syngas cooling.

In the desulfurization section, organic sulfur compounds are first hydrogenated to H_2S, which is then removed from the natural gas feed by absorption. Before being reformed, the desulfurized natural gas is saturated with process water coming from the FT reactor. Higher hydrocarbons in the desulfurized and saturated natural gas are reformed in the prereformer. In the autothermal reforming section, the prereformed natural gas is mixed with external recycle gas from the FT unit and is partly combusted with pure oxygen (99.5 vol%) and then reformed with steam into synthesis gas.

In the CTL or BTL processes, when coal or biomass are the syngas feedstocks, a more complicated sequence of gas cleaning steps is required, namely, cooling, filtering, scrubbing, reforming, WGSR, acid gas, and CO_2 removal. The first steps in gas conditioning are cooling and filtering to remove particulates and tars. The concentration of organic compounds must be below the dew point at working pressure to avoid condensation of the tar in the pipeline, and all solids must be completely removed to avoid obstruction of the pipelines and of the catalyst bed.

Organic chlorine compounds, present in biomass and in coal, generate hydrogen chloride during gasification, which must be brought down to a very low level since it can cause catalyst poisoning and reactor corrosion. A water scrubber or a solid adsorbent in a packed bed is used as it has a marginal effect on the investment

Table 2.4 Typical syngas impurities and their concentrations.

	H_2S (ppm)	COS (ppm)	NH_3 + HCN (ppm)	HCl (ppm)	Tar (g/Nm^3)
GTL	$5–15 \times 10^{-3}$		—	—	—
BTL	180–350	20–40	2100–3000	130–250	2–5
CTL	2000–5000	500–1000	1000	800–1000	0–2

cost; nitrogen compounds such as NH_3 and HCN are also removed in the water scrubber.

The syngas usually still contains considerable amounts of methane and light hydrocarbons, representing a significant part of the heating value of the gas. The reforming processes then convert these compounds into CO and H_2.

The typical processes for desulfurization entail chemisorption and/or physisorption. In the *chemisorption* process, a solvent reacts with the acid gases to form chemical complexes that, with change in pressure and temperature, dissociate to release the acid gases. The most used solvents are alkanolamines. Although MEA (monoethanolamine, $NH_2CH_2CH_2OH$) has been extensively used in the past, currently MDEA (methyldiethanolamine, $NMe(CH_2CH_2OH)_2$) is the solvent most in use, as it has a high H_2S/CO_2 selectivity and is very stable and less corrosive than primary or secondary amines.

The *physisorption* process uses a solvent to dissolve acid gases: they can again be released from the solvent by pressure reduction or temperature change. The physical processes require more energy for refrigeration than the chemical processes, but the solutions are more stable and can be configured to use both their high selectivity for H_2S and COS over CO_2 and to give high levels of CO_2 recovery. Rectisol^{TM} and Selexol^{TM} are two of the most widely used physical solvent processes. They are licensed by Lurgi/Linde and UOP, respectively. Selexol uses the dimethyl ether of polyethylene glycol (DMPEG) ($Me(OCH_2CH_2)_nOMe$), while Rectisol uses methanol, which clearly has some operating cost advantage over the UOP process. About 75% of the world's syngas produced from oil residue, coal, and waste is purified by the Rectisol process [17]. Solubilities of H_2S and COS in methanol, at process conditions, allow sulfur removal to below 0.1 ppmv. Under the same process conditions, due to its relatively lower solubility, CO_2 can mostly be removed. Figure 2.9 shows the adsorption coefficients in methanol of various gases typically present in syngas as functions of temperature, at 1 atm partial pressure [18].

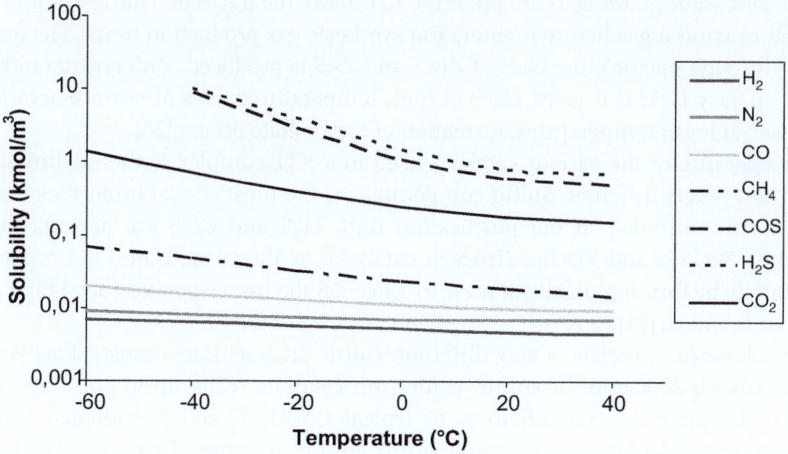

Figure 2.9 Solubility of various gases in methanol.

Table 2.5 Chemisorption and physisorption processes.

Process	MDEA	Rectisol	Selexol
Solvent	MDEA	Methanol	DMPEG
Temperature	43 °C	−40 °C to −60 °C	−20 °C to 50 °C
H_2S–CO_2 selectivity	Good	High	High
H_2S removal limit	4 ppmv	<0.1 ppmv	4 ppmv
Hydrocarbon adsorption	Low	High	High
Main advantages	• CO_2 removal if required	• Very low H_2S removal limit • Effective for COS and CS_2	• Effective for COS and CS_2 removal
Main disadvantages	• Limited removal selectivity	• Higher capital investment • Higher power required	• Higher capital investment • Higher power required

Methanol-based physisorption process can operate selectively to recover hydrogen sulfide and carbon dioxide as separate streams, so that the hydrogen sulfide can be sent to a desulfurizing Claus unit for conversion into elemental sulfur, while the carbon dioxide can be sequestered or used. The solvent from the acid gas absorbed is regenerated by flashing the clean fuel gas and boiling off the methanol. Table 2.5 summarizes the main characteristics of the chemical and physical processes [19].

Solvent methods are not commonly used to treat natural gas streams in GTL plants; zinc oxide guard beds are preferred to remove the traces of hydrogen sulfide present in natural gas before it enters the synthesis gas production units. The gas passes through one or more beds of ZnO, and ZnS is produced. ZnS regeneration is not an easy task: if it takes place at high temperatures, loss of surface area is observed; at lower temperatures, formation of zinc sulfate occurs [20].

The tolerance of the various catalysts used in a XTL complex to the mixture of impurities is very different. Sulfur compounds are the most critical impurities that need to be controlled in the purification unit. H_2S and COS are poisons for reformer, WGSR, and Fischer–Tropsch catalysts, as they deactivate Fe, Co, and Ni catalysts by forming metal sulfides. In Table 2.6 the impurities tolerance for FT catalysts is shown [21].

The tolerance of metals is very different: cobalt catalysts lose almost all activity if they adsorb 2000 ppm of sulfur, while iron catalysts resist up to 20 000 ppm [22–24]. In Table 2.7, the lifetimes of typical Co/Al_2O_3 and precipitated iron catalysts were calculated for a syngas sulfur level of 0.1 ppm; the necessary H_2S concentration for a year's catalyst life is also shown.

Table 2.6 Impurities tolerance for FT catalysts.

Impurity	Tolerance level
Sulfur	<4 ppb for cobalt catalyst <1 ppm for iron catalyst
Halides	<10 ppb
Nitrogen (NH_3, NO_x, HCN)	<1 ppm
Tars	Below dewpoint at FT pressure
Particulate	Absent

Table 2.7 Fischer–Tropsch catalyst lifetimes.

	Lifetime (days) with H_2S 0.1 ppm	H_2S concentration for 1 year catalyst lifetime
Co/Al_2O_3	83	0.02 ppm
Fe precipitated	830	0.2 ppm

Ideally, the sulfur content in the syngas must be zero, but as gas cleaning is very expensive, there is a trade-off between the catalyst cost and the investment and operating cost of the gas cleaning facility. Usually for FT cobalt catalysts, a very efficient sulfur removal is justified by the cost and their sensitivity to sulfur poisoning. Nevertheless, this is not the only element to consider, for instance, the interaction between sulfur and catalyst also depends on the fluidodynamics in the FT synthesis reactor. In slurry bubble column reactors, sulfur is deposited on all catalyst particles, while in the fixed bed reactor, sulfur is preferentially adsorbed on the catalyst at the reactor entrance in the first section of the catalyst bed [25]. Thus, the FT synthesis reactor must play a role in the size of the cleaning facilities.

2.5 Thermal and Carbon Efficiency

In the FT synthesis for making liquid fuels, both the thermal efficiency and the proportion of carbon atoms that are retained are important. The FT reaction (Equation 2.3) where all the carbon atoms are preserved is highly exothermic [26].

$$CO + 2H_2 = -CH_2- + H_2O, \quad \Delta H^\circ = -154.1\,kJ/mol \tag{2.3}$$

$$CO + 2H_2 + {}^3\!/_2O_2 = CO_2 + 2H_2O, \quad \Delta H^\circ = -766.5\,kJ/mol \tag{2.14}$$

$$-CH_2- + {}^3\!/_2O_2 = CO_2 + H_2O, \quad \Delta H^\circ = -612.4\,kJ/mol \tag{2.15}$$

If the energy contained in the initial components (Equation 2.14) is compared with that in the final products (Equation 2.15), we find that if the reaction were quantitative, a carbon yield of 100% and thermal output of 79.9% would be achieved. Although the calculation is at the standard state for the ideal gases, the values in real conditions, 200–250 °C and 20–25 atm, are not significantly different. The heat produced in the FT reactor is exploited by raising medium pressure steam that is utilized in the GTL complex for electrical power and feeding the steam network; 80% is the theoretical thermal efficiency limit for the FT reaction. In an overall balance for the production of liquid fuels from other carbon sources, two other factors must be taken into account: the efficiency in the transformation of primary fuels, that is, natural gas, biomass, or coal, into synthesis gas and the selectivity to useful hydrocarbons within the FT reactor where, inevitably, other products are formed. However, it is worth remembering that despite the FT reaction requiring a H_2/CO ratio of 2, even in reactors with cobalt catalysts (where the water-gas shift reaction (WGSR) is minimal), the ratio needs to be around 2.1 to allow other reactions such as methane formation.

The theoretically simpler and more efficient way to obtain syngas from natural gas is by partial oxidation (Equation 2.16) [6, 27]:

$$CH_4 + 0.5O_2 = CO + 2H_2, \quad \Delta H^\circ = -36.0\,\text{kJ/mol} \tag{2.16}$$

As this is slightly exothermic, to keep the carbon to a theoretical thermal efficiency of 95%, the reaction requires a stoichiometric excess of oxygen, which makes it less attractive [28]. Thus, to feed the FT reaction, autothermal or steam reforming is preferred.

In autothermal reforming [29], the energy needed to reach the temperature necessary for the reaction (1000–1100 °C) is provided by combustion of part of the natural gas with a slight excess of oxygen needed to sustain the reaction at these temperatures. The reactions involved are as follows:

$$CH_4 + 1.5O_2 = CO + 2H_2O, \quad \Delta H^\circ = -519.7\,\text{kJ/mol} \tag{2.17}$$

$$CH_4 + H_2O = CO + 2H_2, \quad \Delta H^\circ = +205.8\,\text{kJ/mol} \tag{2.10}$$

$$CO + H_2O = CO_2 + H_2, \quad \Delta H^\circ = -41.0\,\text{kJ/mol} \tag{2.6}$$

These reactions are suitably adjusted to obtain the desired H_2/CO ratio. For example, with steam and oxygen to carbon feed ratios $H_2O/C = 0.36$ and $O_2/C = 0.57$, a ratio of $H_2/CO = 2.14$ and a conversion of methane higher than 90% are obtained (Equations 2.6, 2.10, and 2.17).

The overall reaction can be written as Equation 2.18:

$$8CH_4 + 5O_2 = 7CO + CO_2 + 15H_2 + H_2O, \quad \Delta H^\circ = -812\,\text{kJ/mol} \tag{2.18}$$

Thus, the heat developed per mole of methane is 102 kJ/mol, sufficient to support the process. Unreacted methane and that produced by the FT reaction can be separated downstream and recycled to the ATR reactor.

Starting from coal, the carbon yield as well as the thermal efficiency are lower [30]. The main reaction of coal gasification (Equation 2.19) [31]

$$C + H_2O = CO + H_2, \quad \Delta H^\circ = +131.4\,\text{kJ/mol} \tag{2.19}$$

being endothermic, must be thermally supported, and produces a hydrogen-poor syngas. This is not suitable for all FT reactions and additional hydrogen has to be provided. The main reactions then are the combustion of coal (Equation 2.20) and the WGSR (Equation 2.6):

$$C + O_2 = CO_2, \quad \Delta H^\circ = -393.3\,\text{kJ/mol} \tag{2.20}$$

$$CO + H_2O = CO_2 + H_2, \quad \Delta H^\circ = -41.0\,\text{kJ/mol} \tag{2.6}$$

These reactions can be combined in different ways depending on the type of gasifier (fixed bed, fluidized bed, or entrained flow). While the combustion reaction always takes place in the gasifier section to provide the necessary heat, the WGSR may also occur in a subsequent reactor to increase the H_2/CO ratio or, in the case of iron catalysts, in the FT reactor itself. In this case, we can consider the FT reaction as the sum of Equations (2.8) and (2.14):

$$2CO + H_2 = -CH_2- + CO_2, \quad \Delta H^\circ = -195.1\,\text{kJ/mol} \tag{2.21}$$

The required H_2/CO ratio becomes 0.5. Anyway if a H_2/CO ratio of 2.1 is desired, the overall production of syngas from carbon can be written as Equation 2.22:

$$5C + 2.05O_2 + 4H_2O = 1.9CO + 3.1CO_2 + 4H_2, \quad \Delta H^\circ = -461.9\,\text{kJ/mol} \tag{2.22}$$

Therefore, each carbon atom develops 92.4 kJ of energy with an overall thermal efficiency around 77%, calculated as the calorific value of the produced gas compared to that of the incoming coal. If the heat recovery from cooling the exit gases is included (since the reaction is carried out between 800 and 1200 °C), a total thermal efficiency of 90% or more can be reached [32]. Coal gasification also produces a considerable quantity of methane and light hydrocarbons (5–10 vol%) [33], which can be converted into hydrogen and carbon monoxide in a steam reforming reactor that is also used to transform the light hydrocarbons coming from the FT synthesis.

Biomass can produce results similar to coal: attention must again be paid to methane reforming, production of light hydrocarbons, and recovery of the heat of the gases. That heat can, for instance, be used to reduce water in the biomass from around 50% to 10–15%. The reactions involved are similar, but with a different starting material composition [34], where the overall reaction can be written as Equation 2.23:

$$5CH_{1.5}O_{0.7} + 1.75O_2 + 0.75H_2O = 2.15CO + 2.85CO_2 + 4.5H_2, \quad \Delta H^\circ = -492.9\,\text{kJ/mol} \tag{2.23}$$

This gives the energy developed per mole of biomass as 98.6 kJ/mol, a thermal efficiency calculated on the calorific value of gas and dry biomass of approximately 77% and a carbon efficiency of 43%.

The overall yield of the XTL process via FT synthesis is in turn influenced by the different reactions that can occur within the reactor. Since the aim is usually to produce heavier hydrocarbons, the CO_2, methanation, formation of C_2–C_5

hydrocarbons, and the formation of light alcohols (Equations 2.21–2.25) are frequently neglected:

$$CO + 3H_2 = CH_4 + H_2O, \quad \Delta H^\circ = -205.8\,\text{kJ/mol} \tag{2.24}$$

$$nCO + 2nH_2 = C_nH_{2(n+1)}O + (n-1)H_2O, \quad \Delta H^\circ/n = -147.0\,\text{kJ/mol} \tag{2.25}$$

Carbon dioxide production by the WGSR is accompanied by production of hydrogen; that, if organized properly, does not affect the overall performance of the process but allows a lower concentration of H_2 at the inlet. In low-temperature reactors, it can reach a concentration of 1–2 vol% [35].

The formation of methane and light hydrocarbons is much greater (10–15 vol%) in a cobalt reactor at low temperature, but can reach 35–45 vol% using an iron catalyst at high temperature [35]. The hydrocarbon products are not completely lost because they are recycled to the reforming reactor; the thermal efficiency of their recovery, as discussed above, is about 80%.

The production of alcohols influences the effective yield of fuel only because the light alcohols, water soluble, leave the process with the wastewater. They make up about 1–2 vol% for cobalt-based FT technology and 4–5 vol% for iron catalysts.

The thermal efficiency of FT synthesis thus falls from 80% to 75% in the case of cobalt catalysts and to 70% for iron catalysts with carbon yields of 95% and 87%, respectively. The overall yield of the process, excluding upgrading operations for waxes and condensates (discussed in later chapters), is summarized in Table 2.8.

When limited only to the syngas generation and FT reaction, these figures lead to overall thermal and carbon efficiencies for a typical GTL process in the range of 70% and 80–85%, respectively. The carbon efficiencies of CTL and BTL are considerably lower (35–40%) due to the lower thermal content of a carbon atom in coal and biomass compared to natural gas and hydrocarbons (Equations 2.22 and 2.23). For this reason, it is necessary to burn a part of the coal or biomass to provide the energy that is stored in the final product. Consequently, CO_2 produced in the formation of syngas from coal and biomass (Equations 2.22 and 2.15) is more than that produced in its formation from natural gas (Equation 2.18).

In complex plants, other ancillary units are necessary: for air separation, water treatment and desalination, steam and power generation, product upgrading (hydrocracking and hydrotreating), pumps and compressors, and so on. In part

Table 2.8 Thermal and carbon efficiencies of selected XTL processes.

	Thermal efficiency	Carbon efficiency
Methane ATR	95	90
Coal gasification	80	40
Biomass gasification	80	43
LTFT – Co catalyst	76	95
HTFT – Fe catalyst	70	87

they use the excess energy produced as steam by the FT reaction together with supplementary energy obtained by the combustion of light hydrocarbon products or fresh natural gas. In this way, the overall efficiency of a GTL complex is decreased by about 10% and a carbon efficiency of 70–75% is found.

2.6
The XTL Gas Loop

All FT applications involve cooling the gases and vapor at the FT reactor outlet to condense and separate the hydrocarbons and water products from the tail gas. Part of the tail gas is usually recycled back to the FT reactor: this stream is known as *internal gas recycle*. The remaining gas can be recycled back to syngas production or can be used as fuel for the production of hydrogen or to generate electrical power with a gas turbine. This overall gas processing scheme is known as the *gas loop*.

A *closed gas loop* design is used when the tail gas at the outlet is kept at the minimum required for inert gas control; in contrast, an *open gas loop* design is when large amounts of tail gas are used as fuel for hydrogen or power production. A purge from the gas loop is always necessary to prevent a buildup of inert gases.

An open loop design is simpler than a closed loop, because there are fewer elementary units due to the absence of external recycle loops that can complicate design and operation. A closed gas loop design increases the total conversion, but an open loop configuration may be chosen if a cogeneration facility is considered part of the design and, specifically, if the FT tail gas has to be used for electricity generation. In such a case, the tail gas may be treated to remove some CO_2 (optionally preceded by a CO shift to increase the hydrogen content) before it is sent to a gas turbine. In addition, a heavy ends recovery unit could be added to separate higher hydrocarbon compounds from the FT tail gas prior to its use as fuel.

When the FT tail gas is recycled, a cryogenic unit is normally included to separate valuable products; H_2O and CO_2 must of course be removed prior to cryogenic separation to avoid problems with solids formation.

In the following section, we describe the gas loop for a HTFT synthesis with a coal bed gasifier and with a natural gas feed, using an iron catalyst, and a LTFT synthesis with natural gas using a cobalt catalyst.

2.6.1
Gas Loop for HTFT Synthesis with a Coal Gasifier

As an example of a gas loop, we consider the one used by Sasol in South Africa for both open and closed loops [36]; the latter is illustrated in Figure 2.10. In the Secunda plant, the syngas is produced by a Sasol–Lurgi fixed bed coal gasifier, and after cleanup it is fed to the Synthol reactor, a transported fluidized reactor, with a H_2/CO ratio as low as 1.7 and a methane content of about 12 vol% [37]. Since HTFT synthesis has a high methane selectivity, it is more appropriate to use an autothermal methane reformer (ATR) that can share the same air separation unit (ASU) used to produce the oxygen necessary for the coal gasifier.

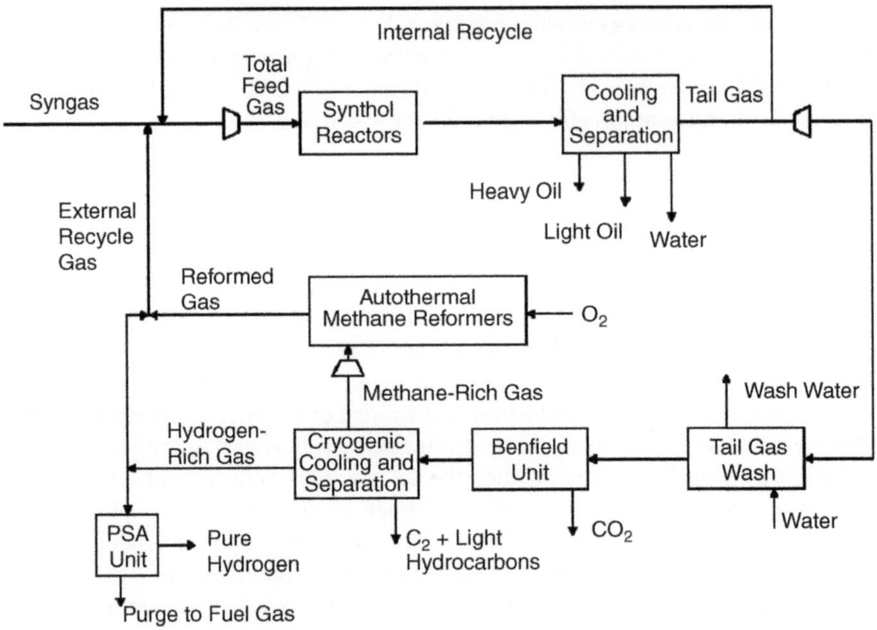

Figure 2.10 Sasol–Lurgi gasifier HTFT gas loop.

Total removal of CO_2 is required in the Benfield unit before the cryogenic unit (cold box). The inert gases are purged via a pressure swing adsorption unit (PSA) and some methane-rich gas is used as fuel. In case of bottlenecks in any of the units in the gas loop, it is possible to vary the gas to fuel ratio or the hydrogen exit.

The gas fed to the FT reactor comes from several streams (syngas, reformed gas, hydrogen-rich gas, and internal recycle), and to ensure optimum performance, the fresh feed must be kept in stoichiometric balance.

If carbon is in excess, it is possible to correct it by decreasing the conversion in the FT reactor, with the effect of increasing the carbon dioxide that is removed in the Benfield unit. If the ratio $P_{CO}/P_{H_2}^2$ increases in the reactor, with a rapid deterioration of the Fe catalyst due to carbon deposition [37, 38], the internal recycle can be decreased; in contrast, if the hydrogen is in excess, the syngas rate can be decreased or some hydrogen-rich gas can be purged from the loop.

2.6.2
Gas Loop for HTFT Synthesis with a Natural Gas Feed

In the Mossel Bay, South Africa, GTL plant, PetroSA, convert natural gas into liquid hydrocarbon fuels using an iron-based catalyst technology with a circulating fluidized bed reactor (Figures 2.10 and 2.11).

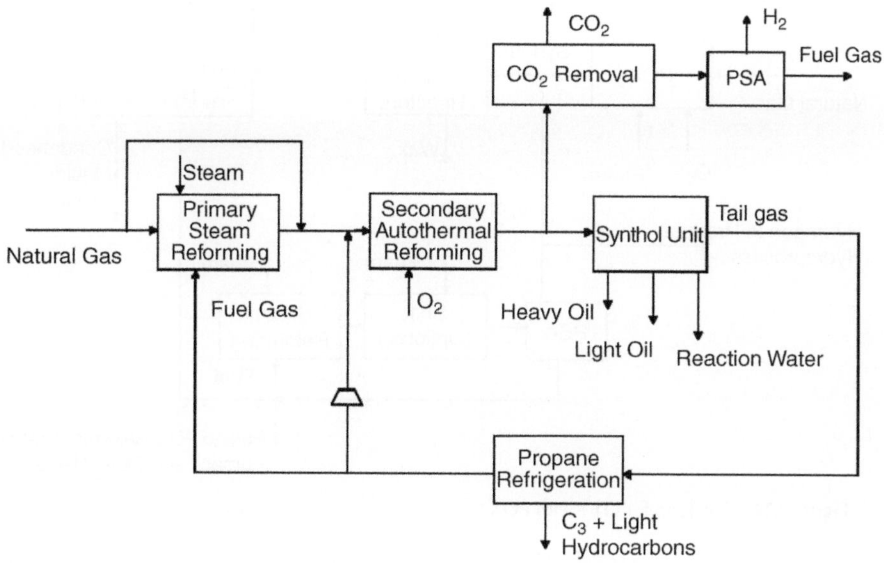

Figure 2.11 Mossel bay gas loop block low diagram.

It would be possible to use autothermal reforming only with low steam to carbon ratios (below 0.6) and to use the tail gas, after CO_2 removal, mixed with the feed to adjust the stoichiometric balance.

With the HTFT process, there is a moderately high selectivity to C_2 hydrocarbons that could be worth recovering for ethylene production. Cryogenic cooling, after CO_2 removal, is the normal way to remove C_2 fractions: this favors an open loop plant design since the compression to recycle the tail gas could be expensive. Without recycling the unconverted reactants, the use of a second reactor in series, after water separation, may be appropriate.

2.6.3
Gas Loop for LTFT Cobalt Catalyst with Natural Gas Feed

Since the cobalt FT catalyst has negligible activity for the water-gas shift reaction, the stoichiometric consumption of reactants depends mostly on the FT reaction itself. The main methane reforming processes for producing syngas in a GTL plant are POX and ATR (Section 2.2.1). With a natural gas feedstock, POX produces a syngas with a H_2/CO ratio below that required, while the ATR can produce a syngas with more hydrogen than necessary, so the H_2/CO ratio is corrected by a recycle stream or a gas separation unit. The LTFT process has a low C_2 hydrocarbons production and only the C_{3+} hydrocarbons can be retrieved from the tail gas. With POX, a parallel steam reforming can be used to adjust the syngas composition. A possible gas loop scheme is illustrated in Figure 2.12.

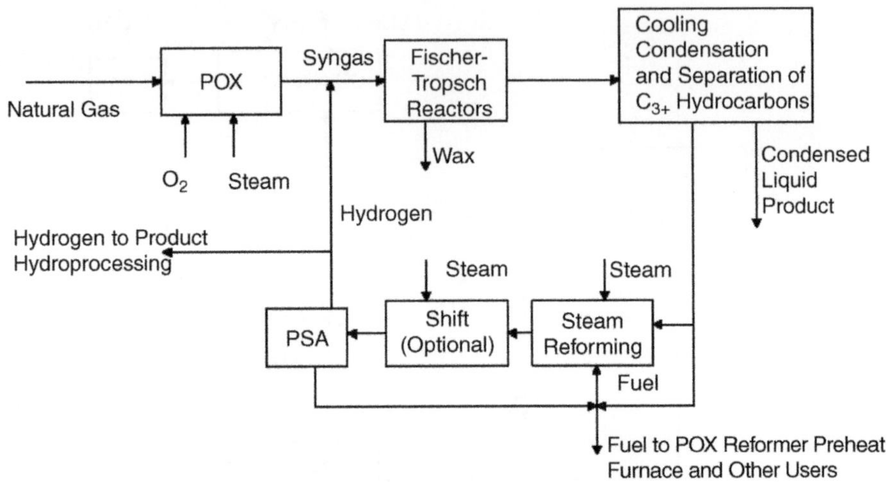

Figure 2.12 Gas loop for LTFT with POX.

In the ATR FT process, the external recycle is primarily based on carbon dioxide and methane, and the carbon dioxide reacts in the ATR unit decreasing the syngas H_2/CO ratio. In this technology, the fuel gas is usually in excess and can be used to heat the furnace of a conventional steam reformer reducing, at the same time, the external recycle: the optimum overall design is reached when there is no excess of fuel gas.

Eni has developed proprietary FT and upgrading technology jointly with IFPEN (*IFP Energies nouvelles*) and AXENS, its process and licensing branch, in a close collaboration, which started in the beginning of 1996 [39]. Several options were considered for syngas production, the one preferred by Eni is based on the ATR integrated in the gas loop scheme shown in Figure 2.13:

The gases leaving the FT reactor are cooled to condense water and heavy hydrocarbons, and they are then further cooled to separate the C_{4+} hydrocarbons ("light ends" for upgrading) and some additional water. The uncondensed gas mainly contains C_1, C_2, C_3 hydrocarbons, H_2, CO, and CO_2. Such a stream is partially recycled to the ATR in an external recycle, while the rest is purged for fuel usage.

Particular attention has been focused on optimizing the efficiency of the whole process. The overall energy production comes from three main sources:

- fuel produced by the process as off-gas,
- very high-pressure steam from synthesis gas production, and
- medium-pressure steam from FT synthesis.

The GTL plant also consumes energy for the following uses:

- as fuel, mainly for synthesis gas production,
- as motive steam (for turbines) mainly for air separation, and
- power generation for plant requirements.

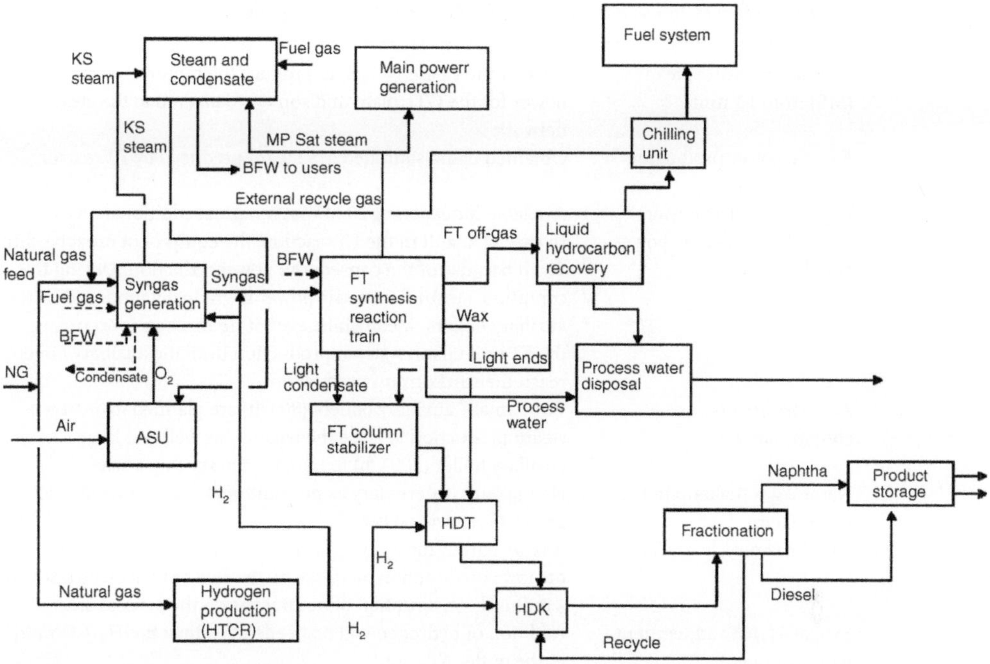

Figure 2.13 Eni gas loop integration.

To optimize the overall GTL plant efficiency, there is close integration in three main units, the so-called gas loop integration:

- syngas generation unit (e.g., ATR),
- FT section, and
- utility system (e.g., fuel and steam system).

The integration has focused on the selection of the best configuration for efficiency, fuel consumption, reliability of operation, and investment costs, by analyzing the different possibilities for using the process off-gas; this is for process heating, steam generation, and recycle to ATR, producing medium-pressure steam (10–12 atm) and very high-pressure steam (107–113 atm). The main configuration is chosen depending on how these energy sources can be used in the above three units.

For the hydrogen production unit, an optimal solution is provided by a Haldor–Topsøe convective reformer (HTCR). This technology is recommended for medium-scale hydrogen production (5000–30 000 Nm^3/h), as its main characteristic is that it absorbs almost all the heat released by the burner in the process so that the intake of natural gas as fuel is minimized [40].

Table 2.9 lists the main technical characteristics for the Eni GTL plant for 17 000 bpd of liquid products (naphtha and diesel) for steam and power generation. For the same plant, Figure 2.14 shows a simplified steam flowchart representative of the integration between the process and the utility systems.

Table 2.9 Main technical features of the Eni GTL "gas loop" integration.

Medium pressure steam (MS) from FT unit	Most of the steam is used to produce the required electrical power for the GTL plant and some is delivered to the steam network
Electric power production for internal use	Obtained using saturated MS steam produced by FT reactor
Export of electric power	Available for export of 20 MW at marginal additional investment
Reliability of electric power supply	In case of a fault in the FT reaction, the equipment must be shut down because of the consequent power reduction. During this condition, steam for electric power is gradually produced by the auxiliary boilers. Meanwhile. even if no reaction is occurring, the FT unit ensures steam production until the auxiliary boilers reach their maximum production
Auxiliary boilers configuration	Two "small" auxiliary boilers (80 t/h) are planned to ensure steam production to close the balance, as well as a large auxiliary boiler (200 t/h) to manage the start-up phase
Natural gas makeup in the fuel system	No makeup is necessary as no additional steam is needed to generate electric power
FT off-gas recycle to ATR	A large part of the FT off-gas can be recycled to the ATR due to the low consumption of off-gas in the fuel system. This leads to a reduced consumption of natural gas on the process side
Syngas H_2/CO adjustment with hydrogen	Addition of hydrogen will be needed to adjust the H_2/CO ratio in the outlet ATR unit

2.7 CO_2 Production and CO_2 as Feedstock

Although the precise details are still under discussion (see Chapter 12), it is generally agreed that the FT reactions start by adsorption of CO on the catalyst and its subsequent deoxygenation/hydrogenation by reaction with chemisorbed hydrogen, affording water together with hydrocarbons [41]. In principle, however, the deoxygenation of carbon monoxide might occur by reaction with a second molecule of CO, affording carbon dioxide. Formally, therefore, both water and carbon dioxide could be regarded as primary products of the FT reaction [42].

Under real FT conditions, however, CO_2 production is quite limited unless the catalyst is also active for the WGSR. As this is not the case for cobalt, CO_2 yields are quite low, typically 1–2 vol% in cobalt-catalyzed FT processes. In contrast, iron can be a very good catalyst for the WGSR and CO_2 yields are significantly higher (up to 50% based on reacted CO) for iron-based FT catalysts. In fact, the iron catalysis of the WGSR is exploited to increase the H_2/CO ratio when a FT process is fed with hydrogen-poor syngas.

CO_2 can also be a significant component of the gas fed to FT plants, particularly when the syngas is obtained by biomass gasification or by reaction of carbon dioxide with methane (dry reforming) (Equation 2.26):

$$CO_2 + CH_4 \rightarrow 2CO + 2H_2 \tag{2.26}$$

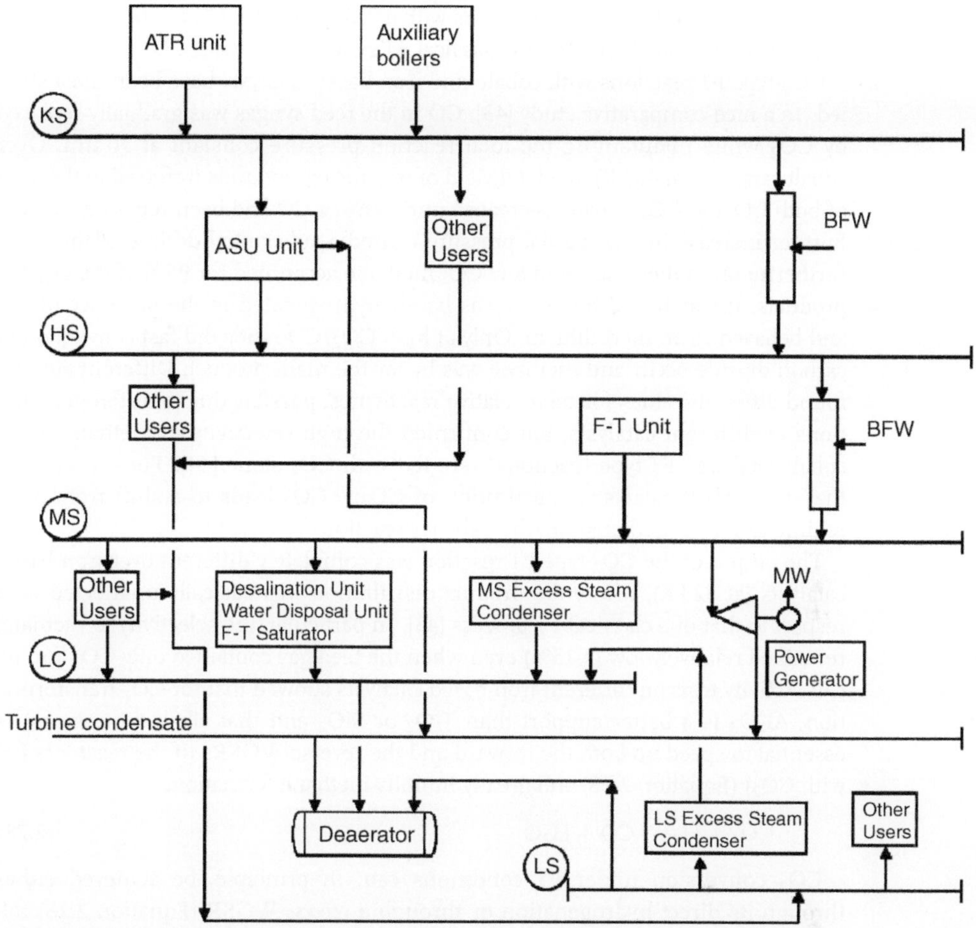

Figure 2.14 Simplified steam flowchart of a typical Eni GTL plant of 850 000 t/a (17 000 bpd) capacity. KS: steam at 119.5 atm, 510 °C; HS: steam at 47 atm, 368 °C; MS: steam at 12.3 atm, 180 °C; LS: steam at 4.5 atm, 200 °C; LC: condensate.

As a result of the increasing concern about the role of CO_2 in environment pollution and global warming, considerable effort has been devoted to the study of FT-like reactions of CO_2. Such reactions fed with CO_2 instead of CO (Equation 2.27) may provide a route to recycle the CO_2 produced in some anthropogenic processes:

$$CO_2 + 3H_2 \rightarrow -CH_2- + 2H_2O \tag{2.27}$$

From the thermodynamic viewpoint, although a CO_2-based FT (Equation 2.27) is less favorable than the CO-based FT process, it is still favorable as formation of additional water provides the chemical energy for the conversion of the very stable CO_2 molecule. Preliminary calculations suggest that at 500 K, the transformation of a CO_2 molecule into a CH_2 unit occurs with a ΔG of about -17 kJ/mol compared to a ΔG of about -38 kJ/mol for the classical FT synthesis fed with CO.

CO_2-type FT reactions with cobalt- and iron-based catalysts have been most studied. In a nice comparative study [43], CO in the feed syngas was gradually replaced by CO_2 while maintaining the total reaction pressure constant at 10 atm. Over cobalt catalysts (at 463 K), the total yield of organic compounds (referred to the sum of both CO and CO_2) slowly decreased until 50% of CO had been replaced by CO_2. Further increase in CO_2 partial pressure caused a substantial decline of the yield; furthermore, in the absence of any CO, methane accounted for 95% of the organic products. It was found that CO_2 was barely hydrogenated in the presence of CO and behaved as an inert diluent. Only at high CO_2/CO ratios did fast conversion of carbon dioxide occur and methane was by far the main product. Different authors found different values for their relative reactivities, possibly due to different conditions or different catalysts, but confirmed the high selectivity to methane when cobalt-catalyzed FT-type reactions were fed with CO_2 alone [44]. Thus, it appears that over cobalt catalysts, substitution of CO by CO_2 leads to a shift from a FT regime to a CO_2 methanation, the Sabatier reaction.

The output of the CO_2-type FT reaction was completely different over iron-based catalysts (at 523 K), where the product distribution was basically unaffected with respect to that of a classical FT process [43]. In particular, the selectivity to methane remained relatively low (<15%) even when the feed gas contained only CO_2 and no CO. Activity tests on different iron-based catalysts showed that for CO_2 transformation, Al_2O_3 is a better support than TiO_2 or SiO_2 and that alkali (potassium) is essential to speed up both the forward and the reverse WGSRs (if the reactor is fed with CO_2) (Equation 2.28) and greatly inhibits methane formation.

$$CO_2 + H_2 \rightarrow CO + H_2O \tag{2.28}$$

CO_2 conversion under FT conditions can, in principle, be achieved either through its direct hydrogenation or through a *reverse* WGSR (Equation 2.28) followed by conventional FT conversion of CO.

The latter pathway is supported by, for instance, the elegant isotopic experiments, already carried out in 1957, where mixtures of CO and radioactive carbon dioxide ($^{14}CO_2$) were fed over iron-based catalysts. The results suggested that carbon dioxide did not build appreciably into hydrocarbons [45]. Despite other more controversial data, the results of new experiments in which $^{13}CO_2$ has been added to syngas supported the earlier work and showed that the hydrocarbon products had negligible ^{13}C content [42]. We may therefore summarize that under conditions in which CO and CO_2 compete as the hydrocarbon precursor, CO_2 is much less reactive than CO, except when close to the WGSR equilibrium, where CO and CO_2 interconvert at a much faster rate than that of the FT process and become kinetically indistinguishable from each other.

Interestingly, it has been suggested that naturally occurring CO_2-type FT reactions might explain the formation of hydrocarbons at the Lost City hydrothermal field, on the seafloor of the Atlantic Ocean at a water depth of ~780 m [46]. Hydration of olivine (a major component of the surrounding peridotite rock) could provide hydrogen, which is indeed found in the hydrothermal fluids. Iron and nickel, both common in the surrounding rocks, could act as catalysts. The relative amounts of C_2–C_4 alkanes appear to follow an Anderson–Schulz–Flory distribution, with the usual excess of methane. Ethylene, acetylene, propylene, and propyne were also detected.

From a process viewpoint, Equation 2.27 requires more hydrogen than the classical FT process and it is therefore less attractive, unless (i) a cheap source of hydrogen is readily available, preferably not involving CO_2 co-production (e.g., solar water splitting or by hydroelectric, wind, or nuclear power), or (ii) the reverse WGSR (Equation 2.28) is substituted for a thermal dissociation of CO_2 to CO at high temperatures (e.g., by a thermochemical solar approach), followed by a standard FT process.

On the other hand, compared to the classical FT process, reaction (Equation 2.27) is less exothermic, thus making temperature control of the reactor easier, even if the CO_2 transformation is likely to require high temperatures that favor the reverse WGSR (Equation 2.28).

A challenging possibility to improve the performance of CO_2-based FT reactions is the use of membranes able to selectively remove water from the reaction medium, thus forcing the reverse WGSR (Equation 2.28) and, consequently, the overall process [47].

In conclusion, an industrial FT-type process fed with CO_2 appears technically feasible. An iron-based catalyst would probably be the catalyst of choice for such a process. Much optimization, however, is still to be done on the reaction conditions (particularly temperature), the catalyst composition, and the reactor configuration.

References

1 Paggini, A., Sanfilippo, D., Pecci, G., and Dybkjaer, I. (1986) Implementation of the methanol plus higher alcohols process by Snamprogetti, Enichem, H.Topsoe A/S "MAS Technology". Proceedings of the VII Symposium on Alcohols Fuels, October 20–23, Paris, Editions Technip, pp. 62–67.

2 Fattore, V., Notari, B., Paggini, A., and Laganà, V. (1985) Catalytic system for producing mixture of methanol and higher alcohols. US Patent 4,513,100, filed October 28, 1982 and issued April 23, 1985.

3 Tijmensen, M.J.A., Faaij, A.P.C., Hamelinck, C.N., and van Hardeveld, M. R.M. (2002) *Biomass Bioenergy*, **23**, 129–152.

4 Iandoli, C.L. and Kielstrup, S. (2007) *Energy Fuels*, **21**, 2317–2324.

5 Bakkerund, P.K., Gol, J.N., Aasberg-Petersen, K., and Dybkjaer, I. (2004) *Stud. Surf. Sci. Catal.*, **147**, 13–18.

6 Basini, L., Aasberg-Petersen, K., Guarinoni, A., and østberg, M. (2001) *Catal. Today*, **64** (1–2), 9–20.

7 Wilhelm, D.J., Simbeck, D.R., Karp, A.D., and Dickenson, R.L. (2001) *Fuel Process. Technol.*, **71**, 139–148.

8 Prins, M.J., Ptasinski, K.J., and Janssen, F.J.J.G. (2007) *Energy*, **32**, 1248–1259.

9 McKendry, P. (2002) *Bioresour. Technol.*, **83**, 55–63.

10 Olofsson, I., Nordin, A., and Soderlind, U. (2005) Initial Review and Evaluation of Process Technologies and Systems Suitable for Cost-Efficient Medium-Scale Gasification for Biomass to Liquid Fuels. Report 05-02, ISSN 1653-0551 ETPC Report.

11 Higman, C. and van der Burgt, M. (eds) (2008) *Gasification*, Elsevier.
12 International Energy Agency (2007) Fossil fuel-fired power generation: case studies of recently constructed coal- and gas-fired power plants.
13 Ratnasamy, C. and Wagner, J.P. (2009) *Catal. Rev.*, **51**, 325–440.
14 Graaf, G.H., Sijtsema, P.J.J.M., Stamhuis, E.J., and Joosten, G.E.H. (1996) *Chem. Eng. Sci.*, **41**, 2883.
15 van der Laan, G.P. and Beenackers, A.A.C.M. (1999) *Catal. Rev. Sci. Eng.*, **41**, 255.
16 Zhang, R.Q., Cummer, K., Suby, A., and Brown, R.C. (2005) *Fuel Process. Technol.*, **86**, 861.
17 National Energy Technology Laboratory (2007) Industrial size gasification for syngas, substitute natural gas and power production. DOE/NETL 401/040607.
18 Echt, W.I. (1997) Chemical solvent-based processes for acid gas removal in gasification application. Presented at the IchemE Conference on Gasification Technology in Practice, February 1997, Milan, Italy.
19 Ranke, G. (1972) *Chem. Econ. Eng. Rev.*, **4**, 25–31.
20 Hassan, K.H. (2010) *Eur. J. Sci. Res.*, **39**, 289.
21 Boerrigter, H., Den Uil, H., and Clis, H. (2002) Green diesel from biomass via Fischer–Tropsch synthesis: new insights in gas cleaning and process data. Presented at the Expert Meeting on Pyrolysis and Gasification of Biomass and Waste, September 30–October 1, 2002, Strasbourg, France.
22 Rankin, J. (1982) Investigation of sulfur-tolerant catalysts for selective synthesis of hydrocarbon liquids from coal-derived gases. Proceedings of the DOE Contractors Conference on Indirect Liquefaction, September 8–9, 1982, Pittsburgh, PA.
23 Visconti, C.G., Lietti, L., Forzatti, P., and Zennaro, R. (2007) *Appl. Catal. A: General*, **330**, 49.
24 Visconti, C.G., Lietti, L., Forzatti, P., Zennaro, R., and Rossini, S. (2010) *Catal. Today*, **154**, 202–209.
25 Dry, M.E. (2004) (eds J.R. Anderson and M. Boudart), *Catalysis Science and Technology*, vol. **1**, Springer, New York.
26 Rowley, R.L., Wilding, W.V., Oscarson, J.L., Yang, Y., Daubert, T.E., and Danner, R.P. (2006) *DIPPR® Data Compilation of Pure Chemical Properties*, Design Institute for Physical Properties, AIChE, New York, NY.
27 Bodke, A.S., Bharadwaj, S.S., and Schmidt, L.D. (1998) *J. Catal.*, **179** (1), 138–149.
28 Aasberg-Petersen, K., Christensen, T.S., Nielsen, C.S., and Dybkjaer, I. (2003) *Fuel Process. Technol.*, **83**, 253–261.
29 Aasberg-Petersen, K., Bak Hansen, J.H., Christensen, T.S., Dybkjaer, I., Seier Christensen, P., Stub Nielsen, C., Winter Madsen, S.E.L., and Rostrup-Nielsen, J.R. (2001) *Appl. Catal. A: General*, **221** (1–2), 379–387.
30 van Vliet, O.P.R., Faaij, A.P.C., and Turkenburg, W.C. (2009) *Energy Convers. Manage.*, **50** (4), 855–876.
31 Molina, A. and Mondragón, F. (1998) *Fuel*, **77** (15), 1831–1839.
32 Ge-wen, Yu, Yuan-yuan, Xu, Xu, Hao, Yong-wang, Li, and Guang-qi, Liu, (2010) *Fuel*, **89** (5), 1070–1076.
33 Hofmockel, J., Liebner, W., and Ulber, D., *Lurgi's multi purpose gasification (mpg) application and further development*. Proceedings of the IChemE Conference, Gasification for the Future, 4 April 2000, Noordwijk, The Netherlands.
34 Buragohain, B., Mahanta, P., and Moholkar, V.S. (2010) *Energy*, **35**, 2557–2579.
35 Dry, M.E. (2004) Chemical concepts used for engineering purposes, in *Fischer Tropsch Technology* (eds A. Steynberg and M. Dry), Studies in Surface Science and Catalysis, Elsevier, Amsterdam, pp. 196–257.
36 Dry, M.E. (1982) *Chem. Technol.*, 744–750.
37 Dry, M.E. and Steynberg, A. (2004) Commercial FT process application, in *Fischer Tropsch Technology* (eds A. Steynberg and M. Dry), Studies in Surface Science and Catalysis, Elsevier, Amsterdam, pp. 406–481.
38 Dry, M.E. (1980) *Hydrocarb. Process.*, **59** (2), 92–94.
39 Perego, C., Bortolo, R., and Zennaro, R. (2007) *Catal. Today*, **142**, 9–16.

40 Winter-Madsen, S. and Olsson, H. (2007) *Hydrocarb. Process.*, **12** (7), 37–40.

41 Maitlis, P.M. and Zanotti, V. (2009) *Chem. Commun.*, 1619–1634.

42 Krishnamoorthy, S., Li, A., and Iglesia, E. (2002) *Catal. Lett.*, **80**, 77.

43 Riedel, T., Clayes, M., Schulz, H., Schaub, G., Nam, S.S., Jun, K.W., Choi, M.J., Kishan, G., and Lee, K.W. (1999) *Appl. Catal. A*, **186**, 201.

44 (a) Visconti, C.G., Lietti, L., Tronconi, E., Forzatti, P., Zennaro, R., and Finocchio, E. (2009) *Appl. Catal. A*, **355** (1–2), 61–68; (b)Zhang, Y., Jacobs, G., Sparks, D.E., Dry, M.E., and Davis, B.H. (2002) *Catal. Today*, **71**, 411.

45 Hall, W.K., Kokes, R.J., and Emmett, P.H. (1957) *J. Am. Chem. Soc.*, **79**, 2983.

46 Proskurowski, G., Lilley, M.D., Seewald, J.S., Früh-Green, G.L., Olson, E.J., Lupton, J.E., Sylva, S.P., and Kelley, D.S. (2008) *Science*, **319**, 604.

47 Rohde, M.P., Unruh, D., and Schaub, G. (2005) *Ind. Eng. Chem. Res.*, **44**, 9653.

3
Fischer–Tropsch Technology

Arno de Klerk, Yong-Wang Li, and Roberto Zennaro

Synopsis

The term *Fischer–Tropsch* (FT) *technology* encompasses the FT catalyst, the operating conditions, and the reactor types that are employed to conduct FT syntheses in industrial practice. An overview of industrially applied FT technologies describes the catalyst, operating conditions, and reactor combinations, pointing out the motivations that underlie specific developments. The requirements for industrial FT catalysts and the main FT reactor types are discussed in detail and guidelines for the selection of an appropriate FT technology for a specific task are proposed.

3.1
Introduction

We have given an overview of FT-based conversion in Chapter 2 that explained how different raw materials can be converted into syngas ($H_2 + CO$) and how the syngas can be cleaned, conditioned, and converted into a synthetic crude oil (syncrude).

FT technology uses many elements. The principles of reaction engineering apply to FT conversion as they do to other conversion processes, and to create a successful industrial FT technology, the catalyst, the operating conditions, and the reactor must be matched.

The technology to be developed must be decided early on, and the final products from the FT-based facility must be agreed upon in advance. Are the products to be syncrude, transportation fuels, lubricants, petrochemicals, or a combination of some of these? The preferred properties of the syncrude are determined by the composition requirements of the final products [1]. An important advantage of FT over crude oil is that the syncrude composition is tunable, within limits. This tunability is possible since the FT synthesis is a conversion process and the various parameters can be chosen to achieve a predetermined outcome.

Greener Fischer-Tropsch Processes for Fuels and Feedstocks, First Edition. Edited by Peter M. Maitlis and Arno de Klerk.

3.1.1
FT Catalyst

There is a distinction between FT catalysts and materials that are catalytically active for the FT reaction. An FT catalyst contains the material that is catalytically active for the FT reaction, but aspects such as the particle size distribution, particle morphology, intraparticle heat and mass transfer resistances, mechanical strength, chemical stability, and large-scale catalyst production must all be considered in the creation of an FT catalyst.

3.1.2
Operating Conditions

The FT catalysis is sensitive to the operating conditions; thus, the operating temperature affects desorption and hydrogenation rates, which in turn affect the chain growth probability and the degree of product hydrogenation (Chapter 4). Temperature also affects the reaction phase and steam quality (pressure) that can be produced. The operating temperature is used to classify the synthesis into high-temperature FT (HTFT), medium-temperature FT (MTFT), and low-temperature FT (LTFT) technologies. Typical operating temperatures for each are >320, ~ 270, and $<250\,°C$, respectively. Pressure mainly affects productivity, while changes in the composition of the syngas during synthesis affect the products and are also involved in the deactivation of the catalyst.

3.1.3
FT Reactor Types

Since the FT reaction involves the very exothermic CO hydrogenation reactions (heat released $\sim$140–160 kJ/mol CO converted) and since the product selectivity is temperature sensitive, the reactor types that can be considered are severely restricted by the heat management requirements. Different reactor types are required due to the differing number of phases present at reaction conditions. Catalyst deactivation rate and catalyst replacement strategy are also important in determining the reactor type, which in turn imposes further requirements on the catalyst.

3.2
Industrially Applied FT Technologies

There is considerable diversity in the technologies that are employed (Table 3.1), and we do not suggest that any one of these technologies is superior to the rest, as most have a proven record of successful and stable industrial operation.

To understand the thinking behind the different technologies, we can examine their origins and some of the thinking that went into their development. More

Table 3.1 Industrially applied FT technologies (as of 2012).

FT technology	FT metal	FT type	Reactor type	Year first commissioned	Still applied?
German normal-pressure synthesis	Co	LTFT	Fixed bed (tube cooled)	1936	No
German medium-pressure synthesis	Co	LTFT	Fixed bed (tube-in-tube)	1937	No
Hydrocol	Fe	HTFT	Fixed fluidized bed	1951	No
Arbeitsgemeinschaft Ruhrchemie-Lurgi (Arge)	Fe	LTFT	Fixed bed (multitubular)	1955	Yes
Kellogg Synthol	Fe	HTFT	Circulating fluidized bed	1955	No
Sasol Synthol	Fe	HTFT	Circulating fluidized bed	1980	Yes
Shell middle distillate synthesis (SMDS)	Co	LTFT	Fixed bed (multitubular)	1993	Yes
Sasol Advanced Synthol (SAS)	Fe	HTFT	Fixed fluidized bed	1993	Yes
Iron Sasol slurry bed process (Fe-SSBP)	Fe	LTFT	Slurry bubble column	1995[a)]	Yes
Statoil cobalt slurry bubble column process	Co	LTFT	Slurry bubble column	2005[a)]	Yes
Cobalt Sasol slurry bed process (Co-SSBP)	Co	LTFT	Slurry bubble column	2007	Yes
High-temperature slurry FT process (HTSFTP)	Fe	MTFT	Slurry bubble column	2008[a)]	Yes

a) The technologies can be arguably called demonstration scale for modern facilities, but at 1000–4000 bbl/day it is on the same order of magnitude as the pre-1970 industrial Fischer–Tropsch facilities in Germany, Japan, France, China, the United States, and South Africa.

details of industrial FT developments are given in Chapter 5, while this chapter focuses on the decisions that shaped FT technology.

3.2.1
German Normal-Pressure Synthesis

The normal pressure process operated at 100 kPa (1 atm) and it employed a kieselguhr-supported Co-LTFT catalyst. Kieselguhr is a natural high surface area silica, but it is not mechanically very strong. A fixed bed reactor design overcame this limitation, while the cooling tubes provided the necessary heat removal capacity; excellent temperature control to within 1 °C was achieved in this way. However, the CO conversion was low and the product yield was only 0.1 kg/h per kg catalyst. Two or three reactors in series with intermediate product removal were employed to improve volumetric productivity and reduce deactivation by oxidation [2].

3.2.2
German Medium-Pressure Synthesis

The medium pressure 1 MPa (∼10 atm) process gave a higher volumetric reactor productivity and more liquid products [2, 3]. Furthermore, the higher H_2 partial pressure gave a two to three times longer catalyst lifetime and the increased liquid production helped to remove deposits from the catalyst surface. Since the same catalyst was used as in the normal-pressure synthesis, the same mechanical limitations applied and a fixed bed reactor design was employed. However, with the increased reactor productivity, the heat release per unit volume increased, and a tube-in-tube fixed bed design was employed to provide a larger heat exchange surface per volume of catalyst.

3.2.3
Hydrocol

The change from Co-LTFT to Fe-HTFT technology arose for several reasons, the most important being the higher quality of the Fe-HTFT syncrude, which produced a much better motor gasoline than the Co-LTFT syncrude [1, 4]. Advances in fluid catalytic cracking (FCC) technology increased confidence in the design of the appropriate reactors [2]. The fixed fluidized bed (FFB) reactor design was elegant in its simplicity and was less complicated to construct than the circulating fluidized bed (CFB) reactors employed for FCC. Operation at high temperatures and pressures compensated for the potentially lower volumetric productivity of the less-dense catalyst bed and >90% CO conversion could be achieved. Since such systems are in principle capable of online catalyst replacement that allows operation with an "equilibrium" catalyst analogous to FCC operation, the shorter anticipated lifetime of an Fe-based catalyst compared to a Co-based catalyst was not a concern. Furthermore, since iron was considerably cheaper than cobalt, interest in iron-based FT synthesis increased even in Germany due to the high cost and limited availability of cobalt. The fluidized bed reactor imposed limitations on the catalyst design. A finely divided robust catalyst with good attrition resistance was required for fluidized bed operation. Fused iron-based FT catalysts had the required attrition resistance and they could also be operated at high temperatures without significant methane production.

3.2.4
Arbeitsgemeinschaft Ruhrchemie-Lurgi (Arge)

The robustness of fixed bed LTFT technology was demonstrated industrially by the German normal and medium pressure processes. However, the tube-in-tube reactor design of the German medium pressure process was mechanically complicated. The Arge multitubular fixed bed reactor, which resembles a vertical shell-and-tube heat exchanger, was a more practical design that achieved the same objective. At the time of writing (2012), the original Arge reactors that were commissioned in 1955 are still in commercial operation, which is testimony to the success of this technology; the main drawback is that catalyst replacement is more onerous.

3.2.5
Kellogg Synthol and Sasol Synthol

The original Kellogg Synthol circulating fluidized bed design was developed based on the industrial experience that was gained with the Hydrocol fixed fluidized bed technology. The change from a fixed fluidized bed to a circulating fluidized bed was not related to FT synthesis requirements, as, in fact, the circulating fluidized bed design was more complicated to construct and placed more demands on the catalyst. The change was motivated by Kellogg experience with fluid catalytic cracking, which employs a CFB design. The Sasol Synthol modification ironed out some of the industrial operating problems with the original reactor design [5].

3.2.6
Shell Middle Distillate Synthesis (SMDS)

Several considerations led to the selection of a multitubular fixed bed reactor design (analogous to Arge) in combination with a supported Co-LTFT catalyst [6]: Cobalt was preferred to iron, because it is easier to develop a high α-value FT catalyst with good lifetime and high activity, while the higher cost of Co was offset by the longer catalyst lifetime and the subsequent recovery of Co from the spent catalyst. The main factor favoring the selection of a multitubular fixed bed reactor was that this technology is more robust and easier to develop and scale-up; even though catalyst loading and unloading is more onerous, this is of little consequence if the catalyst has a lifetime of several years.

3.2.7
Sasol Advanced Synthol (SAS)

The same design principles were employed as for the original Hydrocol technology and for similar reasons, as there is no reason related to FT synthesis for specifically employing a circulating fluidized bed instead of a fixed fluidized bed. Fixed fluidized bed reactors are simpler to construct and have a higher volumetric productivity. In fact, the advantages of fixed compared to a circulating fluidized bed technology led to the replacement of all the Synthol reactors by Sasol Advanced Synthol (SAS) reactors at the Sasol Synfuels facility [7]. The 10.7 m diameter SAS reactor is at present the highest capacity FT reactor in industrial use, with a production capacity per reactor equivalent to 1 000 000 t/a (20 000 bbl/day; 130 m^3/h).

3.2.8
Iron Sasol Slurry Bed Process (Fe-SSBP)

Despite the robustness and advantages of multitubular fixed bed operation, a number of drawbacks noted in the literature have served as justification for the development of a slurry bubble column reactor (SBCR) [8]. Advantages of the Fe-Sasol slurry bed process (SSBP) include the easier replacement of deactivated catalyst leading to better product consistency, a lower pressure drop (and hence less need for recompression), smaller temperature gradients, and the potential for capacity increase by increasing the unit dimensions. The hope is that these advantages can outweigh the main disadvantages of slurry-phase operation, namely, the higher poisoning sensitivity, the need for an attrition-resistant catalyst, and the challenge of separating small catalyst particles from the liquid product. However, it should be pointed out that this technology is arguably only at demonstration scale in industrial production.

3.2.9
Cobalt Sasol Slurry Bed Process (Co-SSBP)

The justification for the development of the Co-SSBP is similar to that of the Fe-SSBP [7]. However, sufficient catalyst attrition resistance is important for the success of this technology [9]. This was highlighted by catalyst–product separation problems due to fines produced by attrition during commissioning of the first industrial application of this technology [10].

3.2.10
Statoil Cobalt-Based Slurry Bubble Column

A single 1000 bbl/day unit is in industrial operation at the PetroSA facility. The industrial commissioning of this unit actually predates that of the Co-SSBP. As in the case of the Fe-SSBP process, this is arguably only a demonstration-scale unit that is employed in industrial production.

3.2.11
High-Temperature Slurry Fischer–Tropsch Process (HTSFTP)

The same advantages previously cited for the Fe-SSBP motivated the development of this technology. The main difference is that it operates at higher temperatures (~270 °C) in order to increase the steam quality that can be produced. The product spectrum is typical of an Fe-LTFT syncrude.

3.3
FT Catalysts

There are essentially three different FT catalyst types that are industrially employed, although there are variations within each type. Further details are given in Chapters 8 and 9.

Fused iron catalysts are used in all the Fe-HTFT technologies (Table 3.1), which were the robust "workhorse" catalysts that accounted for most of the global FT syncrude production before the commissioning of Pearl GTL in 2011. This catalyst was extensively studied in the past [11, 12], but has attracted less research effort recently.

Precipitated iron catalysts are employed for fixed bed and slurry bubble column Fe-LTFT technologies (Table 3.1). Unlike Fe-HTFT, Fe-LTFT is in direct competition with Co-LTFT, because both processes have similar aims. However, there have been fewer investigations into precipitated iron catalysts [11–13]. In fixed bed operation, precipitated Fe-LTFT catalysts are able to achieve very high α-values ($\sim$0.95) and somewhat less in slurry bubble column operation [1]. These differences are due to the reactor design (plug flow versus continuous stirred tank) and are not intrinsic to the catalyst.

The supported cobalt catalysts employed in Co-LTFT technologies (Table 3.1) have evolved considerably from the Co-ThO_2-kieselguhr catalysts employed in the original German work [6, 12, 14], and have been very successful in industrial fixed bed operation, as demonstrated by the SMDS process.

3.4 Requirements for Industrial Catalysts

As industrial experience is not always reflected in the open literature research publications, hindering a comprehensive understanding of the demands for an ideal industrial FT catalyst, some lessons from industrial FT practice are summarized.

3.4.1 Activity

Industrial FT processes use higher activity catalysts, as higher activity leads to lower catalyst consumption. This in turn lowers the direct and the indirect costs of catalyst replacement and waste catalyst handling, as a smaller volume of catalyst is used in the reactors. Higher syngas conversion is possible, leading to easier recycling and (especially for fluidized and slurry reactors) the easier separation of catalyst from product streams. Fixed bed reactor operation benefits less from more active catalysts due to internal transportation limitations in catalyst pellets.

3.4.2 Selectivity

The product spectrum obtained from FT synthesis ranges from methane to heavier hydrocarbons. For a practical FT catalyst using a gas loop without cryogenic separation, it is desirable to suppress as much as possible the formation of the less valuable products such as methane (C_1) and the C_2 hydrocarbons. For HTFT processes, keeping methane selectivity below 10 wt% seems possible for long-term operation; however, this value is still quite high. For LTFT synthesis in general, the methane selectivity can be kept at a level of 3–8 wt%. Synfuels China high-temperature slurry Fischer–Tropsch process technology (HTSFTP) operates at a methane selectivity level of 2.5 wt% by employing a Fe-based catalyst. For Co-based catalysts, methane selectivities on the order of 5–6 wt% are typical. It has been shown that expensive tail gas recovery procedures can be replaced by simple recovery of hydrogen from tail gas if the methane selectivity can be suppressed to 1.0–1.5 wt%. However, such low levels of methane selectivity have not yet been reached by current industrial FT catalysts.

3.4.3 Stability

Catalyst stability has long been an industrial problem. Iron catalysts are expected to maintain "stable" operation only over less than 3000 h, while cobalt catalysts are expected to remain stable for a couple of years. However, there are indications that iron catalysts may be capable of longer term stability [15, 16]. More stable FT catalysts for industrial applications are desirable; furthermore, the nature of the deactivation and how it affects selectivity also has an impact on downstream refining [1].

3.4.4 Other Factors

As catalyst consumption is still high in modern FT-based industries, there is an appreciable catalyst cost. However, higher productivity has been sought during catalyst development, and currently the best FT catalysts may reach 2000 ton C_3 and heavier products per ton of catalyst during the life cycle of the catalyst. Even a catalyst with a productivity of a few hundred tons of $\geq C_3$ products per ton catalyst can still be viable if the cost is low, as is often the case with Fe-based catalysts. However, it should be noted that catalyst cost becomes insignificant for a typical CTL plant only when the productivity of the catalyst is $\geq$5000 tons C_3 and heavier products per ton catalyst (Li, Y.W., personal experience.).

In addition to the activity, stability, and selectivity of industrial FT catalysts, the overall performance of the FT process is also determined by the energy evolution during FT synthesis. Based on past industrial experience, we may expect that the future direction for a good FT process is that of most efficient energy conversion. It is also necessary to incorporate coal, natural gas, and biomass into the supply chain for syngas conversion with high selectivity for C_5 and heavier products that can efficiently be upgraded/refined into liquid fuels and chemicals. Alternatively, the gas loop and the FT refinery design should enable the conversion of the normally gaseous products.

However, the FT technologies producing heavy products with high selectivity are often operated at low LTFT synthesis temperatures. However this means that the FT reaction heat can only produce low-grade steam, carrying about 15–20% energy of the syngas converted. It is thus desirable to push industrial operation to higher temperatures to enable the production of more valuable steam for powering the process, at the same time maintaining the selectivity to heavy hydrocarbons with high material conversion efficiency. This aspect has been considered in the latest development of Synfuels China HTSFTP process, which targets the operation of a slurry bubble column reactor at a temperature between traditional LTFT and HTFT processes. The potential of this type of medium-temperature FT (MTFT) synthesis has not yet been fully realized. Cobalt-based catalysts are not a suitable choice for MTFT applications, but iron catalysts may be successfully adapted to produce high liquid yield under MTFT conditions, thereby meeting the targets.

We also need to consider the water-gas shift reactivity (WGSR) of the catalyst. Since cobalt catalysts are not very active for WGSR conversion, this leads to less gas recycling and tail gas treatment in a cobalt-based FT process than in those using iron catalysts. Thus, it could be advantageous to suppress the WGSR over iron catalysts, if that were possible. The aim to have good WGSR activity on FT catalysts for the conversion of syngas with low $H_2:CO$ ratios appears frequently in the literature, but it must be remembered that this is not necessarily a good way for the industrial process to be developed as that would generate large amounts of CO_2 in the FT gas loop, which will increase the processing cost.

It has been found in industrial practice that both cobalt and iron catalysts deactivate over time, largely due to the high partial pressure of water caused by high single pass conversion during FT synthesis. Cobalt catalysts are generally resistant to slightly higher water vapor pressures than iron catalysts, implying that a higher syngas recycling ratio is needed for iron than for cobalt catalysts in order to keep the deactivation rate low. There is a clear incentive to develop FT catalysts that have a tolerance for higher water vapor pressure.

3.5 FT Reactors

The overview of the main industrially applied FT technologies (Section 3.2) showed that the selection of the reactor type in combination with the FT catalyst is central to each technology. Detailed and critical comparisons of reactor types have been given [11, 17–19], while the reactor types actually used industrially are described in Sections 3.5.1–3.5.4, which is followed by a discussion of important selection criteria (Sections 3.6–3.9).

3.5.1 Tube-Cooled Fixed Bed Reactors

The structures of the FT reactors used in Germany in the 1930s and 1940s were described by Stranges [20]. A cobalt-based FT catalyst was used in an early version of the fixed bed reactor (Figure 3.1), which was a rectangular sheet steel box (5 m long, 2.5 m high, and 1.5 m wide) containing about 600 horizontal water cooling tubes interlaced at right angles with 555 vertical steel sheets. This grid-like arrangement of the reactor, through which the syngas (some 650–750 m^3/h) was fed from

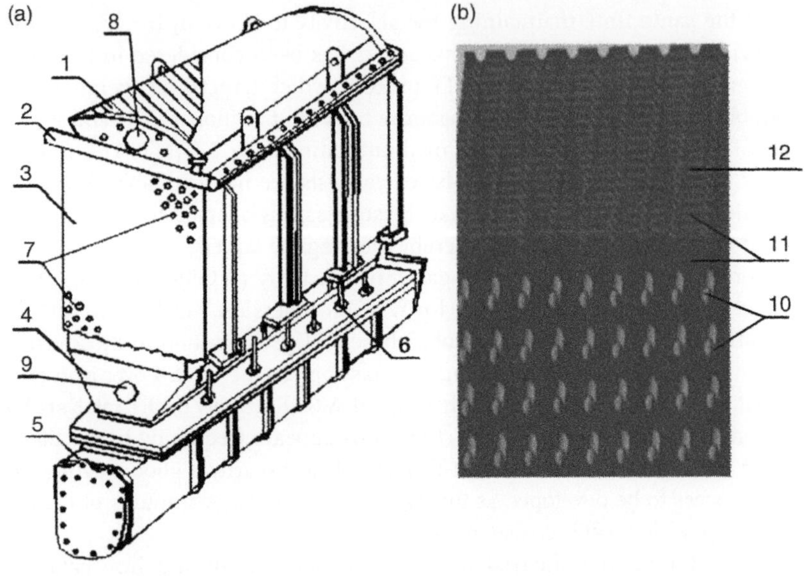

Figure 3.1 A sketch of an early version of a normal-pressure fixed FT reactor using a Co catalyst. (a) Overview of the reactor. (b) The internal sheets plus tubes structure: (1) upper cover of the reactor body, (2) header for cooling water distribution, (3) reactor body, (4) down space for wax–gas separation, (5) wax collector, (6) condensate conduits, (7 and 10) internal cooling tubes, (8) syngas inlet, (9) syngas outlet, (11) internal steel sheets, and (12) spaces for catalysts between the sheets.

the top, eliminated localized heat buildup in the catalyst bed. Each steel sheet was 1.6 mm thick, with a space of 7.4 mm separating adjacent plates. The cooling tubes were 40 mm in diameter, 40 mm apart, and led to a boiler to recover heat released in the FT synthesis. As documentation was lost during the Cultural Revolution in the 1960s, the diagram is based on descriptions by former staff in Jinzhou, China, of an FT plant brought from Germany in 1938 by the Japanese military. This plant operated only during 1952–1962.

From the above information, it can be estimated that the heat transfer area of horizontally arranged cooling tubes was about 370 m^2 and that the vertically arranged plates fixed to the tubes provided another 2000 m^2 heat exchange area. The plates may not have been as effective as the tube walls, but provided significant additional heat transfer area. It may be inferred that the very good heat removal allowed good temperature control during operation.

The typical designs for these normal-pressure FT facilities were for a scale of 30 000–200 000 t/a (600–4000 bbl/day) liquids [20, 21]. An early FT plant was imported from Germany by the Japanese Military into Jinzhou, China, in which 74 such reactors with a catalyst load of 13 tons each were used to produce 30 000 t/a liquids. Catalyst performance was 0.004–0.007 kg liquid oil/kg cat per hour. We can thus infer that current FT technologies are between 20 and 200 times more efficient than the original German normal-pressure technology.

3.5.2 Multitubular Fixed Bed Reactors

The German medium-pressure FT technology made use of a tube-in-tube design within a pressure vessel. This was a complicated way of constructing a multitubular reactor, because the catalyst was loaded into the annular space between the inner and outer tube walls, giving an annular thickness of catalyst bed of only 9 mm [3]. This configuration provided heat transfer on both sides, but today we see that the design was an overkill, even though it enabled excellent near-isothermal temperature control.

The next generation of multitubular fixed bed reactors designed after the Second World War used the more practical approach provided by the Arbeitsgemeinschaft Ruhrchemie-Lurgi (Arge) for the Sasol 1 facility in South Africa. This employed a multitubular design with single tubes, which made loading and unloading of the catalyst easier and the construction simpler. Each multitubular reactor had a diameter of 3 m and contained 2050 tubes of 50 mm internal diameter and 12 m length. The iron FT catalyst was loaded in the tubes. The reactors operate at about 230 °C and 27 atm, with a production capacity of 500 bbl/day (25 000 t/a) for each reactor. In 1987, a new unit was installed in Sasolburg, which operated at 45 atm, with a higher space velocity and 700 bbl/day (35 000 t/a) production capacity [19]. A typical Arge multitubular fixed bed reactor is depicted in Figure 3.2.

The main advance in this type of technology was made by Shell in the scale-up during the development of the SMDS process. This resulted in the design of larger multitubular fixed bed reactors around 7 m in diameter, which are currently used in the Shell Pearl GTL project, with a total capacity of 140 000 bbl/day (6 000 000 t/a) liquids. This GTL project is actually combined with a 120 000 bbl/day (5 000 000 t/a) production of LPG condensates and ethane piped in from the Qatar North field. The total estimated cost for the combined project, including the upstream offshore development (platforms and sea lines), is estimated at US$18–19 billion [22]. This cost excludes the associated 750 000 t/a liquefied natural gas (LNG) plant.

As can be seen from Figure 3.2, the multitubular fixed bed reactor is actually a shell-and-tube type heat exchanger, in which the tubes in the bundle are filled with catalyst pellets. The external walls of these catalyst filled tubes are immersed in boiling water, which enables heat removal on the shell side of the reactor. The temperature in the reactor is controlled by regulating the steam pressure on the shell side. The syngas introduced from the top of the reactor passes through the catalyst bed in the tubes. FT synthesis is conducted in the tubes and the unreacted syngas and FT syncrude product mixture leaves the tubes at the lower end and enters into the reactor bottom where the heavy wax and gaseous streams are separated. The gaseous product mixture leaves the reactor at the gas outlet, situated at the top of the reactor base, while the wax leaves the reactor at the bottom of the reactor. The FT catalyst remains fixed in the reactor.

In a multitubular fixed bed reactor, the main design requirement, essential for the application to FT synthesis, is again the efficient removal of the reaction heat in

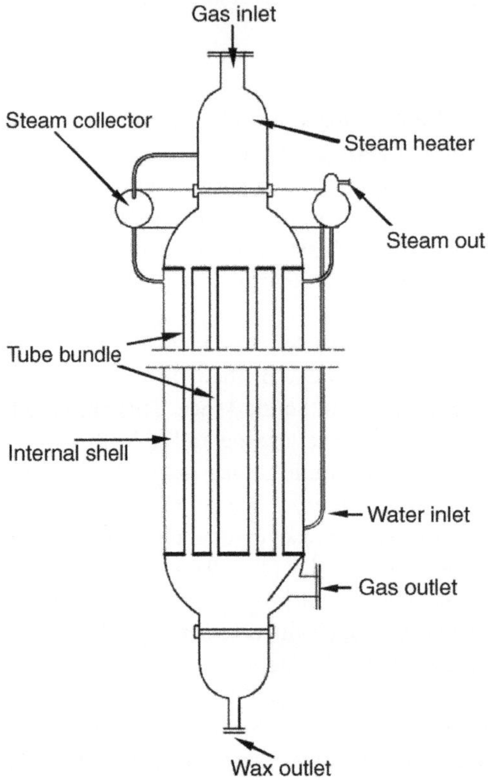

Figure 3.2 Multitubular fixed bed FT reactor (original Arge reactor).

order to maintain a good temperature control in the catalyst beds. For this reason, the maximum diameter of the tubes is 50 mm; indeed, thinner tubes should be considered when more active catalysts are used. Tubes with an internal diameter of 20 mm can meet the most severe conditions for heat transfer, but cause the cost to increase due to the decreased throughput compared to using larger diameter tubes. Another way to improve the heat transfer in multitubular fixed bed reactors is to recycle liquid wax back to the reactor inlet. In this way the complete reactor is in trickle bed operation with high liquid content. In the enhanced trickle bed operation, efficient heat transfer can be achieved by increasing the heat transfer coefficient due to the large liquid flow. Using a high gas recycling ratio can also improve the heat transfer, but this is less effective and consumes more power.

A drawback of using a fixed bed reactor is that the catalyst particles are large. Large catalyst particles have a lower efficiency due to the transport limitations imposed on reactions in the catalyst pellets. It has been estimated that the effectiveness factor in a typical fixed bed FT catalyst pellet of 2–4 mm is much lower than that in the fine catalyst particles of <0.2 mm employed in slurry bubble column reactors [17, 18, 23–25]. For fixed bed FT catalyst pellets 3 mm in diameter, the effectiveness factor is estimated to be between 0.1 and 0.3, implying that between 10 and 3 times more catalyst is required in a the fixed bed reactor compared to a slurry bubble column reactor when using similar but smaller diameter catalyst particles with the same intrinsic activity. To overcome this drawback, the catalyst pellets for fixed bed FT synthesis need to be carefully shaped to reduce the diffusion distance through the catalyst pellet. The use of eggshell catalyst pellets to approach unit effectiveness has also been suggested; however, eggshell-type pellets have other drawbacks, for example, they are more easily deactivated by catalyst poisons, because they have less active metal, which is concentrated in the outer surface of the catalyst particles.

Furthermore, as charging and discharging catalyst in a multitubular fixed bed FT reactor is time consuming and costly, when a multitubular fixed bed reactor is selected, it is important that the catalyst should have a long lifetime during industrial operation.

To summarize, multitubular fixed bed reactors have intrinsic drawbacks, only some of which can be overcome by appropriate design:

i) It is more difficult to control the temperature inside the reactor axially and radially.
ii) The complexity of large multitubular reactors leads to high construction costs.
iii) A large pressure drop occurs across the fixed bed.
iv) The efficiency of the catalyst is low due to internal transfer limitations in 2–5 mm catalyst particles.
v) The need for labor-intensive catalyst replacement.

In spite of these drawbacks, multitubular fixed bed technology is the dominant reactor technology for LTFT synthesis based on industrial production capacity. Multitubular fixed bed reactors also represent a promising technology for processing syngas from biomass with relatively small production capacity. The recent development of microchannel reactor technology [26] is just a more compact way of improving heat transfer and reducing mass transfer resistance compared to the traditionally larger multitubular fixed bed reactors.

On the other hand, multitubular fixed bed reactors have advantages that are equally applicable to large- and small-scale reactors:

i) They are robust in operation, as demonstrated industrially over many decades.
ii) They are resistant to syngas contaminants like H_2S, since the H_2S is adsorbed by the top layer of catalyst, which serves as a guard bed to the rest of the catalyst bed.
iii) There are no wax and catalyst separation problems.
iv) Scale-up based on pilot plant data obtained with a single tube reactor is straightforward and robust.
v) Attrition resistance is not a key catalyst design requirement.

3.5.3
Circulating and Fixed Fluidized Bed Reactors

Circulating fluidized bed and fixed fluidized bed reactors are only used for HTFT synthesis. Under operating conditions, typically >320 °C and 25 atm, the reaction system contains only gas-phase and solid-phase (catalyst) material, and the FT catalyst must be designed to have an α-value of 0.7 or lower in order to avoid the formation of liquid products.

The CFB FT reactor technology was initially developed by Kellogg (in the United States) in the 1950s. At that time Kellogg was actively developing fluid catalytic cracker technology, which employs a CFB reactor; the same design principles were applied to FT, where it was also seen as a way to solve problems that were encountered in the first FFB FT reactors in the Hydrocol HTFT facility (in Brownsville, TX). CFB reactors were chosen for the Sasol 1 plant (in South Africa) due to the perceived operating problems of the Hydrocol FFB compared to the Kellogg CFB reactors [19, 27]. However, FFB is now the dominant reactor type for HTFT synthesis.

The CFB reactors in the Sasol 1 plant were scaled up from the original 100 mm internal diameter to 2.3 m. The initial production capacity for the first two Kellogg reactors was about 75 000 t/a (1500 bbl/day) each, which was later increased to 125 000 t/a (2500 bbl/day) as problems with the Kellogg CFB technology were ironed out. The Synthol CFB reactors were modified versions of the Kellogg CFB reactors, and were used in the Sasol 2 and 3 (Secunda, South Africa) and had a capacity of 365 000–400 000 t/a (7300–8000 bbl/day), which was three times higher than that of the Sasol 1 reactor. Three 40 000 t/a (8000 bbl/day) Synthol CFB reactors were also used in the PetroSA "Mossgas" GTL plant (Mossel Bay, South Africa). The main features of the CFB FT reactor are shown in Figure 3.3.

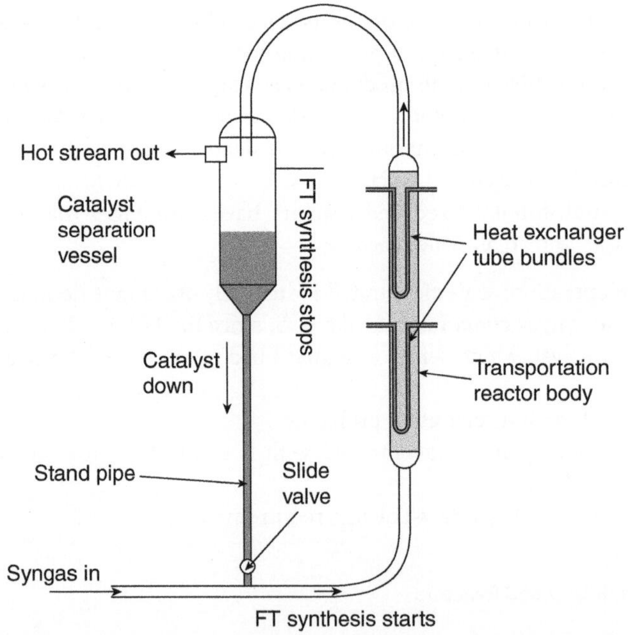

Figure 3.3 Gas–solid circulating fluidized bed (CFB) Fischer–Tropsch reactor operating at 320–350 °C and 2.5 MPa with catalyst particles in the range of 50–200 μm in size.

The CFB reactor operates in a fluidized transportation (entrained flow) mode. The hot syngas is mixed with the catalyst moving down in the stand pipe and flow is controlled through a slide valve. At this mixing point FT synthesis starts. The catalyst and gas mixture passes through the pipe and enters into the main section of the transportation reactor, where reaction heat is removed by the heat exchanger tube bundles. High-quality steam (>50 atm can be achieved) is produced and the reaction temperature is controlled by regulating the steam pressure. The reacting gaseous mixture containing the catalyst passes through the reactor body and leaves it at the top; it then enters the catalyst separation cyclones where the catalyst is separated from the gas stream. After catalyst separation the gas mixture leaves the FT reactor system. The CFB reactor operation entails a very large catalyst recycle and the high gas flow rate causes catalyst attrition, as well as erosion. In addition to the severe operation conditions for both the catalyst and the equipment, the pressure drop balance between the CFB reactor and the standpipe have to be carefully maintained in order to achieve safe operation. Finally, the reactor contains a large amount of nonreacting catalyst in the catalyst separation vessel and standpipe, which lowers the overall volumetric reactor productivity and the catalyst efficiency.

Despite the variable track record of operation in the Hydrocol plant, a FFB reactor in principle has a higher volumetric reactor productivity and greater catalyst usage efficiency than a CFB reactor. The operation of a FFB reactor is also less erosive than that of a CFB reactor due to the lower gas velocity. In collaboration with Badger, a 1 m diameter FFB reactor was built in 1984 at Sasol 1, which was scaled up to 5 m diameter and a capacity of 175 000 t/a (3500 bbl/day) in 1989. Further scale-up work on FFB reactors was carried out at Sasol from 1995 to 1999 and 16 Synthol CFB reactors at Sasol 2 and 3 were replaced with FFB units, called Sasol Advanced Synthol reactors (Figure 3.4). Four 8 m diameter FFB reactors with a capacity of 550 000 t/a (11 000 bbl/day) each and four 10.7 m diameter FFB reactors with a capacity of 1 million t/a (20 000 bbl/day) each were constructed.

The FFB FT reactor operates in the dense fluidization mode, allowing all the catalyst to stay inside the reactor body without the need for external catalyst recycling as in the CFB reactor. The syngas introduced at the bottom of the reactor passes through a gas distributor located at the bottom of the catalyst bed and then enters the fluidized bed for FT synthesis. The product gas leaves the catalyst bed at the top. In the upper catalyst-free space of the reactor, cyclones are used to separate entrained catalyst from the exiting gas. The heat exchanger tube bundles that are immersed in the fluidized catalyst bed remove the reaction heat.

The following are the major advantages of FFB reactors over CFB reactors for FT synthesis [19, 27]:

i) Higher throughput for a single reactor.
ii) Lower rate of online catalyst replacement and lower overall catalyst consumption.
iii) Forty percent lower construction cost because of the simplified structure of the FFB reactor.
iv) Lower operating and maintenance cost.

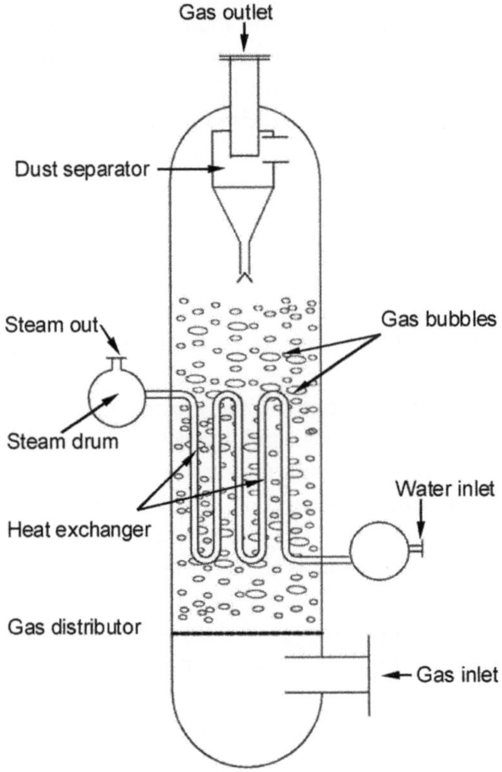

Figure 3.4 Gas–solid fixed fluidized bed (FFB) FT reactor operating at 320–350 °C and 2.5 MPa with catalyst particles in the range of 50–200 μm in size.

An FFB reactor approximates to a continuous stirred tank reactor (CSTR) more than to a CFB reactor that better approximates to a plug flow reactor (PFR), which is a fundamentally more efficient reactor type.

3.5.4 Slurry Bed Reactors

Large-scale application of slurry bubble column reactors for iron-based FT synthesis has been successfully practiced by Sasol since 1993, Synfules China since 2008 (Statoil has also practised Co-LTFT slurry reactor operation at industrial scale.). Although Sasol also applied slurry bubble column reactors industrially for Co-LTFT, some significant problems were experienced with catalyst fines [10].

The earliest slurry bed studies were conducted by Fischer in 1932. In the early 1950s, Rheinpreussen AG and Koppers GmbH developed a semicommercial slurry reactor operated with a low $H_2:CO$ ratio at a superficial gas velocity of

0.1 m/s. A high (~90%) per pass carbon monoxide conversion was reported. The Rheinpreussen company also started work on a slurry bed reactor in 1937 under the direction of Köbel. The catalysts used were mostly iron based, as opposed to the cobalt-based catalysts used for fixed bed operation at that time. A slurry reactor with a diameter of 1.5 m and a bed height of 7.7 m was developed, with a working volume of around 10 m^3 [19]. Early versions of slurry bubble column reactors for FT synthesis were designed with insufficient understanding of the hydrodynamics and of FT catalysis in a slurry phase. Separation of waxy products from the small catalyst particles was difficult and made slurry bed technology unattractive for industrial application.

In the late 1980s and the early 1990s, Sasol converted the 1 m diameter FFB reactor that was employed to develop the SAS technology as a pilot-scale slurry reactor. The catalyst used was similar to that used in the multitubular fixed bed Arge reactor, but instead of extrudates the catalyst was produced as finely divided particles. Central to the development was a better understanding of slurry-phase hydrodynamics and a method for separating the fine catalyst particles from the liquid product. The process was scaled up in the FFB demonstration reactor used for the development of the SAS technology, which has an inside diameter of 5 m and is 22 m high. After the successful conversion and recommissioning of the FFB reactor as a slurry bubble column reactor in 1993, it was adopted for commercial production at Sasol 1. The reactor has a production capacity of 12 500 t/a (2500 bbl/day) and the process is known as the Sasol slurry bed process. The technology was later adapted for use with a supported cobalt FT catalyst, instead of a precipitated iron catalyst, and the reactor was scaled up to an internal diameter of 10 m and a height of 45 m. The design capacity of this unit is 750 000–850 000 t/a (15 000–17 000 bbl/day). The first industrial use of the Co-SSBP technology was in the Oryx GTL plant (Ras Laffan, Qatar). Although some difficulties were originally reported, the new Co-SSBP-based LTFT configuration seems to be Sasol's standard technology for XTL projects, with the same technology being employed in the Escravos GTL plant (under construction in Nigeria).

Synfuels China started the scale up of slurry-phase FT synthesis in 2000, when its 1000 t/a (20 bbl/day) pilot plant was built in Taiyuan China. Through 2002–2008, systematic pilot tests were conducted using iron catalysts in this slurry reactor (0.35 m internal diameter and 45 m high). In 2005, three demonstration projects were started with designed capacities of 200 000–225 000 t/a (4000–4500 bbl/day) using 5.3–5.8 m internal diameter and 57 m high slurry reactors. These demonstration plants were successfully commissioned during 2008–2009, and have been in operation ever since. They form the basis for the HTSFTP technology and will be the standard for planned projects with capacities between 2 000 000 and 7 500 000 t/a (40 000–150 000 bbl/day) both in China and outside.

A typical slurry bubble column reactor is shown in Figure 3.5. The syngas is introduced via the gas distributor at the bottom of the slurry bed and enters the slurry phase where the FT synthesis reactions take place. The reacting gas goes through the bed and leaves the slurry bed interface at its top. In the upper slurry-free space of the reactor, provision is made to separate any mist carried over by the

(a)

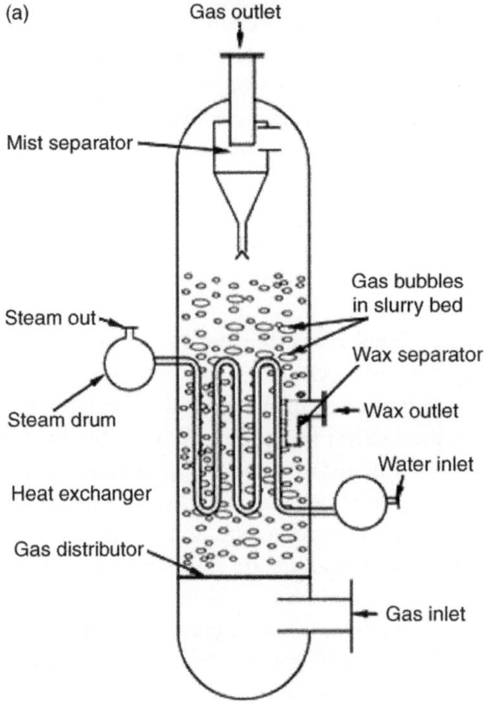

(b)

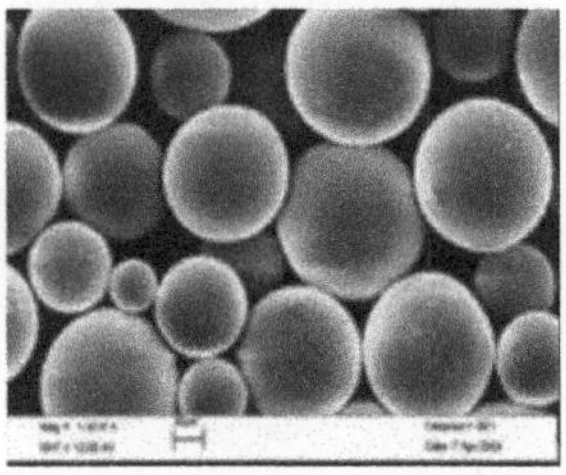

Figure 3.5 (a) A FT slurry bubble column reactor. (b) A scanning electron microscope image of the spherical catalyst particles ranging in size from 50 to 200 μm.

gaseous stream, which leaves the reactor vessel at the top after demisting. The light FT syncrude fractions in the gaseous stream are recovered downstream of the reactor and the heavy waxy products remaining in the slurry are extracted by *in situ* filtration. The reaction heat is removed with high efficiency by the heat exchangers immersed in the slurry bed.

There are many advantages of slurry bed reactors over multitubular fixed bed reactors for iron-based FT processes, but some of them are mute for Co-LTFT:

i) Catalyst consumption per ton of product from an Fe-FT slurry bed reactor is about 25–10% of that from a fixed bed multitubular reactor, because of the smaller catalyst particle size and improved mass and heat transfer in the suspended slurry state.
ii) The operating temperature is more isothermal.
iii) The pressure drop in the reactor is about a quarter of that in fixed bed reactors.
iv) Online removal and addition of catalyst are possible, which allow longer reactor runs if the catalyst lifetime is short.
v) Higher selectivity to heavier products and low methane selectivity can easily be achieved using an Fe-FT catalyst.
vi) The cost of a slurry reactor is much lower than that of a multitubular fixed bed reactor with the same capacity.

However, there are also some disadvantages of slurry bed reactors for example:

i) Syngas contaminants, such as H_2S, can instantly spread over the whole catalyst inventory in the reactor and thus lead to fast catalyst deactivation.
ii) A slurry bed approximates to a continuous stirred tank reactor (CSTR), which is less efficient than a plug flow reactor.
iii) The vigorous movement and collision of catalyst particles that occur in slurry bubble column reactors ultimately lead to catalyst erosion and attrition. This wear and tear produces micrometer-sized catalyst particles that significantly increase the viscosity of the slurry phase and also make the separation of the catalyst from the wax product extremely difficult, which in turn leads to considerably increased downstream processing costs.

The liquid/solid separation design is one of the main issues of a slurry bubble column reactor, if we consider the hydrodynamic regime, coupled with the Fischer–Tropsch environment. Fines formed during initial transient conditions with fresh catalyst, as well as potential submicrometer fines produced as a result of mechanical and chemical stresses, can lead to side effects, such as failure of the solid–liquid separation device or foam formation inside the reactor, and can have a severe impact on the downstream upgrading section [9].

Eni IFPEN (IFP Energies Nouvelles), has carried out several studies to identify the optimal solution for minimizing the operating risks, especially when dealing with large column reactors (10 m diameter and >45 m height). In the Gasel^{TM} technology (XTL technology suite for the conversion of synGAS to diesEL) the preferred reactor configuration is one that employs an external filtration unit downstream of the wax outlet. The catalyst separated from the liquid hydrocarbon product is circulated back to the bottom of the reactor, below the gas distributor, in order to keep the catalyst concentration controlled.

3.6 Selecting the Right FT Technology

The selection of a specific FT technology for industrial application has a ripple effect on many of the design decisions that will have to be taken in the overall facility. Some of the aspects that are directly affected by the selection of the FT technology include the following requirements.

3.6.1 Syngas Composition

The FT catalyst will either be WGSR active or not. When the catalyst is not WGSR active, CO_2 and H_2O are final products, or can be considered "inert" if present in the feed. When the catalyst is not WGSR active, the H_2 : CO ratio in the feed will change through the reactor depending on the usage ratio of the FT technology. It is best to provide syngas with a H_2 : CO ratio at or close to the usage ratio. However, when the catalyst is WGSR active, it is the overall composition of H_2, CO, CO_2, and H_2O that matters, because the catalyst is able to change the ratios. The rate of WGSR equilibration is determined by the operating conditions, whereas the local H_2 : CO ratio will be determined by the reactor type (PFR or CSTR behavior) and the reaction kinetics. The FT technology therefore directly impacts the design of the gas loop (see Chapter 2).

3.6.2
Syngas Purity

When the FT technology employs a fixed bed reactor (PFR configuration), syngas contaminants that are FT catalyst poisons will deactivate the top of the catalyst bed and the deactivation front will over time progress through the bed. The catalyst downstream of the deactivation front caused by the poisoning remains unaffected and active for FT synthesis. When the FT technology employs a slurry bed or a fluidized bed reactor (CSTR configuration), syngas contaminants will affect all the catalyst. The operating robustness of fixed bed FT technology is much better than that of slurry bed or fluidized bed technology. The syngas cleaning requirements (Chapter 2) for fixed bed reactors are less stringent and fixed bed technology can weather operating upsets in syngas production, cleaning and conditioning better than either slurry bed or fluidized bed technologies.

3.6.3
Impact of Catalyst Deactivation

All catalysts deactivate over time. The nature of the FT technology will determine the type and rate of deactivation and how the yield and selectivity are influenced by deactivation. However, of all the FT catalyst types, Fe-LTFT is the only one where the selectivity change due to deactivation has a beneficial effect on the FT refinery [1]. Usually FT catalyst deactivation is accompanied by an increase in methane selectivity, which is particularly detrimental to the productivity of the facility. The catalyst replacement cost is affected by both the rate of catalyst replacement and the unit cost per mass of catalyst. Cobalt-based FT catalysts are more expensive than iron-based FT catalysts and must therefore have a longer lifetime (and/or better yield) to justify the cost. The causes for catalyst deactivation and the limitations it places on per pass conversion are directly related to the FT catalyst and operating conditions. The life cycle of FT catalysts is discussed in detail in Chapter 13.

3.6.4
Catalyst Replacement Strategy

The reactor type will determine whether it is possible to remove and replace catalyst online. This is possible with slurry bed and fluidized bed reactors and the FT technology can in principle be operated with an "equilibrium" catalyst composition that has a stable activity and selectivity profile. Although this is advantageous from an operability viewpoint, the removal of catalyst from an "equilibrium" mixture implies that almost fresh and very old catalyst are withdrawn together, while some very old catalyst remains in the reactor. Online catalyst replacement is not possible with fixed bed reactors and thus far no FT moving bed reactors have been commercialized. Catalyst replacement in multitubular fixed bed reactors is labor intensive and must be carefully scheduled. When a single or small number of reactors are operating in parallel, the conversion and selectivity profile will vary somewhat over time.

3.6.5 Turndown Ratio and Robustness

Turndown ratio is the ratio of actual feed rate to design feed rate. The ability to operate the facility at lower syngas feed rate may or may not be important and depends on the robustness of the facility. If it is important to make provision for turndown, fixed bed reactors are better than slurry or fluidized bed reactors, where a minimum slurrying or fluidization velocity must be maintained. The same is true for overall robustness, with smaller capacity fixed bed reactors being more robust than larger capacity slurry bed or fluidized bed reactors. If a reactor has to be taken offline, the impact is less for smaller than for larger units. However, reducing the operating risk comes at a cost, since multitubular fixed bed reactors are more expensive than slurry bed or fluidized bed reactors for the same total production capacity.

3.6.6 Steam Quality

FT synthesis is very exothermic and the heat of reaction can be beneficially used within the facility or for power generation. The steam quality (i.e., the pressure) determines the usefulness of the energy. The operating conditions of the FT technology affects the temperature (and thereby the maximum pressure) at which the steam can be generated, thus FT technologies operating at a higher temperature have an advantage.

3.6.7 Syncrude Composition

The syncrude composition affects the ability to refine the products to the targets for which the facility is designed. There are significant differences in composition between LTFT and HTFT syncrudes, as well as important differences between technologies, which affect the type of products that can be efficiently produced by refining [1, 28]. The carbon number distribution and relative concentrations of various compound classes are both important (see Chapter 4). The syncrude composition also affects the design of the syncrude recovery section and the possible inclusion of cryogenic separation in the gas loop (Chapters 2 and 4). If the yield of gaseous products is high, or increases substantially due to catalyst deactivation, it becomes inefficient not to recover and refine these lighter products. The same is true of the water-soluble oxygenates found in the aqueous products: when their yield is significant, it becomes inefficient not to recover and refine this fraction.

3.6.8 Syncrude Quality

The downstream recovery and refining of syncrude is affected by dissolved or suspended material from the FT synthesis. Slurry bed and fluidized bed reactors both require catalyst–syncrude separation steps as part of the design. Depending on the efficiency of the catalyst separation from the syncrude, catalyst particles produced by attrition in the FT reactor may find their way into the product [9], which may also contain dissolved metals. The design of downstream equipment must make provision for such material when it is present or it will lead to lower efficiency and reduced run lengths caused by plugging and fouling [1, 29].

3.7 Selecting the FT Operating Conditions

In order to help guide the technology selection process (Section 3.6), we discuss the impact of selecting FT operating conditions. However, it is important to state clearly that there is not a simply best FT technology and operating condition combination; some technologies are more successful than others, but technology selection must be performed on a case-by-case basis. The saying, "horses for courses," applies very much to FT technology selection.

Selecting the operating conditions and specifically the operating temperature range of the process is the first important decision to be taken. This decision must be based on the products that are desired from the facility, since the FT-based facility should be designed to produce specific products, which should be implicit in the business case that justifies the facility. The operating temperature directly affects the syncrude composition and a high operating temperature (e.g., HTFT) generally leads to a lighter syncrude (lower FT α-value). Guidelines for an efficient matching of products to syncrude type are given in Chapter 4, as well as in the literature [1, 28].

Once the FT operating conditions have been selected, there is less choice for the following other aspects of the FT technology:

i) The extent of light gas recovery needed in the gas loop.
ii) The steam pressure that can be produced for internal use or power generation.
iii) The FT metal catalyst that can be employed (higher temperatures rule out cobalt due to increased methane selectivity).
iv) The reactor configurations that are viable (Note: It has not been shown in the current FT technology domain that the FT catalyst, reactor and gas loop configuration can be selected and combined from different sources into an integrated plant.).

3.8 Selecting the FT Catalyst Type

The FT operating conditions chosen place some limitations on the carbon number distribution and compound selectivity that can be achieved. However, although they have a significant effect, syncrude composition is not determined by operating conditions alone. The hydrogenating nature of the metal, the choice of promoters to alter selectivity, and the impact of deactivation over time are all catalyst properties that also influence the syncrude composition. The FT catalyst can therefore be selected or designed to best match the product requirements for the facility.

The selection of the catalyst type is unfortunately not independent of the selection of the reactor type. Nevertheless, catalyst selection will be presented as if it precedes reactor selection or, as suggested by Krishna and Sie [30], is the first step in multiphase reactor selection. Some of the following aspects should be considered during catalyst selection or the design of a new catalyst.

3.8.1 Active Metal

Industrial FT catalysts use only Fe (Chapter 8) or Co (Chapter 9). Other metals are also FT active (Chapter 10), but a very good reason is needed for selecting a different metal, especially if it is more expensive than Co. Even Co is quite costly, and it is necessary to recover the metal from the spent FT catalyst for both environmental and economic reasons (Chapter 13). However, even more expensive metals are sometimes added to the main FT-active metal in industrial catalysts to essentially improve specific property of the catalyst, for example, Sasol employs 0.05% Pt with 20% Co in their new industrial FT catalyst [31]. The selection of the main active metal may be constrained by the operating conditions. For example, higher operating temperatures preclude the use of Co-based catalysts since Co gives increased methane selectivity at higher temperatures; since Co is more hydrogenating than Fe, it also influences product selectivity. When FT is considered as a strategic technology, availability of the active metal may play a role in the decision-making process.

3.8.2 Catalyst Complexity

It is possible to develop very sophisticated FT catalysts with good control over surface area, metal dispersion, and promotion. However, as the complexity of the catalyst increases, it becomes more difficult to scale up and potentially more costly to manufacture. Furthermore, there is usually a trade-off between mechanical strength (e.g., attrition resistance) and level of complexity that can be engineered into the catalyst design. A complex catalyst may be better designed for fixed bed application, where fewer mechanical demands will be placed on the working catalyst.

3.8.3 Catalyst Particle Size

For the same catalytically active material, the particle size of the actual catalyst affects the activity, heat transfer, and mass transfer. There are generic trade-offs to be made in catalyst design [30]. Viewed in isolation, the highest possible activity per unit catalyst volume is desirable, because it maximizes the volumetric productivity of the FT reactor, that is, production rate per unit reactor volume. In practice, a balance must be achieved between the volumetric catalyst activity and the ability to overcome transport resistance. It is very important to efficiently remove reaction heat and to ensure adequate supply of syngas, while removing product from the catalyst. Poor heat or mass transfer will lead to lower selectivity to desirable products and/or an increase in catalyst deactivation rate. Smaller catalyst particles have lower heat and mass transfer resistance and such catalysts can be designed with higher volumetric activity. Conversely, if larger catalyst particles are employed, out of necessity the volumetric activity must be lower and the catalyst should be designed for better selectivity or longer lifetime to compensate for the loss in volumetric activity.

3.9 Other Factors That Affect FT Technology Selection

Once the FT operating conditions and catalyst have been decided, the choices remaining for reactor selection are significantly reduced [30]. Among the other factors to be considered are the following.

3.9.1 Particle Size

The particle size that was selected for the FT catalyst determines what reactor types may be considered. When the particle size is in the millimeter range, a fixed bed reactor can be appropriate, but it becomes impractical for smaller catalyst particle sizes due to the increased pressure drop. Nevertheless, smaller diameter catalyst particles can be employed in microchannel fixed bed technology, where the order of magnitude increase in transport coefficients allows the use of much shorter bed lengths. However, in general, when the particle size is smaller, around 100 μm, a fluid bed reactor is more appropriate. As the particle size becomes smaller, other hydrodynamic considerations may affect the specific type of reactor that can be employed, for example, in gas–solid systems, the fluidization velocity versus transport velocity.

3.9.2 Reaction Phase

The combination of the FT operating conditions and the α-value of the catalyst determines whether the reaction system is two phase, gas–solid, or three phase, gas–liquid–solid. In the case of the fluid bed reactor type, the number of reaction phases dictates whether it is a fluidized bed (gas–solid) or slurry bubble column (gas–liquid–solid) reactor.

3.9.3 Catalyst Lifetime

When the lifetime of the FT catalyst is only months, rather than years, it may be advantageous to consider a reactor type that allows online addition and removal of catalyst. Slurry bed and fluidized bed technologies are therefore preferable for shorter lifetime FT catalysts, whereas fixed bed technology is better suited to catalysts with a longer lifetimes. For example, it has been reported that the Co-LTFT catalyst used in the SMDS process has an overall catalyst lifetime of 5 years [32], which is better than most fixed bed catalysts routinely employed in conversion processes. The use of multitubular fixed bed reactors for Co-LTFT is consequently a logical decision based on lifetime. However, in industrial practice, catalyst lifetime does not seem to have been a major consideration in reactor selection. For example, multitubular fixed bed reactor technology has been employed industrially since the 1950s with Fe-LTFT catalysts, despite the perceived "short" catalyst lifetime of Fe-LTFT compared to Co-LTFT catalysts. Conversely, new FT technologies were developed employing slurry bubble column reactor technology with Co-LTFT catalysts, despite claims of a longer lifetime for Co-LTFT than for Fe-LTFT catalysts.

3.9.4 Volumetric Reactor Productivity

The volumetric reactor productivity is one of the primary economic driving forces in the selection of the FT reactor. The elegant pilot plant study reported by Dry [11] gives a good comparative indication of the efficiency of the main industrially employed reactor types. In this study, the same iron-based FT catalysts but with different, though appropriate, particle sizes were evaluated at the same operating conditions (Table 3.2). The volumetric reactor productivity increased in the

Table 3.2 Comparison of the catalyst and volumetric reactor productivity of iron-based FT catalysts in different reactor types operated at similar conditions.

Description	Precipitated Fe-LTFT (230–240 °C)		Fused Fe-HTFT (320–330 °C)	
	Fixed bed	Slurry bed	Slurry bed	Fixed fluidized bed
Catalyst loading (kg Fe)	2.7	0.8	1.0	4.2
Bed volume during synthesis (l)	7.5	7.5	7.5	3.9
Syngas conversion (%)	46	49	79	93
Catalyst productivity (l/(kg s)[a]	0.1	0.4	0.7	0.2
Reactor productivity (s^{-1})[b]	0.044	0.046	0.094	0.21

All tests were conducted at pilot plant scale in a 50 mm diameter reactor.
a) Catalyst productivity = volume of syngas converted per time (l/s) per mass of catalyst (kg Fe).
b) Reactor productivity = volume of syngas converted per time (l/s) per reactor bed volume (l).

following order: fixed bed < slurry bed < fixed fluidized bed. The catalyst productivity is different, with slurry bed reactor technology making the best use of the catalyst, even though it does not have the highest volumetric reactor productivity. It can also be seen that better productivity can be obtained during HTFT synthesis than LTFT synthesis. Since the results in Table 3.2 cannot be directly extrapolated for use in industrial reactors, one should compensate for the additional volume required by cooling. For example, in an Arge-type multitubular fixed bed reactor, 43% of the volume of the actual reaction section is occupied by the cooling medium. In slurry bed and fluidized bed reactors, the volume occupied by the cooling coils is less, but still substantial. At industrial scale, it is also possible to optimize each reactor type hydrodynamically and otherwise. Nevertheless, the trends remain the same even though a direct numerical comparison is not possible. The 5 m diameter and 22 m high reactor at the Sasol 1 site, which was first employed to scale up the SAS technology and later the SSBP technology, provides comparative data at an industrial scale. When it was employed as fixed fluidized bed reactor, the production capacity was 145 000 t/a (3500 bbl/day) and when it was employed as a slurry bed reactor, the production capacity was 100 000 t/a (2500 bbl/day) [19]; however, see also Ref. [33] for different capacity values. The volumetric reactor productivity of FFB Fe-HTFT synthesis is therefore more than that of slurry bed Fe-LTFT synthesis under industrial operating conditions.

3.9.5
Other Considerations

Issues such as robustness, turn down ratio, unit size, capital cost, and steam pressure can all contribute to the decision to select one reactor type over another. There are also other practical considerations, for example, road transportation restrictions to move equipment to an inland location may limit the maximum reactor size (diameter) that can be specified. In such cases, the incentive to select cheaper fluid bed reactor technologies over fixed bed reactor technology is diminished and this may influence the reactor selection based on other criteria.

References

1 De Klerk, A. (2011) *Fischer–Tropsch Refining*, Wiley-VCH Verlag GmbH, Weinheim.
2 Asinger, F. (1968) *Paraffins Chemistry and Technology*, Pergamon Press, Oxford.
3 Weil, B.H. and Lane, J.C. (1948) *Synthetic Petroleum from the SYNTHINE Process*, Remsen Press, Brooklyn.
4 De Klerk, A. (2009) *Advances in Fischer–Tropsch Synthesis, Catalysts, and Catalysis* (eds B.H. Davis and M.L. Occelli), Taylor & Francis, Boca Raton, pp. 331–364.
5 Holtkamp, W.C.A, Kelly, F.T., and Shingles, T. (1977) *ChemSA*, **3** (3), 44–45.
6 Sie, S.T. (1998) *Rev. Chem. Eng.*, **14** (2), 109–157.
7 Collings, J. (2002) *Mind Over Matter. The Sasol Story: A Half-Century of Technological Innovation*, Sasol, Johannesburg.
8 Jager, B. and Espinoza, R. (1995) *Catal. Today*, **23**, 17–28.
9 Perego, C., Bortolo, R., and Zennaro, R. (2009) *Catal. Today*, **142**, 9–16.

10 Anonymous, *Petrol. Econ.* (2008) **75** (6), 36–38.

11 Dry, M.E. (1981) *Catalysis Science and Technology*, vol. **1** (eds J.R. Anderson and M. Boudart), Springer, Berlin, pp. 159–255.

12 Dry, M.E. (2004) *Stud. Surf. Sci. Catal.*, **152**, 533–600.

13 De Smit, E. and Weckhuysen, B.M. (2008) *Chem. Soc. Rev.*, **37**, 2758–2781.

14 Bezemer, G.L., Bitter, J.H., Kuipers, H.P. C.E., Oosterbeek, H., Holewijn, J.E., Xu, X., Kapteijn, F., Van Dillen, A.J., and De Jong, K.P. (2006) *J. Am. Chem. Soc.*, **128**, 3956–3964.

15 Storch, H.H., Golumbic, N., and Anderson, R.B. (1951) *The Fischer–Tropsch and Related Syntheses*, John Wiley & Sons, Inc., New York.

16 Janse van Vuuren, M.J., Huyser, J., Grobler, T., and Kupi, G. (2009) *Advances in FT Synthesis, Catalysts and Catalysis* (eds B.H. Davis and M.L. Occelli), Taylor & Francis, Boca Raton, pp. 229–241.

17 De Swartz, J.W.A., Krishna, R., and Sie, S.T. (1997) *Stud. Surf. Sci. Catal.*, **107**, 213–218.

18 Sie, S.T. and Krishna, R. (1999) *Appl. Catal. A*, **186**, 55–70.

19 Steynberg, A.P., Dry, M.E., Davis, B.H., and Breman, B.B. (2004) *Stud. Surf. Sci. Catal.*, **152**, 64–195.

20 Stranges, A.N. (2003) Germany's synthetic fuel industry 1927–45. Proceedings of the AIChE 3rd Topical Conference on Natural Gas Utilization, New Orleans, March 30–April 3, 2003, pp. 635–646.

21 Stranges, A.N. (2007) *Stud. Surf. Sci. Catal.*, **163**, 1–27.

22 Overtoom, R., Fabricius, N., and Leemhouts, W. (2009) Shell GTL, from bench scale to world scale, advances in gas processing. Proceedings of the 1st Annual Gas Processing Symposium, Doha, Qatar, January 10–12, 2009, Elsevier, Amsterdam, pp. 378–380.

23 Wang, Y.N., Xu, Y.Y., Xiang, H.W., Li, Y. W., and Zhang, B.J. (2001) *Ind. Eng. Chem. Res.*, **40**, 4324–4335.

24 Wang, Y.N., Xu, Y.Y., Li, Y.W., Zhao, Y.L., and Zhang, B.J. (2003) *Chem. Eng. Sci.*, **58**, 867–875.

25 Krishna, R. and Sie, S.T. (1994) *Chem. Eng. Sci.*, **49**, 4029–4065.

26 Lerou, J.J., Tonkovich, A.L., Silva, L., Perry, S., and McDaniel, J. (2010) *Chem. Eng. Sci.*, **65**, 380–385.

27 Steynberg, A.P. (2004) *Stud. Surf. Sci. Catal.*, **152**, 1–63.

28 De Klerk, A. (2011) *Energy Environ. Sci.*, **4**, 1177–1205.

29 De Klerk, A. (2008) *Catal. Today*, **130**, 439–445.

30 Krishna, R. and Sie, S.T. (1994) *Chem. Eng. Sci.*, **49**, 4029–4065.

31 Botes, F.G., Van Dyk, B., and McGregor, C. (2009) *Ind. Eng. Chem. Res.*, **48**, 10439–10447.

32 Schrauwen, F.J.M. (2004) *Handbook of Petroleum Refining Processes*, 3rd edn (ed. R.A. Meyers), McGraw-Hill, New York, pp. 1525–1540.

33 Duvenhage, D.J and Shingles, T. (2002) *Catal. Today*, **71**, 301–305.

4
What Can We Do with Fischer–Tropsch Products?

Arno de Klerk and Peter M. Maitlis

Synopsis

Two main types of product, low-temperature FT (LTFT) *syncrude* and high-temperature FT (HTFT) *syncrude*, are produced industrially from Fischer–Tropsch synthesis. Although the chief products from the FT reaction, linear alkenes, linear alkanes, methane and water, some branched chain hydrocarbons, and oxygenates are also formed. Typical compositions of the main syncrude types are reviewed. Syncrude has both similarities and some notable differences to crude oil. One fundamental difference is that the composition of syncrude can be manipulated during the FT reaction and the composition of the syncrude fractions obtained can be changed after synthesis by adjusting the cooling and separation stages during recovery. The fuel, lubricant, and petrochemical products that can efficiently be produced from FT syncrudes are discussed.

4.1
Introduction

The products from a Fischer–Tropsch process form a synthetic crude or *syncrude*. What can we do with this FT syncrude?

It is useful to compare FT syncrude with the more familiar *crude oil*. The latter conjures up an image of a viscous dark liquid that has little use in its raw form and that needs refining into useful products such as the transportation fuels, petrochemicals, and lubricants that we encounter daily. In this respect, FT syncrude is no different. In its raw form, the syncrude is of little value and must be refined in order to become useful. We can choose what we want the syncrude to become and adapt the FT refinery design accordingly [1]. However, not all the products are accessible by efficient refining pathways and they must also be selected in relation to both environmental and economic constraints.

Unlike crude oil, syncrude is not a viscous dark liquid: in fact, it is not dark, viscous, or a single-phase liquid. There is no such thing as a single-standard syncrude composition. Depending on the FT technology used to make it, syncrude at

Greener Fischer-Tropsch Processes for Fuels and Feedstocks, First Edition. Edited by Peter M. Maitlis and Arno de Klerk.

standard conditions exists in three to four different phases: gaseous, organic liquid (oil), aqueous liquid, and solid. This has considerable consequences. First, the syncrude composition depends on the FT technology used, which involves the FT catalyst, the synthesis reactor, the operating conditions, and the level of catalyst deactivation (Chapter 3).

Second, since the syncrude composition is not a geological given, as in the case of crude oil, the composition can be manipulated and tailored to facilitate refining to specific products.

Third, since the composition of the different syncrude phases depends on the design of the primary syncrude cooling and separation steps after the FT process, the conditions of separation, the phase equilibria, and the approach to equilibrium determine how the compounds distribute between the different product phases. This may affect the ability to efficiently produce specific products.

Finally, the availability of light hydrocarbons from the gaseous product phase is determined by the design of the syncrude recovery after the FT process and not just the FT synthesis.

The path to useful products from FT syncrude starts with the design of the FT process. The composition of the syncrude and how it is recovered after synthesis are as important as its refining. Syncrude composition, the manipulation of the syncrude composition and syncrude recovery will be considered first. Then we will consider which products can be produced by refining in an efficient manner: "efficient" here is not a literary embellishment, but a statement of purpose. The Green Chemistry principles of *avoiding waste, maximizing atom economy, increasing the energy efficiency,* and *designing less hazardous chemical syntheses* all require that refining pathways and technologies should be compatible with the molecular composition of the syncrude. Syncrude can in principle be converted into the same products as crude oil, but with very different degrees of refining effort [2]. The aim here is not to provide an anthology of all possible products, but to highlight those that can most efficiently be produced from a FT process.

4.2
Composition of Fischer–Tropsch Syncrude

The composition and carbon number distribution of syncrude is a direct result of the desorption kinetics and the FT mechanism (Chapter 12). The carbon number distribution can be mathematically described reasonably well [3], but a molecular description of the selectivity is more elusive.

The main syncrude types produced industrially can be broadly classified into iron-based high-temperature FT (Fe-HTFT), iron-based low-temperature FT (Fe-LTFT), and cobalt-based low-temperature Fischer–Tropsch (Co-LTFT) syncrudes. Generic compositions for each of these syncrudes are given in Table 4.1 [4]. These compositions are not representative of any specific technology, and within each syncrude type, considerable variation can be found. For example, fixed bed

Table 4.1 Generic compositions of the main industrially produced FT syncrude types.

Product fraction	Carbon range	Compound class	Syncrude composition (mass%)[a]		
			Fe-HTFT	Fe-LTFT	Co-LTFT
Tail gas	C_1	Alkane	12.7	4.3	5.6
	C_2	Alkene	5.6	1.0	0.1
		Alkane	4.5	1.0	1.0
LPG	C_3–C_4	Alkene	21.2	6.0	3.4
		Alkane	3.0	1.8	1.8
Naphtha	C_5–C_{10}	Alkene	25.8	7.7	7.8
		Alkane	4.3	3.3	12.0
		Aromatic	1.7	0	0
		Oxygenate	1.6	1.3	0.2
Distillate	C_{11}–C_{22}	Alkene	4.8	5.7	1.1
		Alkane	0.9	13.5	20.8
		Aromatic	0.8	0	0
		Oxygenate	0.5	0.3	0
Residue/wax	$>C_{22}$	Alkene	1.6	0.7	0
		Alkane	0.4	49.2	44.6
		Aromatic	0.7	0	0
		Oxygenate	0.2	0	0
Aqueous product	C_1–C_5	Alcohol	4.5	3.9	1.4
		Carbonyl	3.9	0	0
		Carboxylic acid	1.3	0.3	0.2

a) The syncrude composition is based on the total mass of product from FT synthesis, *excluding* inert gases (N_2 and Ar) and water-gas shift products (H_2O, CO, CO_2, and H_2). Zero indicates low concentration and not necessarily the total absence of such compounds.

Co-LTFT synthesis as employed in the Shell middle distillate synthesis (SMDS) process produces a heavy paraffinic syncrude [5], which is close in composition to the generic Co-LTFT syncrude composition listed in Table 4.1. In contrast, fixed bed Co-LTFT synthesis as employed by the original German normal pressure process produced a lighter more olefinic syncrude, which contained less than 10% wax and around 20% alkenes [6]. Furthermore, the syncrude composition changes over time during synthesis due to deactivation of the FT catalyst, even though some FT technologies approximate steady-state operation. In this respect, the FT process is not different from any other catalyzed conversion process.

In addition to the usual FT products (namely, linear hydrocarbons, *n*- alkenes, and *n*-alkanes), a typical syncrude also contains some branched chain hydrocarbons (largely monomethyl-substituted alkenes and alkanes), some aromatics, and some oxygenates. The oxygenates are mainly 1-alkanols, but aldehydes, ketones, and carboxylic acids are also found. There has been considerable speculation concerning the exact origins of these other materials and whether they should be regarded as primary products from the FT reaction, or whether they are produced in later steps by secondary reactions.

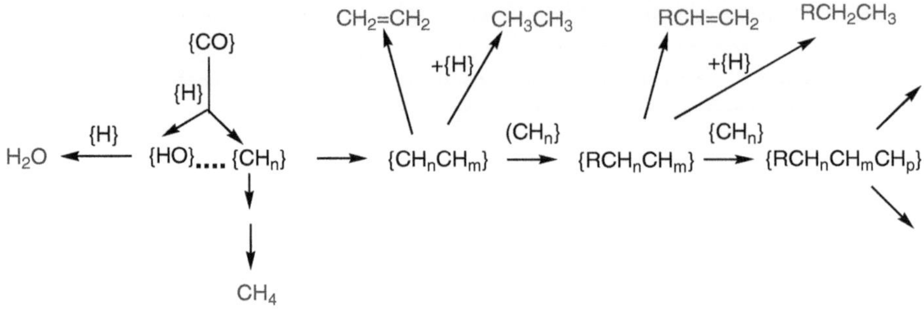

Figure 4.1 A diagrammatic flowchart showing possible relationships between the FT reactants ($CO + H_2$) and the hydrocarbon products (methane, *n*-alkenes, and *n*-alkanes) and water. No detailed mechanisms or intermediates are implied, but some surface hydrocarbyl species that may be relevant are shown, within curly brackets: $\{CH_n\}$, $\{RCH_nCH_m\}$, and so on as their structures and binding to the surfaces are not well defined (see Chapter 12 for more details).

To illustrate how the products *may* be related, we have constructed the flowchart diagrams in Figures 4.1 and 4.2 linking the starting materials ($CO + H_2$) and the various products. *However, no detailed mechanisms are given or should be implied;* Chapter 12 *gives a summary of present-day mechanistic thinking based on recent results.*

In this pictorial representation, the first steps involve the adsorption of carbon monoxide and hydrogen on the catalyst surface and their conversion via adsorbed monohydrocarbyl species $\{CH_x,\}$ ($x = 1$–4) into methane. These $\{C_1H_x\}$ species can also join together to make longer chain hydrocarbyls $\{C_nH_y\}$, which can then undergo various modes of reaction,

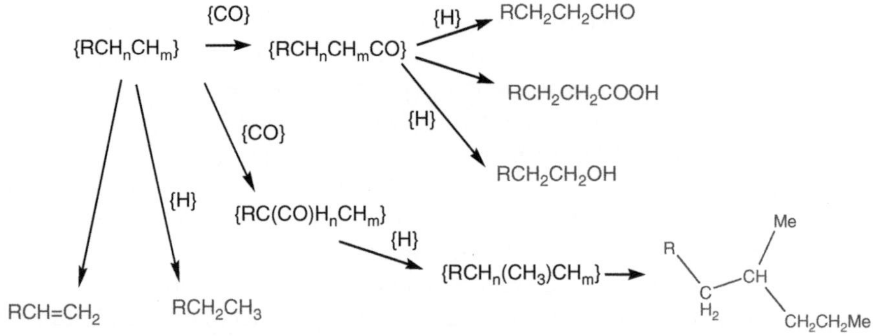

Figure 4.2 A diagrammatic flowchart showing the possible relationships between putative FT hydrocarbyl intermediates and major FT products (linear *n*-alkenes and *n*-alkanes) as well as minor products (alcohols, aldehydes, and carboxylic acids) and branched chain hydrocarbons. Some surface species that may be relevant are indicated within curly brackets, for example, $\{CH_n\}$, $\{RCH_nCH_m\}$, and so on as their structures and binding to the surfaces are not well defined (see Chapter 12 for more details).

(i) The $\{C_nH_y\}$ species may desorb, leaving the surface as alkenes C_nH_{2n}. The probability of desorption will depend on the temperature and the strength of the interaction with the catalytic surface. This principle is often applied to determine site strengths experimentally through temperature-programmed desorption (TPD). On a molecular level, this process may correspond to the *β-elimination* reactions of alkyl–metal complexes where an olefin and a metal hydride are formed (Equation 4.1), in a step that is reversible: M is a generic surface metal.

$$RCH_2CH_2{-}M \rightarrow RCH{=}CH_2 + M{-}H \tag{4.1}$$

(ii) Alternatively, the hydrocarbyl may interact with hydrogen to give a product, which is now *physisorbed* rather than *chemisorbed* on the surface and no longer participates in chain growth. This physisorption is a much weaker interaction than chemisorption, and precedes complete desorption. The alkane of chain length C_nH_{2n+2} is then desorbed as a final product, a step that has a counterpart in the *reductive elimination* step of an alkane, in alkyl–metal chemistry (Equation 4.2):

$$RCH_2CH_2{-}M(H) \rightarrow RCH_2CH_3 + M \tag{4.2}$$

(iii) A further possibility, is that the hydrocarbyl can react with CO, an olefin, another unsaturated molecule, or a species derived from one of those to form new functionalized entities. Such steps can then lead to chain growth (polymerization), as illustrated in Figure 4.2. Since the FT end products are largely linear or monomethyl branched molecules, with only a low incidence of ethyl and longer chain branching, any coupling of two hydrocarbyl chains probably occurs fairly specifically, head-to-head or head-to-tail.

Chain growth seems to take place very largely by interaction with CO or a C_1 species derived from CO. This chain growth probability is referred to as the α-value, where α denotes the probability that the adsorbed C_n species will grow to chain length C_{n+1}, rather than be desorbed as a hydrocarbon of chain length C_n. Once a C—C bond is formed, desorption may again take place as previously described, but with some additional possibilities.

In the phenomenological discussion thus far, all surface species are hydrocarbyls, leading to the production of alkenes or alkanes, as well as water. However, if the $\{C_n\}$ reacts with CO, the chain grows to give an oxy-species that can produce an oxygenate such as an aldehyde on desorption. Hydrogen transfers can also occur to give an alcohol or a carboxylic acid.

For the special case where $n = 1$, reactive desorption by partial hydrogenation might be expected to lead to methanol (CH_3OH). However, although methanol is also made by CO hydrogenation, a completely different (and presumably much lower energy) route is preferred (see Section 6.2.3).

Some details of our present understanding of what happens on surfaces and how that relates to the mechanism of FT hydrocarbon synthesis and other reactions are presented in Chapters 11 and 12. In fact, it is now clear that one single mechanism cannot explain all the FT products, and we support the view that even for

FT hydrocarbon synthesis, two types of mechanism occur, one involving *electrophilic* species on highly polar surfaces, while the other involves largely neutral species and occurs on nonpolar surfaces (see Section 12.6).

4.2.1
Carbon Number Distribution: Anderson–Schulz–Flory (ASF) Plots

The carbon number distribution from the products of an FT hydrocarbon synthesis is regular and is mathematically similar to that seen for an alkene polymerization. The chain growth probability (α-value) in the Anderson–Schulz–Flory (ASF) description is a constant. The ratio of the molar fraction (x) of any two carbon numbers in the product is related through the probability of chain growth (Equation 4.3):

$$x_n/x_m = a^{(n-m)} \tag{4.3}$$

In practice, three deviations from the ideal ASF formulation are found (Figure 12.3): the methane yield is usually higher, the C_2 yield is lower than predicted, and there is a slight curvature to the yield versus carbon number plot indicating a higher α-value for higher carbon numbers. The latter is only noticeable in LTFT synthesis.

The deviations of C_1 and C_2 from the ASF distribution can be described using the LTFT model proposed by Botes [3]. The model recognizes the main hydrocarbon desorption pathways outlined above, as well as the special case of the surface hydrodicarbyl C_2H_m where chain growth can take place at either carbon.

The chain length-dependent α-value gives rise to an ASF distribution in LTFT syncrude that exhibits two different regions, with a transition in the region C_8–C_{20} [7]. Different approaches have been suggested to describe these deviations. The change can be described mathematically by two different α-values, one for the light products (α_1) and one for the heavier products (α_2). It can also be described in terms of a chain length-dependent probability [3]. However, the apparent "transition" in chain growth probability observed in LTFT syncrude may indicate a vapor–liquid equilibrium effect, rather than an intrinsic change in the FT hydrocarbon synthesis mechanism.

4.2.2
Hydrocarbon Composition

Aliphatic hydrocarbons are primary products from FT hydrocarbon synthesis and the alkanes and alkenes are also the two most abundant compound classes (Table 4.1). Aromatic hydrocarbons are secondary products that are produced at higher operating temperatures; thus, LTFT syncrudes contain almost no aromatic compounds, whereas HTFT syncrudes contain a significant fraction of aromatics.

Identification of individual isomers present in the syncrude becomes increasingly difficult as the molecular mass increases. Even using techniques such as two-dimensional gas chromatography coupled with mass spectrometry (GC–GC and GC–MS), an isomer-by-isomer identification of even a comparatively "simple" naphtha-cut is a daunting task [8]. However, with improvements in analytical techniques, greater resolution is becoming possible. One approach to simplify the composition is to hydrogenate the syncrude, thereby converting the unsaturated compounds and oxygenates into the corresponding alkanes. This then makes it possible to define the skeletal structures of the isomers. Analysis of the unhydrogenated straight run syncrude also allows some unsaturated hydrocarbon functionalities, for example, of the *n*-1-alkenes, to be determined.

a) The alkane to alkene ratio shows a monotonic increase from C_3 upward and this is a universal trend in FT hydrocarbon synthesis. The alkene content of LTFT waxes is consequently low in comparison to that of the oily liquids. Thus, we can say that the probability of reactive desorption by hydrogenation as an alk*a*ne relative to desorption as an alk*e*ne increases with increasing molecular mass.
b) The alkane to alkene ratio is also affected by the FT metal, operating conditions, and reactor type. The FT metal determines how hydrogenating the catalyst is. When comparing Fe- and Co-based syntheses at similar conditions, the syncrude derived from Co-based FT has a higher alkane to alkene ratio because Co is more hydrogenating. The conditions determine the degree of desorption, thus operating a FT catalyst at higher temperatures leads to a decrease in the alkane to alkene ratio (and to a lower α-value), because the relative contribution of thermal desorption as opposed to reactive desorption is increased. In general, HTFT syncrude is more olefinic than LTFT syncrude.

   The reactor type affects the alkane to alkene ratio by changing the probability that alkenes will be hydrogenated in secondary reactions. The extent of further hydrogenation at comparable conversion is higher in reactor types that approach ideal plug flow reactor behavior than in reactors that approach continuous stirred tank reactor behavior. The syncrude from LTFT fixed bed operation is consequently more paraffinic than that from LTFT slurry bed operation [9].
c) While branched chain hydrocarbons are generally minor products, some are produced at percentage levels. The degree of branching in the hydrocarbons depends on the metal and is a function of the carbon number. Anderson and coworkers devoted considerable effort to quantify and describe branching [10]. In the naphtha range material, where the isomers of the hydrogenated syncrude could be identified, it was found that there was a fixed probability of branching (f) during chain growth. Each time the carbon chain grows by one carbon atom, that carbon can either be inserted into the chain for linear growth, or added to the growing chain as a branch (Figure 4.1). For Fe-based FT, $f = 0.115$, and for Co-based FT, $f = 0.035$. The ratio of linear to branched material therefore decreases with increasing carbon number and based on this formalism reasonable agreement was found between experimental and predicted values (Table 4.2). However, although the formalism holds for the lighter hydrocarbons, the predicted decrease in the ratio of linear to branched material with increase in carbon number does not hold true for the heavier hydrocarbons [11]. The branching probability (f) is therefore not constant, but starts to decrease at some point with increasing chain length.

Table 4.2 Prediction of branching in the products from FTS employing the formalism of Anderson [10].

Carbon number	Isomer	Relative abundance	Fe-based FTS (%)		Co-based FTS (%)	
			Predicted (f=0.115)	Observed	Predicted (f=0.035)	Observed
C_4	*n*-Butane	1	89.7	89.4	96.6	—
	2-Methylpropane	f	10.3	10.6	3.4	—
C_5	*n*-Pentane	1	81.3	81.2	93.5	95.0
	2-Methylbutane	$2f$	18.7	18.8	6.5	5.0
C_6	*n*-Hexane	1	73.6	78.8	90.4	89.6
	2-Methylpentane	$2f$	16.9	11.2	6.3	5.7
	3-Methylpentane	f	8.5	9.5	3.2	4.7
	2,3-Dimethylbutane	f^2	1.0	0.4	0.1	0
C_7	*n*-Heptane	1	66.7	66.0	87.5	87.7
	2-Methylhexane	$2f$	15.3	13.1	6.1	4.6
	3-Methylhexane	$2f$	15.3	19.1	6.1	7.7
	2,3-Dimethylpentane	$2f^2$	1.8	1.6	0.2	0
	2,4-Dimethylpentane	f^2	0.9	0.3	0.1	0

d) Although 1-alkenes dominate the olefinic products, some internal alkenes are also formed by isomerization. The extent of double bond isomerization depends on the catalyst, since it is a secondary reaction that requires the competitive readsorption of the alkenes on the catalyst. Two possibilities for the formation of internal alkenes exist. At low pressures, the hydrogenation–dehydrogenation equilibrium may result in internal alkene formation by dehydrogenation. This can explain the low 1-alkene to internal alkene ratio found in the syncrude from the German normal pressure Co-LTFT process: for example, the 1-alkene fraction was 36%, 28% and 18%, respectively, for the C_6, C_7, and C_8 alkenes [6]. Double bond isomerization through the action of {H} is the dominant pathway at higher pressures [12]. While the mechanisms are not established, we may hypothesize that the alkene is adsorbed on the metal through the carbons of the C=C bond. The adsorbed surface intermediate can then react with {H} in a second step to give a hydrocarbyl C_nH_m. Subsequently, this can either be hydrogenated to yield an alkane or it can desorb as an alkene by returning a hydrogen atom to metal surface (Figure 4.3). This is not an acid-catalyzed migration and does not involve skeletal isomerization. Cobalt is more active than iron for this type of double bond isomerization and different degrees of isomerization at constant degree of branching have been reported [13, 14].

e) Cyclic hydrocarbons can be formed by dehydrocyclization in another secondary reaction, related to double bond isomerization. In this case, the orientation of the adsorbed hydrocarbyl intermediate must be such that {H} desorbs

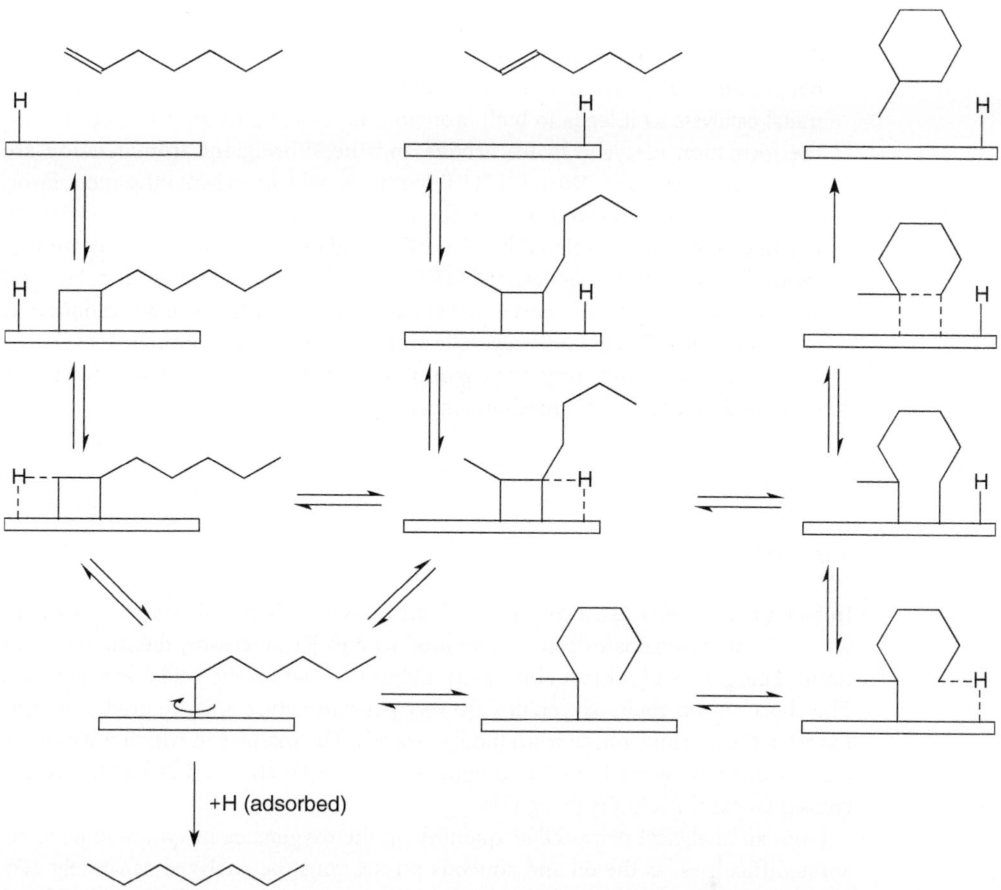

Figure 4.3 Pictorial representations of possible alkene interconversions on metal surfaces. Alkene adsorption on the FT metal may be followed by partial hydrogenation by adsorbed atomic hydrogen. Further hydrogenation results in desorption as an alkane. Hydrogen can be returned from an adjacent carbon, potentially leading to double bond isomerization, or it can be returned from another carbon that can enable the formation of a five- or six-membered ring.

from a carbon in such a way that allows ring closure (Figure 4.3). Under FT synthesis conditions, the likelihood of alkane adsorption and dehydrogenation is low due to the competitive adsorption and the high H_2 partial pressure. Thus, the desorption step for cyclization may be expected to be more demanding than for double bond isomerization. At higher temperatures, where the adsorption of the alkyl chain is less strong, there is sufficient mobility that the alkyl group can more easily "flip over" to enable ring closure. Thus, the cyclic content of HTFT syncrude is consequently higher than that of LTFT syncrude.

f) Aromatics are formed at higher temperatures, and probably arise by dehydrogenation of cyclohexyl species. Aromatization is favored by low H_2 partial pressures and high temperatures and may be related to dehydrogenation on metal catalysts as it leads to both aromatics and coking of the FT catalyst. Both the formation of cyclic hydrocarbons and the subsequent aromatization are demanding reactions. During HTFT synthesis, which presents the most favorable conditions for aromatization, the total aromatics content in the syncrude reaches percentage levels (Table 4.1). The highest aromatics concentration is found in the residue fraction of HTFT syncrude. The combined distillate and residue fraction from industrial Fe-HTFT operation contains 26.3% monocyclic aromatics and 0.7% di- and polycyclic aromatics [15]. This dominance of monocyclic aromatics with long alkyl groups in the heavy HTFT syncrude can be explained in terms of the mechanism in Figure 4.3.

4.2.3
Oxygenate Composition

In fact, the most abundant compound from FT synthesis contains oxygen, namely, water. Organic oxygenates can also be produced in FT processes; the main classes found being alcohols (most abundant), aldehydes, carboxylic acids, and ketones. The short carbon chain oxygenates are very polar in nature and are predominantly found in the aqueous phase from the FT process. The manner in which the oxygenates partition between the oil and aqueous phases (Section 4.3.2) has important consequences for refinery design [1].

From an analytical perspective, quantifying the oxygenates in syncrude presents some difficulties, as the oil and aqueous phases must be analyzed separately. Wet chemical analysis techniques for alcohols and carbonyl compounds have significant uncertainty. Chromatographic analysis employing a flame ionization detector is likewise challenging, because the oxygenates have different response factors depending on their functionality [16]. The oxygenates are also more reactive and susceptible to thermal decomposition, which can further complicate identification and quantification. It is therefore not surprising that many published FT studies ignore oxygenates and focus exclusively on the hydrocarbons. The main oxygenate classes found in syncrude and how the selectivity to each is affected by the FT process are considered in the following [1, 17].

a) Based on the flowchart in Figure 4.2, the ratio between oxygenates and hydrocarbons following reaction with {CO} is governed by the probability that chain growth on the FT catalyst is terminated without deoxygenation (see Chapter 12). Generally speaking, oxygenate selectivity is found to increase with decreasing hydrogenating power of the FT metal, with Co being more hydrogenating than Fe. Oxygenate selectivity is further affected by H_2 partial pressure in the reactor, by the operating conditions, and by the reactor technology. It will also be affected by the operating temperature, since the probability increases that oxygen-containing surface intermediates will be desorbed before they are deoxygenated at higher temperatures. The relatively high concentration of light oxygenates found in HTFT syncrude is presumably because thermal desorption reduces the α-value and increases the probability that the desorbed species retains an oxygenate functionality.

b) It has been shown that oxygenates are readsorbed and hydrogenated, when low molecular mass alcohols, ketones, or carboxylic acids are cofed under FT process conditions and a sizable fraction is converted [18]. The extent of conversion must depend on how effectively the oxygenates can be adsorbed in competition with the other compounds present in the reactor. The conversion also depends on the temperature since the alcohols, aldehydes, and carboxylic acids (Equations 4.4–4.6) are potentially in thermodynamic equilibrium and readily interconvert at typical HTFT reaction temperatures (>320 °C) [19].

$$RCH_2OH + H_2O \rightleftharpoons RCOOH + 2H_2 \tag{4.4}$$

$$RCH_2OH \rightleftharpoons RCHO + H_2 \tag{4.5}$$

$$RCHO + H_2O \rightleftharpoons RCOOH + H_2 \tag{4.6}$$

c) Alcohols comprise about 90% of the oxygenates in LTFT syncrude and about 40–60% of the oxygenates in HTFT, while carboxylic acids and other carbonyl compounds are minor components. Alcohols can be primary products (Figure 4.3), as well as arising by subsequent partial hydrogenation of aldehydes, ketones, and, to a lesser extent, carboxylic acids. The alcohols are also consumed: over Fe-based catalysts, readsorption and further chain growth are possible, whereas this does not seem to occur over Co-based catalysts [20].

d) Aldehydes can also be primary products from the FT process, whereas ketones are secondary products that probably arise by decomposition of carboxylic acids (Equation 4.7) [19, 20]. Since the ketonization reaction requires high temperature, fewer ketones are found in LTFT than in HTFT syncrude. As ethanoic (acetic) acid is the major carboxylic acid, most of the ketones are β-ketones, as would be expected from ethanoic acid decomposition [19]. Both aldehydes and ketones can readily be hydrogenated and part of the 1-alcohols formed in secondary processes arise by aldehyde hydrogenation, whereas all the internal alcohols are derived from partial ketone hydrogenation.

$$RCOOH + {'RCOOH} \rightarrow R(CO)'R + H_2O + CO_2 \tag{4.7}$$

e) Carboxylic acids may also be primary products from FT synthesis. A positive correlation of carboxylic acid selectivity to CO_2 and an inverse correlation to 1-alkene selectivity have been noted [21]. The carboxylic acids can readsorb on the FT catalyst to produce metal carboxylate species that can decompose to yield ketones. However, the metal carboxylates can be remarkably stable under LTFT conditions [22], and those that end up in the syncrude are particularly troublesome and make upgrading of the syncrude more difficult [1].

4.3
Syncrude Recovery after Fischer–Tropsch Synthesis

At reaction conditions, the HTFT reaction product is a two-phase gas–solid mixture, while the product from LTFT synthesis is a three-phase gas–liquid–solid mixture. As the per pass conversion during FT synthesis is usually maintained at an intermediate level to maximize reactor productivity and minimize catalyst deactivation, synthesis gas conversion is not 100%. The lower the conversion per pass, the higher the productivity: this is expressed as kg/s syncrude produced per m^3 reactor volume. However, the cost-efficiency of syngas–syncrude separation, combined with the cost of syngas recycle, presents a trade-off that indicates better efficiency at higher per pass conversion. At high conversion, deactivation of the catalyst becomes an issue, since the hydrogen partial pressure is lower and the water partial pressure is higher, which promote coking and oxidation, respectively. FT catalyst deactivation affects not only the synthesis but also the product yield and refining (see Chapters 8–10) [1].

The efficiency of syngas separation from the reactor product determines the amount of inert material contained in the unconverted syngas. Since inert material builds up over time if it is not purged from the system, the ability to exclude inert material from the syngas recycle is critical in reducing syngas loss to purging. Although purged syngas may still be employed as a fuel gas, it has a much larger environmental footprint than the carbon-based feed before it was turned into syngas.

The same principle applies to the recovery of syncrude and refining. As the total "combined carbon" ("C") moves through the facility, its environmental cost increases. Once the "C" reaches the FT process, it already has an E-factor [23] cost of 1–2 kg C waste per kg C in the syngas [35]. The environmental responsibility associated with syncrude recovery and the recycling of unconverted syngas is consequently significant. When the unconverted syngas or syncrude is employed as a process fuel, it has two to three times the CO_2 footprint of fueling the process by the carbon-based raw material. Although better separation and recovery may add to the complexity and cost of an FT-based facility, the currently increased levels of environmental responsibility suggest that some of the design decisions in present industrial practice should be revisited. In an economic context, it speaks to the value that is attached to "C" and the environment.

4.3.1
Stepwise Syncrude Cooling and Recovery

The principles behind stepwise cooling and recovery of syncrude are the same for different FT technologies, but the nature of the technology affects both the detailed design and what has to be accomplished. The first separation step occurs within the FT reactor and involves the separation of the catalyst from the reaction product. Subsequent cooling and separation take place outside the reactor and result in multiple feed streams for refining. When properly designed, this is one of the

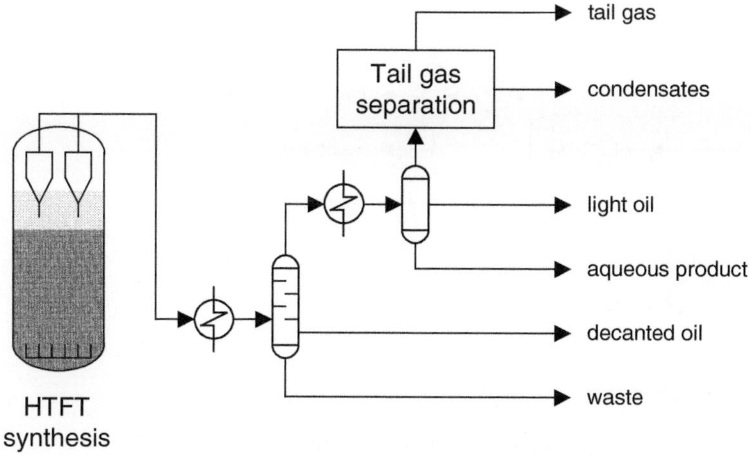

Figure 4.4 Generic syncrude cooling and recovery section typical of industrial HTFT operation.

advantages inherent in FT processes, because it allows the downstream refinery to receive prefractionated cuts as feed, rather than a single mixed feed [1].

In HTFT technology (Figure 4.4), which employs fluidized bed reactors, the gaseous product is separated from the solid catalyst by cyclones in the reactor. In the circulating fluidized bed design, the direction of gas flow must be reversed before passing through the cyclones; this aids gas–solid separation by reducing the solid loading entering the cyclones. In the fixed fluidized bed design, the gas directly enters the cyclones. Irrespective of the design, some catalyst particles are carried over in the gaseous product leaving the reactor. As the gaseous product is cooled down, the heaviest material condenses and much of the remaining solids are trapped by the heavy liquid. The catalyst can be recovered in a mixture with part of the heavy organic liquid, which becomes a waste stream. The "clear" liquid decanted from this heavy oil is the decanted oil, which contains mainly atmospheric residue and distillate range material. Further cooling condenses the remainder of the distillate and naphtha range material as a light oil fraction, together with the water that forms the bulk of the aqueous product. Oxygenates partition between the organic and aqueous phases (Section 4.3.2). The light oil also contains dissolved lighter hydrocarbons that are usually stripped from the oil to produce stabilized light oil (SLO). The remaining gaseous material contains unconverted syngas, some of the light naphtha, and most of the lighter hydrocarbons. A substantial fraction of HTFT syncrude is contained in the normally gaseous product (Table 4.1). The C_3 and heavier material can be recovered by pressure distillation, but the separation and recovery of ethane, ethene, and methane from the unconverted syngas and CO_2 mixture require cryogenic distillation. The advantage of cryogenic distillation is that it improves the carbon efficiency considerably and enables petrochemical opportunities based on C_2 hydrocarbons.

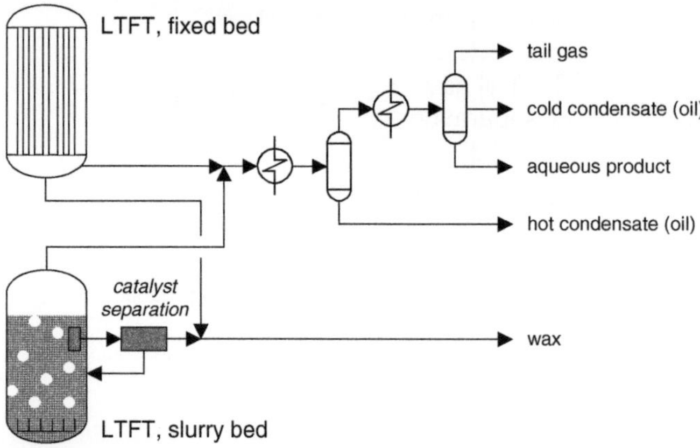

Figure 4.5 Generic syncrude cooling and recovery section typical of industrial LTFT operation, with the difference between fixed bed and slurry bed FTS shown.

The first separation step in LTFT processes depends on the reactor type selected (Figure 4.5). Separating the syncrude (liquid and gas) from the solid FT catalyst presents no problems for fixed bed reactors, where the catalyst is naturally retained as a packed bed. In the slurry-phase operation, the catalyst particles are smaller and suspended in the liquid phase. Depending on the technology, catalyst separation (e.g., filtration) in the slurry reactor and/or downstream from the slurry reactor is required. The wax is a liquid under FT reaction conditions and it does not require cooling; in fact, the opposite is true, cooling may result in the wax congealing with serious operational consequences. The gaseous product, which contains the unconverted syngas, is stepwise cooled and recovered in an analogous way to HTFT syncrude. Unfortunately, the nomenclature employed industrially is confusing, with the LTFT light oil being referred to as "condensate." The combined LTFT hot and cold condensates are equivalent to the HTFT light oil. Oxygenate partitioning takes place between the aqueous product and cold condensate (Section 4.3.2). Although no tail gas separation section is shown in Figure 4.5, the syncrude can be separated by pressure distillation and cryogenic distillation to recover the C_3 and heavier material and C_2 and lighter material from the unconverted syngas. The normally gaseous material from LTFT synthesis constitutes a smaller fraction of the total syncrude than that in the case of HTFT synthesis; but in the LTFT synthesis, the per mass syngas conversion is lower and the need to recycle the unconverted syngas is greater.

4.3.2
Oxygenate Partitioning

Under reaction conditions, the water produced during FT synthesis is in the vapor phase, but when the syncrude is cooled down (Figures 4.4 and 4.5), water is condensed together with the organic products. Water is only sparingly soluble (typically <0.1%) in the apolar hydrocarbon-rich organic liquid and the water forms a separate polar liquid phase.

The oxygenates, alcohols, aldehydes, carboxylic acids, and ketones all have a polar oxygen-containing functional group on an apolar hydrocarbon backbone. The oxygenate functionality is capable of polarizing the neighboring atoms, but this polarization is a short-range effect and it does not extend much beyond the nearest neighbors. As the chain length of the hydrocarbon backbone increases, the molecule becomes increasingly apolar in nature, despite the presence of the oxygenate functionality. This change in polarity affects the phase preference of the molecule. Short-chain oxygenates that are predominantly polar preferentially dissolve in the aqueous phase, whereas the longer chain oxygenates that are more apolar preferentially dissolve in the hydrocarbon-rich phase. Phase preference is governed by thermodynamics through Gibbs free energy minimization and can superficially be described as "like dissolves like." Depending on the design of the syncrude cooling and recovery section, there may be sufficient time for oxygenate partitioning to reach equilibrium. Irrespective of this, partitioning is sensitive to temperature and the compositions of the oil and aqueous products are affected by the design. The composition of the FT aqueous product consequently depends on the FT technology and the design of the phase separation step during syncrude cooling and recovery. Generally speaking, C_4 and the lighter oxygenates preferentially partition into the aqueous product, but there is not a clear point of division; thus, the aqueous product may contain some C_5 and heavier oxygenates and the organic liquid may contain some C_4 and lighter oxygenates [1].

4.3.3
Oxygenate Recovery from the Aqueous Product

On a mass basis, around half of the syngas that is converted during FT synthesis ends up as water. The light oxygenates dissolved in this aqueous phase are present in a dilute solution. Currently, there are no facilities that can extract all the oxygenates from the aqueous product, and carboxylic acids are generally not recovered. The most complete recovery is practiced in conjunction with Fe-HTFT syncrude, because the aqueous product contains around 10% of the total syncrude mass (Table 4.1). Industrially, the least effort is expended on Co-LTFT syncrude and the oxygenates are treated as waste products, although this does not have to be the case.

Industrial designs for oxygenate recovery from the aqueous product all follow the same basic strategy (Figure 4.6) and differ mainly in the extent to which separation is performed [1]. Separation is complicated by numerous oxygenate–water azeotropes. The least "C"-efficient design treats the total aqueous product as a wastewater stream, which avoids separation altogether. Whenever oxygenates are recovered, the first step is a primary distillation to separate most of the nonacid oxygenates from the bulk of the water and carboxylic acids. The oxygenate-rich product still contains around 25% water and can be further separated into carbonyl-rich and alcohol-rich products. Most of the water is retained by the

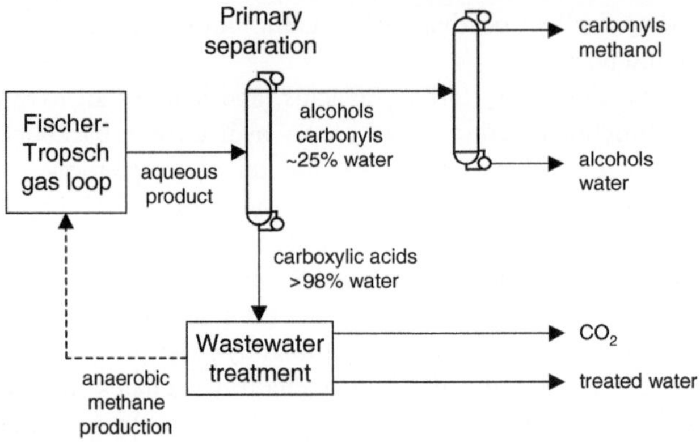

Figure 4.6 Generic flowchart for the recovery of oxygenates from a Fischer–Tropsch aqueous product.

alcohol-rich product. Alternatively, the oxygenate-rich product from primary separation can be partially hydrogenated to convert the ketones and aldehydes into their corresponding alcohols (not shown in Figure 4.6), thereby simplifying subsequent refining. The degree of subsequent separation depends on which products are of interest.

4.4
Fuel Products from Fischer–Tropsch Syncrude

The fuel products that can be obtained from the FT syncrude are determined by the FT refinery design and all the major fuel types can be produced [1, 24]; they are listed in approximate order of increasing boiling points in Section 4.4. In this respect, syncrude is not much different from crude oil. The refinery design and not the FT process determines the types and quality of the fuels. The raw material selected, the syngas production technology, and the FT process affect only the ease and efficiency of refining. Refining advantages can be found for each combination of technologies and these will be pointed out for the main fuel types.

4.4.1
Synthetic Natural Gas

Natural gas is a convenient energy carrier. When appropriate pipeline and consumer infrastructure creates a market for natural gas, there is also an opportunity to produce synthetic natural gas. SNG is a methane-rich gas (Box 4.1) that can be produced and sold like natural gas. This is not a product that is considered in gas-to-liquids facilities, which by definition exploit the lack of a market for natural gas. However, in the case of solid feed materials, such as coal, biomass, or waste, there may be a market for SNG. The Great Plains synfuels plant in North Dakota is a

Box 4.1 Synthetic Natural Gas (SNG) specifications

Pipeline gas specifications vary by region and supplier. A typical specification for synthetic natural gas (SNG) is minimum 75% methane. Limits are usually placed on other compounds: 10% ethane, 4% H_2, 3% N_2, and 2% CO_2. There is likewise a specification on the minimum heating value, with that of pure methane being around 40 MJ/m^3.

prime example of this approach, but without an associated FT process [25]. Coal is gasified and the syngas is cleaned, as described in Chapter 2, before the syngas is converted into methane for sale as SNG (see Section 2.1).

Methane is invariably produced during FT synthesis. In FT facilities employing a solid feed as raw material, the syngas production, cleaning, and conditioning units do not necessarily include a natural gas reformer. Without a gas reformer methane cannot be turned back to syngas unless it can be coprocessed in the gasifier. The methane can be used as fuel gas, but, as explained in Section 4.3, when FT-derived methane is used as fuel gas within the facility, it has a three times higher CO_2 footprint than heating by the raw material. In such instances, SNG is an attractive fuel product, because it takes credit for the conversion of an inconvenient solid carbon carrier into a more convenient and clean energy carrier.

4.4.2
Liquefied Petroleum Gas

Liquefied petroleum gas (LPG) is a generic term that refers to C_2–C_4 alkane mixtures in their liquid state, that is, under pressure at ambient temperature. It is a useful fuel for mobile and remote applications, because it requires only moderate pressure (<2 MPa) to remain in liquid form at ambient temperature and it readily vaporizes when the pressure is released. This is the main advantage of LPG as fuel, because it is liquid for transportation and storage, but gaseous for use. The composition of LPG consequently depends on the season and the location where it is marketed (Box 4.2).

The normally gaseous hydrocarbons are produced as primary products during the FT process and are also generated during some refinery conversion processes,

Box 4.2 Compositions of Liquefied Petroleum Gas (LPG)

Different grades of LPG can be produced. In regions where the winter temperatures drop below 0 °C, the main constituent of LPG is propane, because butanes are liquids at such conditions. In hot climates, LPG can be a butane–propane mixture, but, generally speaking, butanes are more valuable as a blending component for motor gasoline. Ethane is commonly found in large concentration in refinery gas streams and is not as often found in high concentration in LPG mixtures.

such as hydrocracking (Chapter 14). The C_2–C_4 alkenes (olefins) have value as petrochemicals and can be readily converted into transportation fuels, but the conversion of the C_2–C_4 alkanes (paraffins) is more demanding. When the refinery does not include conversion units for the light alkanes, LPG is a convenient fuel product.

4.4.3
Motor Gasoline

Motor gasoline is the most refining intensive transportation fuel to produce. The refining effort is not related to FT syncrude, but to the molecular requirements of the fuel e.g. the alkanes must be highly branched to have a high octane number. It is easier to produce motor gasoline from HTFT than LTFT syncrude and it is easier to produce motor gasoline from Fe-LTFT than Co-LTFT syncrude [1, 24]. The on-specification motor gasoline from FT syncrude is not fundamentally different from that produced from crude oil. It can be distributed using the same infrastructure as employed for crude oil-derived motor gasoline. Motor gasolines produced from FT syncrude have been used in fuel blends and on their own in Germany (1930–1940s), the United States (1950s), and South Africa (1950s until the present) [26]. The composition of motor gasoline changed over time to reflect the limitations imposed on composition and performance criteria by the fuel specifications. The properties in Table 4.3 reflect contemporary specifications [27]. The only two industrial FT refineries that produce motor gasoline at present are in South Africa, the PetroSA (Mossgas) facility in Mossel Bay and the Sasol Synfuels facility in Secunda.

Table 4.3 Motor gasoline properties from industrial FT facilities in relation to the contemporary fuel specifications.

Motor gasoline property	PetroSA unleaded	Sasol Synfuels unleaded	Sasol Synfuels lead replacement[a)]	South African SANS 1598	European EN228:2004
Research octane number	95	93	93	93/95 min	95 min
Motor octane number	85	83	83	83/85 min	85 min
Density at 20 °C (kg/m^3)	748	729	723	710–785	720–775
Reid vapor pressure (kPa)	72	67	66	75 max[b)]	60 max[b)]
Alkene content (mass%)	8	30	30	Not regulated	18 max
Aromatic content (vol%)	37	29	25	50 max	35 max

a) Methylcyclopentadienyl manganese tricarbonyl (MMT) is in some cases added as a lead replacement for antiknock properties.
b) Region- and season-specific specifications.

Motor gasolines with properties superior to that shown in Table 4.3 can be produced from syncrude with appropriate refinery design [1, 24].

4.4.4
Jet Fuel

Aviation turbine fuel, or jet fuel, is the easiest transportation fuel type to refine from FT syncrude. It is easier to produce from LTFT than from HTFT syncrude and it is easier to produce jet fuel from Fe-LTFT than Co-LTFT syncrude [1, 24]. The biggest obstacle to commercial jet fuel production from FT syncrude is the Jet A-1 specifications. Properly refined FT-derived fully synthetic jet fuel has properties well within the range of jet fuel obtained by crude oil refining. Blends of FT- and crude oil-derived kerosene (semisynthetic jet fuel), as well as HTFT-derived kerosene mixtures (fully synthetic jet fuel), have been successfully tested and certified for commercial aviation use [28]. The inclusion of a minimum 8 vol% aromatics requirement and the distillation slope for synthetic jet fuel are sensible technical requirements based on elastomer compatibility in a mixed fuel environment and engine performance. The Jet A-1 specifications was previously written in such a way that it depended not only on the jet fuel properties, but also on the exact refinery configuration employed to produce the jet fuel. This overly restrictive requirement was removed in DEF-STAN 91-91 Issue 7 of 2011 [29].

From a strategic perspective, FT-derived jet fuel is also a desirable military fuel [30]. Military specifications for jet fuel are very similar to that for Jet A-1, but without some of the above limitations imposed on synthetic fuels for commercial aviation.

4.4.5
Diesel Fuel

Considerable marketing effort established diesel fuel as the main product from the FT process. The distillates produced by the refining of HTFT and LTFT syncrudes are quite different (Table 4.4) [27]. This is related to the difference in the straight run properties of the distillate and residue fractions (Table 4.1). The distillate and residue fractions of HTFT syncrude are aromatic, whereas the same fraction of LTFT syncrude is paraffinic. Selective hydroprocessing (hydrotreating and hydrocracking) of HTFT syncrude yields a typical petroleum-like diesel fuel, but hydrocracked LTFT distillate is a very high cetane number, low aromatic, and low-density distillate. These properties are due to the high content of linear and mono methyl branched alkanes.

Depending on the specification of the transportation fuel market, a typical LTFT-derived distillate may not qualify for sale as diesel fuel. The density of LTFT-derived distillate is typically around 770–780 kg/m^3 and in regions where the minimum density of diesel fuel is regulated, FT distillate can only be employed as a blending component. In fact, on a molecular level, FT syncrude is poorly suited for diesel fuel production when the minimum density is regulated [4]. It is difficult to increase the density without sacrificing either quality (cetane number) or yield. Most of the other issues that may be encountered in a mixed fuel environment

Table 4.4 Diesel fuel properties from industrial FT facilities in relation to the contemporary fuel specifications.

Diesel fuel property	PetroSA GTL diesel (HTFT)	Sasol Synfuels CTL diesel (HTFT)	Oryx GTL distillate (LTFT)	South African SANS 342	European EN590:2004
Cetane number	52	55	87	45 min	51 min
Density at 20 °C (kg/m^3)	815	829	771	800 min	820–845
Viscosity at 40 °C (cSt)	2.5	2.2	2.4	2.2–5.3	2.0–4.5
Flash point (°C)	83	77	65	62 min	55 min
Cold filter plugging point (°C)	−4	−6	−7	−4 max[a)]	Regulated[a)]
Aromatic content (mass%)	~15	~25	<1	Not regulated	Not regulated
Polynuclear aromatics (mass%)	<1	<1	<1	Not regulated	11 max

a) Region- and season-specific specifications.

(i.e., when consumers may select between crude oil and FT-derived diesel fuels) can be readily overcome.

It was the original intent to market LTFT distillate as a blending component and not as a diesel fuel (Box 4.3). Yet, the changes in transportation fuel specifications and subsequent changes in crude oil refining practice eroded many of the

Box 4.3 LTFT distillate characteristics

When Shell started marketing LTFT GTL distillate from the Shell middle distillate synthesis process in Europe, the Fischer–Tropsch material was blended into their crude oil-derived diesel fuel. At that stage, Euro-2 (1994) specifications were coming into effect, with a maximum sulfur limitation of 500 μg/g. The zero sulfur and high cetane number LTFT distillate had a significant blending advantage. Time moved on and in many regions the sulfur specification of diesel fuel has been reduced to maximum 10–15 μg/g. The crude oil had to be refined to a low sulfur product and the low sulfur blending advantage of LTFT distillate was lost. There are still cetane number and density blending advantages, but even those have been eroded by the need to produce a more refined diesel fuel directly from crude oil. LTFT distillate can no longer be differentiated on quality, except in specialized applications that require a low aromatic distillate.

advantages of LTFT distillate compared to crude oil-derived diesel fuel. This does not detract from the good quality of LTFT distillate, but illustrates how much the transportation fuel business changed in the past two decades.

4.5
Lubricants from Fischer–Tropsch Syncrude

Lubricant base oils are classified according to the American Petroleum Institute (API) guidelines into five groups. The first three groups (API Group I, II, and III) refer to lubricants of increasing quality that is refined directly from crude oil. FT syncrude can be refined to produce lubricant base oils in API Group III, which have a sulfur content of less than 300 μg/g, aliphatic content of >90%, and viscosity index (VI) >120.

High-viscosity index lubricant base oils in API Group III can be prepared by hydroisomerization (hydrodewaxing) of slack wax from crude oil dewaxing. The same technology can be applied to LTFT waxes to produce isoparaffinic base oils, with viscosity indexes in the range of 130–140. Such good quality lubricant base oils have been commercially produced from LTFT wax with the Shell middle distillate synthesis process since the mid-1990s [31]. The production of these types of lubricant base oils form an integral part of the Shell Pearl GTL facility in Qatar [32]. In order to appreciate the quality of the lubricant base oil that can be produced from LTFT wax, a comparison within the API Group III is provided (Table 4.5) [31].

API Group III lubricant base oils can also be prepared from HTFT residue [33]. The processing that is required is more severe on account of its high aromatic content and the product resembles an average crude oil-derived product. Branched

Table 4.5 Comparison of lubricant base oil commercially prepared from LTFT wax (GTL-5) and a sample of 17 different crude oil-derived API Group III lubricant base oils.

Lubricant base oil property	GTL-5	Crude oil derived		Preference
		Minimum	Maximum	
Kinematic viscosity at 100 °C (cSt)	4.5	4.0	5.0	—
Viscosity index	144	120	141	Higher
Dynamic viscosity at −25 °C (cP)	816	729	2239	Lower
Noack volatility index (mass%)	7.8	10.4	14.8	Lower
Pour point (°C)	−21	−24	−12	Lower
Composition (mass%)				
Alkanes	100	47.3	80.9	Higher[a)]
Cycloalkanes (monocyclic)	0	18.7	28.8	Higher[a)]
Cycloalkanes (polycyclic)	0	5.3	22.2	Lower
Aromatics	0	0	12	Lower

a) Branched alkanes are preferred over monocyclic cycloalkanes, but both make good quality lubricant base oils.

alkanes are the best compound class for API Group III oils and LTFT wax is clearly preferred over HTFT residue as starting material for lubricant base oil refining.

Poly-alpha-olefin (PAO) oils can be prepared from linear 1-alkenes by selective oligomerization. It is also possible to prepare lubricant base oils from the oligomerization of linear internal alkenes, called poly-internal-olefin (PIO) base oils. Collectively, these lubricant base oils are classified as API Group IV and generally have high-viscosity indexes (typically VI >140). The quality of the PAO oil depends on the carbon chain length of the linear 1-alkene from which it is prepared. The best balance of properties is achieved with oils prepared from 1-decene, but good quality oils can be obtained with somewhat both shorter and longer 1-alkenes. These types of oil were prepared from FT-derived 1-alkenes in the past [34]. When the linear 1-alkenes are properly purified from the FT syncrude, there is no difference in the PAO lubricant base oils prepared from the FT- and ethene-derived linear 1-alkenes. Since the linear 1-alkene content is higher in HTFT syncrude than in LTFT syncrude, the former is the more obvious starting material for 1-alkene recovery and purification for PAO lubricant production.

The last of the API lubricant base oil classes (API Group V) contains all the other synthetic and nonhydrocarbon specialty lubricants. Many of these synthetic lubricants are oxygenates, for example, based on polyols, diesters, polyalkylene glycols, and polyphenyl ethers. Oxygenates are produced from the FT process and can in principle serve as feed for the production of API Group V lubricants. Unlike the other lubricant types, little has been reported on the application of FT syncrude for API Group V lubricant base oil production.

4.6 Petrochemical Products from Fischer–Tropsch Syncrude

Since most petrochemicals are produced from crude oil, it is by analogy possible to produce the same selection of petrochemicals from FT syncrude. In addition, there are also chemical products that can be produced in association with syngas production and purification, such as inert gases from air separation, nitrogen-based chemicals via ammonia, and chemicals from pyrolysis liquids obtained during gasification [17].

The following discussion relates to the petrochemicals that are conveniently recovered or prepared from FT syncrude. Once a "platform" chemical is produced, it enables the production of derivative chemicals. This leads to a rapidly expanding network of possibilities.

4.6.1 Alkane-Based Petrochemicals

Two main classes of alkane-based petrochemicals are found: paraffinic solvents and paraffinic waxes. Both products can be readily produced from LTFT syncrude by

Box 4.4 Petrochemicals from LTFT at Sasol 1

The Sasol 1 facility (1955) was originally designed to mainly produce transportation fuels. In the 1990s, the facility was converted into a chemicals production facility. The Fischer–Tropsch synthesis section was converted to LTFT only. The main products are paraffinic solvents and paraffinic waxes. It demonstrates that LTFT syncrude is well suited for the production of alkane-based petrochemicals.

hydrotreating and fractionation (Box 4.4). It is more difficult to produce paraffinic solvents from HTFT syncrude, because the naphtha and distillate contain aromatics. The HTFT residue is very aromatic and it is not a practical feed material for the production of paraffinic waxes. Within the class of paraffinic solvents and paraffinic waxes, there are multiple grades that are differentiated based on boiling range or congealing point.

4.6.2 Alkene-Based Petrochemicals

The three main commodity alkenes based on global market size are ethene, propene, and butadiene. Of these, ethene and propene are primary products from FT hydrocarbon synthesis and can be recovered directly from the syncrude. These two molecules are particularly abundant in HTFT syncrude, constituting around 20% of the straight run product. In general, the attractiveness of different types for the production of alkene-based petrochemicals is Fe-HTFT > Fe-LTFT > Co-LTFT.

Longer carbon chain alkenes are also abundant in HTFT syncrude. The extraction and purification of linear 1-pentene, 1-hexene, 1-heptene, 1-octene, and C_{12}/C_{13} 1-alkenes from HTFT syncrude is practiced industrially [1]. For many applications, there is little difference between the 1-alkenes from ethene oligomerization and the 1-alkenes extracted from FT syncrude. Trace level impurities differ between the two sources of 1-alkenes, and in sensitive applications the trace level impurities may cause performance differences.

Hydroformylation of 1-alkenes is one application that has good synergy with an FT process, since both require syngas as feed material. It is therefore not surprising to find that hydroformylation followed by aldehyde to alcohol hydrogenation is industrially used in conjunction with FT 1-alkenes for the production of 1-butanol, 1-octanol (for dehydration to 1-octene), and detergent alcohols.

Ethene hydration also has good synergy with an FT process, because the purification steps required are similar to those employed for FT aqueous product upgrading [1]. The implementation of ethene hydration in an FT facility is consequently considerably cheaper than its implementation in a conventional petrochemical facility.

4.6.3
Aromatic-Based Petrochemicals

Benzene, toluene, and the xylenes (BTX) are major commodity chemicals. FT syncrude is not generally associated with the production of these aromatic-based petrochemicals, yet it has some advantages over crude oil as feed material for such production.

The naphtha range material can be converted using nonacidic Pt/L-zeolite-based reforming technology, which is more efficient for aromatics production from C_6–C_8 naphtha than conventional catalytic naphtha reforming [17]. The nonacidic Pt/L-zeolite catalyst is extremely sulfur sensitive and the technology requires deep desulfurization of the feed when employed with a crude oil-derived naphtha. The nonacidic Pt/L-zeolite catalyst also yields benzene in high yield from *n*-hexane, reducing dependence on toluene transalkylation to produce benzene.

FT syncrude contains a significant fraction of straight run light alkanes, and some refinery conversion processes, such as hydrocracking, may also produce additional light alkanes. Instead of using these sulfur-free alkanes for fuel production, they can be converted into aromatics over a metal-promoted H-ZSM-5 aromatization catalyst [1, 17]. There is a natural synergy with FT tail gas processing. The separation strategy for ethene and/or propene can be simplified by combining aromatic alkylation with aromatization [1]. A further salient benefit is that it avoids transportation of benzene, which has many restrictions due to its carcinogenicity.

4.6.4
Oxygenate-Based Petrochemicals

The major classes of oxygenates in FT syncrude have applications as petrochemicals. The most abundant and useful species are the short-chain oxygenates that are dissolved in the FT aqueous product. Based on production volume for the recovery and purification of oxygenate-based petrochemicals, the syncrude preference is Fe-HTFT > Fe-LTFT > Co-LTFT.

Industrially, some alcohols and ketones are recovered as pure compounds from the FT aqueous product. These include bulk commodities such as methanol, ethanol, 2-propanol (isopropanol), and acetone, as well as specialty chemicals such as 1-propanol. Despite the value and abundance of the carboxylic acids in the FT aqueous product, there is at present no efficient commercialized technology for the recovery of these compounds from the dilute aqueous solution.

References

1 De Klerk, A. (2011) *Fischer–Tropsch Refining*, Wiley-VCH Verlag GmbH, Weinheim.
2 De Klerk, A. (2008) *Green Chem.*, **10**, 1249–1279.
3 Botes, F.G. (2007) *Energy Fuels*, **21**, 1379–1389.
4 De Klerk, A. (2009) *Energy Fuels*, **23**, 4593–4604.
5 Sie, S.T., Senden, M.M.G., and Van Wechem, H.M.W. (1991) *Catal. Today*, **8**, 371–394.

6 Asinger, F. (1968) *Paraffins Chemistry and Technology*, Pergamon, Oxford.

7 Huyser, J., van Vuuren, M.J., and Kupi, G. (2009) *Advances in Fischer–Tropsch Synthesis, Catalysts, and Catalysis* (eds B.H. Davis and M.L. Occelli), Taylor & Francis, Boca Raton, pp. 185–197.

8 Van der Westhuizen, R., Crouch, A., and Sandra, P. (2008) *J. Sep. Sci.*, **31**, 3423–3428.

9 Satterfield, C.N., Huff, G.A., Jr., Stenger, H.G., Carter, J.L., and Madon, R.J. (1985) *Ind. Eng. Chem. Fundam.*, **24**, 450–454.

10 Anderson, R.B. (1956) Catalysis, in *Hydrocarbon Synthesis, Hydrogenation and Cyclization*, vol. **IV** (ed. P.H. Emmett), Reinhold, New York, pp. 257–371.

11 Le Roux, J.H. and Dry, L.J. (1972) *J. Appl. Chem. Biotechnol.*, **22**, 719–726.

12 Twigg, G.H. (1939) *Trans. Faraday Soc.*, **35**, 934–940.

13 Weitkamp, A.W., Seelig, H.S., Bowman, N.J., and Cady, W.E. (1953) *Ind. Eng. Chem.*, **45**, 343–349.

14 Asinger, F. (1968) *Mono-Olefins Chemistry and Technology*, Pergamon, Oxford.

15 Leckel, D.O. (2009) *Energy Fuels*, **23**, 38–45.

16 Sternberg, J.C., Gallaway, W.S., and Jones, D.T.L. (1962) *Gas Chromatography* (eds N. Brenner, J.E. Callen, and M.D. Weiss), Academic Press, New York, pp. 231–267.

17 De Klerk, A. and Furimsky, E. (2010) *Catalysis in the Refining of Fischer–Tropsch Syncrude*, Royal Society of Chemistry, Cambridge, UK.

18 Dry, M.E. (1981) *Catalysis Science and Technology*, vol. **1** (eds J.R. Anderson and M. Boudart), Springer, Berlin, pp. 159–255.

19 Dry, M.E. (2004) *Stud. Surf. Sci. Catal.*, **152**, 196–257.

20 Weitkamp, A.W. and Frye, C.G. (1953) *Ind. Eng. Chem.*, **45**, 363–367.

21 van Vuuren, M.J., Govender, G.N.S., Kotze, R., Masters, G.J., and Pete, T.P. (2005) *Prepr. Pap.-Am. Chem. Soc., Div. Petrol. Chem.*, **50** (2), 200–202.

22 Blyholder, G., Shihabi, D., Wyatt, W.V., and Bartlett, R. (1976) *J. Catal.*, **43**, 122–130.

23 Sheldon, R.A. (2007) *Green Chem.*, **9**, 1273–1283.

24 De Klerk, A. (2011) *Energy Environ. Sci.* doi: 10.1039/c0ee00692k

25 Stelter, S. (2001) *The New Synfuels Energy Pioneers. A History of Dakota Gasification Company and the Great Plain Synfuels Plant*, Dakota Gasification Company, Bismarck, ND.

26 De Klerk, A. (2009) *Advances in Fischer–Tropsch Synthesis, Catalysts, and Catalysis* (eds B.H. Davis and M.L. Occelli), Taylor & Francis, Boca Raton, pp. 331–364.

27 Kamara, B.I. and Coetzee, J. (2009) *Energy Fuels*, **23**, 2242–2247.

28 UK Ministry of Defence (2008) Turbine Fuel, Aviation Kerosine Type, Jet A-1. NATO Code: F-35. Joint Service Designation: AVTUR. Defence Standard 91-91, Issue 6; UK Ministry of Defence, April 8.

29 UK Ministry of Defence (2011) Turbine Fuel, Kerosine Type, Jet A-1. NATO Code: F-35. Joint Service Designation: AVTUR. Defence Standard 91-91, Issue 7; UK Ministry of Defence, February18.

30 Forest, C.A. and Muzzell, P.A. (2005) SAE Technical Paper Series, 2005-01-1807.

31 Henderson, H.E. (2006) Chapter 19, in *Synthetics, Mineral Oils, and Bio-Based Lubricants. Chemistry and Technology* (ed. L. R. Rudnick), Taylor & Francis, Boca Raton.

32 Fabricius, N. (2005) *Fundamentals of Gas to Liquids*, 2nd edn, Petroleum Economist, London, pp. 12–14.

33 Morgan, P.M., Van der Merwe, D.G., Goosen, R., Leckel, D.O., Saayman, H.M., Loubser, H., and Botha, J.J. (1999) *S. Afr. Mech. Eng.*, **49**, 11–13.

34 Horne, W.A. (1950) *Ind. Eng. Chem.*, **42**, 2428–2436.

35 De Klerk, A. (2011) *ACS Symp. Ser.*, **1084**, 215–235.

5
Industrial Case Studies

Yong-Wang Li and Arno de Klerk

Synopsis

The main technical stages in the world-wide development of industrial Fischer–Tropsch synthesis plants are summarized. The first work started in Germany before and during the Second World War. Work later continued and was expanded in Germany, the USA, other European countries and most notably South Africa. Expertise is now widespread and FT plants are under construction or planned in many countries. The industrial technology, engineering, and overall process integration are discussed. Prospects for future developments in the FT industry are highlighted.

5.1
Introduction

The origin and development of industrial Fischer–Tropsch (FT) facilities can be traced back to a lack of crude oil, not necessarily everywhere, but locally. All economies and indeed life itself depend on energy, and yet only a few countries are self-sufficient; the others have to meet their needs by trade. This chapter starts by briefly reviewing the role that industry played in providing energy in Germany and the circumstances that led to the construction of the first industrial FT facilities [1, 2]. Today, we can see that the pace of industrial development has fostered an "addiction" to oil products globally, and that in the medium term, the high and increasing levels of consumption will lead to a diminishing supply of crude oil and much higher prices everywhere. Despite efforts to develop very large-scale "green low-carbon" industrial processes, based, for example, on solar, wind, or tidal power for sustainable energy production, such aims are unlikely to be realized quickly enough to meet the foreseeable needs. In the 50+ year transition period, we will have to continue to rely on various forms of conventional energy, including crude oil and gas.

Greener Fischer-Tropsch Processes for Fuels and Feedstocks, First Edition. Edited by Peter M. Maitlis and Arno de Klerk.

This situation may be expected to lead to an increase in the industrial use of FT conversion of coal, shale gas, biomass, and carbon containing combustible wastes to supplement conventional oil production [3].

For many decades Fischer–Tropsch based processes, once put into industrial practice, have attracted attention due to the fact that these specific industries are closely connected to local energy trends and policies. The capital-intensive nature of FT facilities also makes the industry sensitive to the local and global economic–political environments.

This chapter discusses industrial processes practiced by many organizations, emphasizing the major achievements and more importantly the experiences and lessons that have been learnt and that form the basis of the current understanding of chemical engineering and energy science principles. The information that is presented here is the best that can be gleaned from the open literature; however, as the fine details are often proprietary information, it is not possible to verify that the practiced technologies are exactly those that are published. In presenting an overview of industrial FT facilities, we have tried to take care not to allow our future thinking to be eclipsed by our past errors, but to rather remember and allow them. The aim is to provide a forward-looking overview that does not erase all the deficiencies in past industrial applications.

5.2 A Brief History of Industrial FT Development

5.2.1 Early Developments

The early industrial Fischer–Tropsch synthesis and coal liquefaction plants were developed in Germany during 1913–1944, as it was the only country with the capability to synthesize fuels from coal and to use that to work toward energy independence [2–6]. A British synthetic fuel program started at about the same time as that of the Germans and in the 1920's research work on FT related syntheses was being carried out as pilot plant tests at the University of Birmingham. ICI planned to construct a coal liquefaction plant at Billingham with a capacity of 1.28 million barrels of oil products in 1935; however, eventually only four small experimental plants were constructed because of lack of support and the high cost of the coal liquefaction process [2–5]. In the United States, the Bureau of Mines began FT research in about 1927, but without significant investment in any commercial developments until the end of the Second World War, mainly because of the lack of government support and for economic reasons [2, 5, 7]. The early developments in Germany are summarized in Table 5.1.

The industrial FT facilities varied in design from location to location. Nevertheless, the same elements can be found, as illustrated by the description of the German-designed FT plant in Jinzhou, China (see below). Syngas preparation was through the reaction between coke (from coal) and superheated steam in a

Table 5.1 Chronology of German developments culminating in the industrial use of FT synthesis.

Year	Technology and process developments	Catalyst
1913	BASF patent for CO hydrogenation	Ce, Co, Mo oxides, or alkali metallic oxides
1923	Fischer and Tropsch observed oxygenates in a tubular reactor at 400–450 °C, 10–15 MPa	Alkali Fe
1932	A small pilot plant in Mülheim	Ni: Mn: Al: Kieselguhr
1934	Ruhrchemie acquired the patents and constructed a large pilot plant in Oberhausen-Holten	Ni: Mn: Al: Kieselguhr initially, then Co catalyst
1935	Four FT commercial size plants under construction with total capacity of 724 000–868 000 barrels per year of motor gasoline, diesel fuel, and lubricating oil and chemicals. 180–200 °C, 0.1 MPa (later 0.5–1.5 MPa)	100 Co: 5 ThO_2: 8 MgO: 200 Kieselguhr
1938–1939	Nine FT plants constructed with total design capacity of 5.4 million barrels (740 000 metric tons) per year	100 Co: 5 ThO_2: 8 MgO: 200 Kieselguhr
1944	Nine FT plants actual production peaked at 4.1 million barrels (576 000 metric tons) per year.	100 Co: 5 ThO_2: 8 MgO: 200 Kieselguhr

vessel containing coke particles piled up as a fixed bed and heated indirectly. Superheated steam was introduced into the heated coke bed to produce a mixture of CO and H_2 (*syngas*, also known as *water gas*) with a H_2 : CO ratio slightly higher than 2 : 1. This dirty raw gas was cooled and washed with water to eliminate dust, and then purified in a series of adsorption columns containing active carbon, as well as iron oxide adsorbents, especially to remove sulfur from the syngas. The syngas feed was then used for FT synthesis. A detailed description of the tube cooled fixed bed reactors used for FT synthesis in Germany during 1930–1940s is given in Section 3.4.1. Downstream refining of the FT syncrude yielded transportation fuels and petrochemicals. The German FT refinery designs exploited the properties of the FT syncrude and were quite different from the crude oil refinery designs of that time [8].

In summary, the major early achievements were that oil (syncrude) was successfully synthesized from coal-derived syngas on an industrial scale and that the syncrude could be refined to useful products. However, the cost of building the early coal-to-liquid plants employing FT synthesis was high considering the limited amounts of product obtained. The justification for these facilities was strategic. While the FT plant designs were acceptable at that time, in the context of the historical development of syngas production, they are not acceptable today. The efficiency indexes of equipment and catalyst performance are typically one or two orders of magnitude lower than those of the present FT technology and the energy efficiency is also several times lower than that of similar modern industries.

5.2.2
Postwar Transfer of FT Technology across Oceans

After the Second World War, much of the FT knowledge and the further development work were gradually transferred from Germany to the United States and then to South Africa. Information on German FT work was collected by the Allied Investigative Team at the end of the Second World War and was sent to the Combined Intelligence Office Subcommittee (CIOS) in London. This investigative work resulted in the public release of a huge amount of invaluable information [2, 5–7].

In addition to the industrial FT plants using a Co catalyst, Fischer and his coworkers also worked on an iron-catalyzed process. This led to the development of the Arge process using a pressurized tubular fixed bed reactor technology that was commercialized by the Ruhrchemie and Lurgi companies (Section 3.4.2). The first industrial use of this technology was in the Sasol 1 plant in South Africa.

Hydrocarbon Research Inc. in the United States developed the Hydrocol process. The industrial implementation consisted of a two-phase fixed fluidized bed reactor, 5 m in diameter, which used a fused iron catalyst and was operated at high temperatures. A large-scale Hydrocol plant in Brownsville, TX, was commercially operated using natural gas as a feedstock during 1951–1957. The plant was designed to make mainly motor gasoline, but it was plagued with start-up problems and eventually ceased operation for economic and technical reasons. However, the Brownsville plant paved the way for later developments in FT fluidized bed reactor technology (Section 3.4.3) [2, 4].

Another important type of FT reactor developed early on in Germany was the slurry reactor. In this reactor design, fine catalyst particles are suspended in the wax generated from FT synthesis. Based on this concept, large-scale tests were conducted in Germany [2–5, 9–12]. However, the difficulty of separating the wax product from the catalyst eventually halted further German development and industrial application of this technology was only achieved much later (Section 3.4.4).

5.2.3
Industrial Developments in South Africa

Early in the 1930s, a South African mining company called Anglovaal had shown interest in coal-to-liquids conversion by coal gasification and FT synthesis. In 1938, Etienne Rousseau was appointed as the research engineer, who was later regarded as the '"father" of Sasol. Franz Fischer visited South Africa around the same time. The first of the FT plants that were constructed in South Africa was commissioned in 1955. It used both German and American technologies for FT synthesis. The state-owned company Sasol was later privatized and it continued to develop technology for FT synthesis in isolation. Over time a further three FT-based facilities were constructed employing the technology developed by Sasol. The main industrial and commercial FT developments in South Africa are presented in Table 5.2.

PetroSA, Lurgi, and Statoil Hydro formed a joint venture in 2004 for the demonstration and commercialization of a cobalt-based low-temperature Fischer–Tropsch

Table 5.2 Chronology of industrial and commercial FT developments in South Africa.

Year	Industrial and commercial developments
1949	License to construct a synthetic fuels facility issued
1950	South African Coal, Oil and Gas Corporation (Sasol) formed
1955	Sasol 1 (Sasolburg) coal-to-liquids (CTL) facility commissioned
1974	Government announced its intent to construct second synthetic fuels facility
1979	Government announced its intent to construct third synthetic fuels facility
1979	Sasol privatized
1980	Sasol 2 (Secunda) CTL facility commissioned
1982	Sasol 3 (Secunda) CTL facility commissioned
1986	Government announced its intent to construct fourth synthetic fuels facility
1992	Mossgas (Mossel Bay) gas-to-liquids (GTL) facility commissioned
1993	Fe-LTFT Sasol slurry bed process (SBP) commercialized
1995	Sasol Advanced Synthol (SAS) fixed fluidized bed commercialized
2002	Petroleum, Oil and Gas Corporation of South Africa (PetroSA) formed[a)]
2004	Sasol 1 facility changed from CTL to GTL
2005	Co-LTFT slurry bed process (1000 bbl/day) commissioned at Mossgas GTL facility

a) PetroSA owns and operates the Mossgas GTL facility.

(LTFT) catalyst and slurry reactor technology and subsequently established the gas-to-liquids (GTL) technology licensing company, GTL.F1. The construction of a 40 000 t/a (1000 bbl/day) demonstration plant was sanctioned in 2002; it has been in operation since 2004 at the PetroSA GTL facility [13]. The $Co/Re/Al_2O_3$ FT catalyst is prepared by incipient wetness impregnation of the alumina support. The cobalt loading is 12 mass% and an α-alumina support is used to provide attrition resistance, which is of critical importance during slurry bubble column operation. The scale-up and production of the Statoil FT catalyst used in commercial operation at the PetroSA facility were carried out by Johnson Matthey Catalysts [14]. More recently, the use of Ni as a substitute for Re was reported [15].

The most recent FT technology developed in South Africa by Sasol is the Co-LTFT slurry bed process (SBP). This technology was commercialized in the Oryx GTL facility in Qatar. Commissioning started in 2006, but some further equipment was later added to reach design capacity [16]. FT catalyst attrition initially led to the formation of fine particles in the syncrude product [17]. A sister facility, in Escravos Nigeria, which uses the same technology as Oryx GTL, is under construction.

Some technology was also developed in South Africa for downstream refining of the FT syncrude. An H-ZSM-5-based olefin oligomerization, the COD (conversion of olefins to distillate) process, was developed by the South African Central Energy Fund (CEF) in collaboration with Süd-Chemie, specifically for the conversion of FT naphtha in the Mossgas GTL refinery [18]. Sasol developed and commercialized a number of processes for the purification of linear α-olefins from FT naphtha and distillate [8].

5.2.4
Industrial Developments by Shell

Shell has worked on the development of its Shell middle distillate synthesis (SMDS) process since 1973 at its research facilities in Amsterdam. It was decided to concentrate the work on FT using a cobalt catalyst in multitubular fixed bed reactors. The engineering decisions leading to the selection of this technology platform have been discussed by Sie *et al.* [19]. The detailed design of the multitubular fixed bed reactors was performed by Lurgi. The supported cobalt catalyst is contained inside the narrow reactor tubes (Section 3.4.2). Originally, each 7 m diameter reactor had a capacity of 130 000 t/a (3000 bbl/day), but this has since been increased to around 400 000 t/a (9000 bbl/day) by an increase in diameter and improved catalyst activity [20].

In 1989, Shell announced the building of a 500 000 t/a (12 500 bbl/day) plant to convert natural gas into FT products, at a capital cost of about $850 million. The plant, which is in Bintulu, Malaysia, produces fuels and petrochemicals [8]. It started operation in 1993, and uses natural gas as feedstock to make distillate, naphtha, and wax. The SMDS technology produces as much clean distillate as possible. The cobalt catalyst used has a high selectivity toward wax, part of which is then hydrocracked or hydroisomerized to produce blending material for diesel fuels.

The success of the first industrial application of the SMDS process led to the Pearl GTL project. Shell partnered with Qatar Petroleum (Doha, Qatar) to invest around $17–19 billion in building a plant using the SMDS process with a capacity of about 6 000 000 t/a (140 000 bbl/day) that is being constructed in two phases. The total plant capacity is about 11 000 000 t/a (260 000 bbl/day), the additional capacity coming from processing natural gas liquids (NGL). The first phase of the Pearl GTL facility was commissioned and started production of both SMDS phases in 2012. When fully commissioned, Pearl GTL will be the largest nameplate capacity industrial FT facility in the world.

5.2.5
Developments in China

In 1938, an early version of the German FT technology and equipment was brought from Germany by the Japanese military. The plan was to construct an FT plant with a capacity of 30 000 t/a in Jinzhou, Northeast China. The plant was not completed during the Second World War, and the construction was finally finished in the beginning of the 1950s by the newly formed Chinese government-owned organization. During 1952–1962, this plant produced a total of 400 000 tons of liquid oil.

Later, from the early 1980s onward, further development of FT technologies was conducted mainly in the Institute of Coal Chemistry (ICC) of the Chinese Academy of Sciences. As shown in Table 5.3, the research and development in the ICC focused on fixed bed FT synthesis using iron catalysts. Slurry-phase FT synthesis had been studied only at a laboratory scale before 2000. The fixed bed FT technology that used precipitated iron catalysts in the first stage and a molecular sieve

Table 5.3 Chronology of FT developments in China.

Year	Technology	Details	Site
1937	German Co-LTFT	Japanese military imported 30 000 t/year plant, but it was never in operation during the Second World War	Jinzhou
1952–1962	German Co-LTFT	Operation of Jinzou plant, produced 400 000 ton oil	Jinzhou, owned by No. 6 Petroleum Plant
1953	Fluidized bed pilot plant	4500 t/year, iron catalyst	Dalian Petroleum Research Institute
1980s	ICC two-stage fixed bed MFT	100 t/year pilot test	Shanxi Dai County Fertilizer Plant
1993–1994	Industrial pilot test	2000 t/year	Shanxi Jincheng Fertilizer Plant
1996	Single tube test	3000 h test run	ICC
1997–1999	ICC-IIA, B	Slurry and kinetics	ICC
2000–2004	Slurry bed reactor pilot test	1000 t/year, 3000 h operation	ICC
2003–2004	Yankuang pilot plant	10 000 t/a, 4706 h test run	Lunan Fertilizer Plant
2005–2009	Synfuels China demonstration plants	160 000 t/year	Yitai Coal Group (Inner Mongolia); Luan Coal Group (Luan)

catalyst in the second stage is known as the modified FT (MFT) technology. The MFT process was scaled up to a capacity of about 2000 t/a (40 bbl/day) in 1993.

In 1997, with a clear strategy to advance FT synthesis in a slurry bed reactor to an industrial scale, the ICC initiated a number of studies. The importance of basic research was realized and topics that were explored included analysis of reaction kinetics and modeling, catalyst design and preparation, aided by advanced fundamental research on understanding the catalysis in the FT reaction, as well as FT process integration and optimization [21–35].

In order to promote the FT research and development and further to commercialize the technology, Synfuels China Ltd., a company that evolved from the ICC, was founded in 2006. Synfuels China Ltd. has carried out further work, including systematic proprietary catalyst development, reactor scale-up, process integration at the pilot plant, and demonstrations under the framework of the China Coal Utilization Program. A pilot plant employing slurry-phase FT synthesis with a capacity of 750–1000 t/a (15–20 bbl/day) was commissioned and operated from 2001 to 2004. Since then, two commercial scale CTL coal-based demonstration plants, each with a capacity of 160 000 t/a (about 4000 bbl/day), have been commissioned and are now running with good product availabilities [34, 35].

The high-temperature slurry Fischer–Tropsch process (HTSFTP) was developed by the Chinese Academy of Sciences, and Synfuels China Ltd. has recently commercialized this process at a scale of 200 000 t/a (5000 bbl/day) at three locations in China. A flow diagram of the HTSFTP is shown in Figure 5.1. Eventually, the FT gas loop will be arranged slightly differently to further improve the overall efficiency of the large-scale CTL plants that are at least 10 times of the demonstration plant size.

This process uses a proprietary iron catalyst in a slurry reactor, which is operated at 270 °C, making it a medium-temperature FT (MTFT) technology. The higher operating temperature leads to easier heat removal from the FT reactor and to better recovery of FT reaction heat for steam production. The product is similar to that from LTFT synthesis.

Purified syngas and recycled syngas are mixed and introduced into the slurry reactor where the FT reactions take place to produce a wide spectrum of hydrocarbons and other by-products, including water. Unconverted syngas together with the gaseous FT products at the FT reaction conditions leave the slurry reactor at the top of the reactor. This hot stream is cooled by the recycled gas and further cooled by boiler water and cooling air. Most of the condensates, heavy oil, light oil, and water are separated from the gas stream, which is then partially recycled (main recycle) back to the FT gas loop and partially released as tail gas. For the demonstration plant design, the tail gas is first sent to a Benfield unit operating at 105 °C to remove CO_2 and the CO_2-free tail gas is split into two parts. One part (20% of the main recycle stream volume in the demonstration plant) is compressed back to FT gas loop conditions and the remainder is sent to a low-temperature (−20 °C) solvent adsorption unit to recover the FT lights, including LPG and some uncondensed C_5 and heavier products. This avoids cryogenic separation. The dry tail gas is then sent to pressure swing absorption (PSA) to recover hydrogen for product workup units and for H_2 : CO ratio balancing of the whole system. Thanks to the extremely low

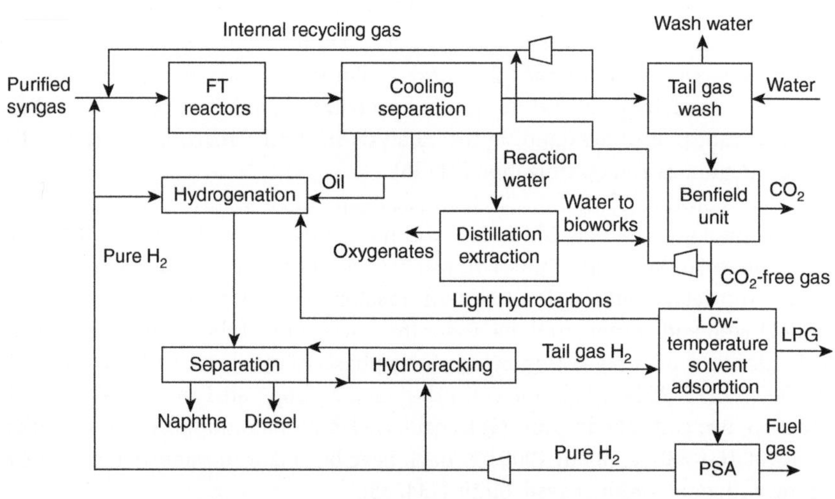

Figure 5.1 Block flow diagram of the HTSFTP demonstration plants in Inner Mongolia, Shanxi, China.

selectivity to methane (about 2.5 mass%) of the catalyst used in HTSFTP, the volume of tail gas is low compared to that from other FT technologies.

The demonstration operation shows that this FT gas loop can be operated at high syngas conversion, up to 93%, which is only slightly lower than the 94% proposed in the original design. The designed capacity of desired products has been reached and the fuel gas is balanced, mainly due to the improved C_{3+} selectivity and very low methane selectivity, which was reduced from 4% in the original design to 2.5% in HTSFTP. The steam from the FT reactor can reach 32 atm (steam drum) under full load operation of the FT reactor, showing a promising improvement for the efficiency of the reaction heat recovery over typical LTFT processes.

In the Synfuels China Ltd. demonstration plants, the product workup unit is composed of hydrogenation followed by hydrocracking. LPG, naphtha, and middle diesel range distillate are the only oil products that are produced by the process.

5.2.6
Other International Developments

Eni and IFP (now IFP-EN), who have collaborated on research and development since 1996, built a 1000 t/a (20 bbl/day) pilot unit in the Eni refinery at Sannazzaro, Italy, which came into operation in 2001. Axens was responsible for the basic engineering design and the design and operation of a 100 kg/day FT catalyst plant in Salindres, France. The slurry bed reactor was operated with cobalt-based catalysts to gain operating experience in slurry handling, reactor design, and product wax–catalyst separation. The catalyst mechanical stability and hydrodynamics inside the slurry reactor were examined in the pilot-scale plant. In addition, special mild hydrocracking and isomerization technology was developed and tested in a hydrocracking pilot plant unit at the IFP Research and Development center in Solaize, France. An engineering study of the overall GTL complex was completed up to the front-end engineering design stage [36, 37].

In 1980, the Mobil Corporation initiated an FT study with a small pilot plant using iron catalysts. The work was stopped in 1983 for economic reasons. Later on Mobil merged with Exxon, and ExxonMobil spent $300 million on the development of its Advanced Gas Conversion for the twenty-first century (AGC-21) technology. A 1.2 m diameter slurry bed reactor with a capacity of 9000 t/a (200 bbl/day) was started up in Baton Rouge, Louisiana, in 1989, which was successfully operated until 1992. The reactor used a cobalt-based catalyst. Exxon holds a large number of FT related patents [38].

BP developed a FT technology with downstream upgrading that was piloted in a 13 000 t/a (300 bbl/day) facility in Nikiski, Alaska. The facility started up in 2003 and included a hydrocracking unit on-site [39]. The FT technology employs a Co-based catalyst in a multitubular fixed bed reactor, analogous to the SMDS technology. The catalyst is prepared by coprecipitation, impregnation, extrusion, and pelletization. During piloting at the Nikiski facility, α-values in the range of 0.92–0.95 were achieved [40].

ConocoPhillips developed a FT technology that was tested in a 17 000 t/a (400 bbl/day) facility, which became operational in 2003; the syngas is produced by catalytic partial oxidation of natural gas [41]. This program was discontinued.

Rentech, a Colorado-based company, was founded by Charles Benham who had research experience in a biomass-to-fuel (BTL) project at the US Solar Energy Research Institute and the Government's China Lake missile research lab. Benham also built an FT pilot plant in 1982 using an iron-based catalyst. Rentech's first commercial plant with a capacity of 10 000 t/a (235 bbl/day) was started up in 1992, and it used landfill gas as a feedstock. However, it was shut down for economic reasons and sold to an Indian company, Donyi Polo Petrochemicals Ltd., in 1994. Currently, Rentech holds several FT patents (http://www.rentechinc.com/tech2.htm). The Rentech technology is based on an Fe-LTFT catalyst that is operated in a slurry bubble column reactor.

Syntroleum was founded in 1984 by Ken Agee to convert natural gas into liquid fuel products. The Syntroleum process is based on a cobalt catalyst and focused on the development of an FT process with adaptability and portability [42]. One of the key features that differentiates the Syntroleum GTL technology from other similar technologies is that it relies on autothermal reforming with air, not with oxygen. Consequently, there is no need for an air separation unit, enabling applications on a smaller scale. The FT reactor is operated at high per pass syngas conversion over a Co/Al_2O_3 catalyst to avoid syngas recycle. The FT technology was developed for two reactor types. The first was a multitubular fixed bed, which was piloted at a 100 t/a (2 bbl/day) scale at Syntroleum's facilities in Tulsa. The second was a moving bed reactor, which was piloted at a 3000 t/a (70 bbl/day) in the Cherry Point, WA refinery [43].

Oxford Catalysts and Velocys are developing a different type of FT technology that employs a microchannel reactor for small-scale applications. This principle is also applied for steam methane reforming [44]. A Co-LTFT catalyst is employed.

A number of other organizations are also now developing FT technologies since FT-based processes are being recognized as practical alternative routes to crude oil for the production of fuels and chemicals. The rest of the chapter focuses on the industrial FT facilities.

5.3
Industrial FT Facilities

Industrial FT facilities are inherently complex, with many production units in each facility. All these units need to work together and the production process is not purely sequential. Generally speaking, four important process transformations can be identified in an industrial FT facility: feed preparation, syngas production/conditioning, FT synthesis, and syncrude refining/upgrading. The heart of any FT facility is the FT gas loop, which determines how the syngas is manipulated (Chapter 2) and how the products are separated and recovered from the unconverted syngas (Chapter 4).

In addition to the process itself, the importance of the utility network should not be underestimated. An industrial FT facility deals with extremes of heat flow and temperature conditions. For example, syngas production is a very endothermic process, whereas FT synthesis is a very exothermic process. Syngas production requires temperatures of 1400–1500 °C, whereas air separation requires temperatures close to −200 °C. The utility network is responsible for efficient energy management between these extremes.

Detailed descriptions of historical and current industrial FT facilities have been published [8], and only an overview of current facilities that highlights some of the main design decisions is given here.

5.3.1 Sasol 1 Facility

Sasol 1 in Sasolburg, South Africa, is the oldest operational FT facility. In its present form, the Sasol 1 facility is an LTFT synthesis-based gas-to-liquids plant (Figure 5.2) [8]. Since its commissioning in 1955, it has seen many changes; the most important changes were the following:

a) The change from a coal-to-liquids (CTL) facility with Lurgi moving bed dry ash coal gasifiers to a gas-to-liquids facility with autothermal reforming of natural gas.
b) The change from having combined high-temperature FT (HTFT) and low-temperature FT (LTFT) syntheses to having mixed LTFT synthesis for syncrude production, with multitubular fixed bed and slurry bubble column Fe-LTFT technologies operated in parallel.
c) The change in refinery output from on-specification transportation fuels and chemicals to the production of only chemicals and intermediates.

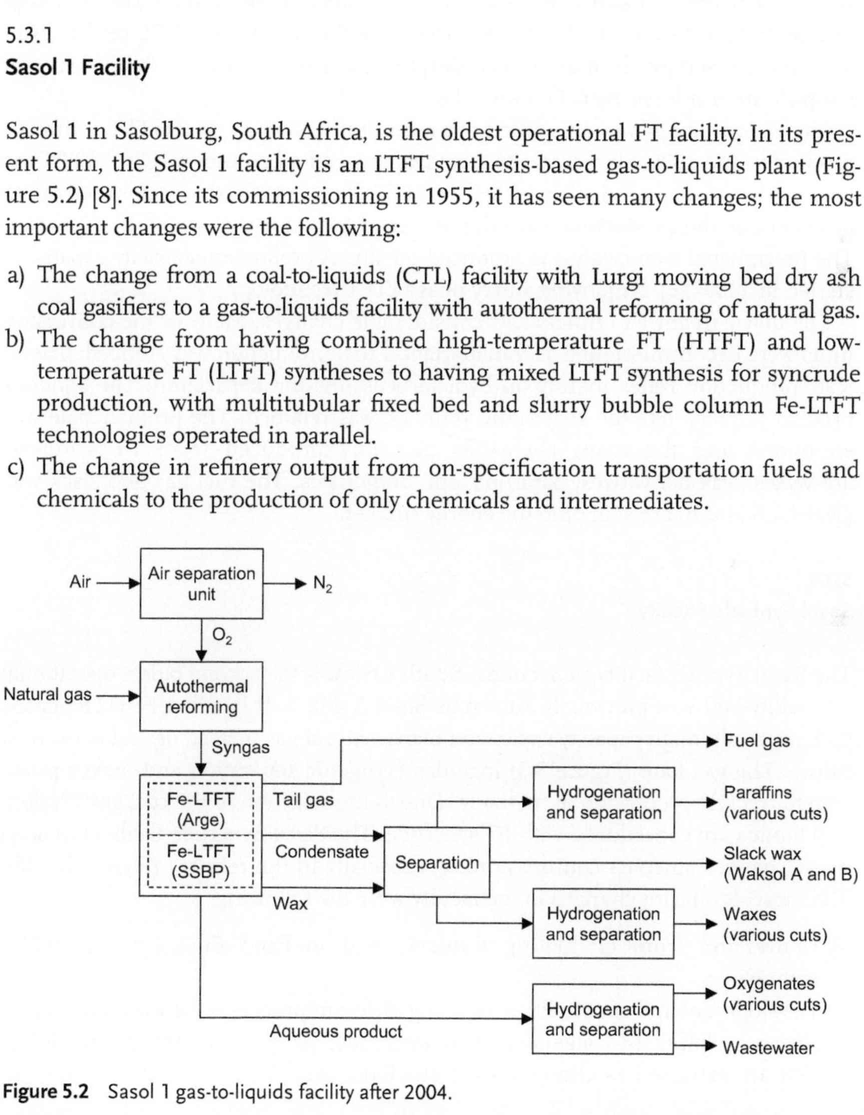

Figure 5.2 Sasol 1 gas-to-liquids facility after 2004.

Feed preparation was simplified with the decision to convert the facility from CTL to GTL. Natural gas can be desulfurized before being converted into syngas avoiding the expensive gas cleanup using Rectisol (cold methanol absorption) used to clean the coal-derived syngas. Natural gas contains much less sulfur than coal and despite many endeavors to reduce sulfur emissions [8, 45], it was only when the feedstock was changed from coal to natural gas in 2004 that the main H_2S emissions from the Sasol 1 facility were fully dealt with. The conversion did not need to improve syngas purity, since coal-derived syngas after Rectisol purification is as clean as or even cleaner than natural gas-derived syngas. Autothermal reforming technology was selected for natural gas to syngas conversion. For other reasons, the coal liquid refinery at the Sasol 1 site was kept in operation with feedstock supplied from the Sasol Synfuels CTL facility even after the conversion to natural gas, despite the loss of some of the benefits of simplification offered by GTL over CTL.

The FT gas loop was simplified after decommissioning the HTFT synthesis reactors in the 1990s, but some of the advantages of a combined HTFT–LTFT gas loop were lost. The original fixed bed Fe-LTFT (Arge) synthesis reactors remained in operation and production was supplemented by a slurry-phase Fe-LTFT reactor. The precipitated iron catalyst is produced on-site. A production capacity expansion started in 2009 [46], employing slurry-phase LTFT technology.

The downstream FT refinery saw considerable changes. Many of the conversion units were decommissioned as transportation fuel production was stopped. Chemicals production relies mainly on hydroprocessing and separation. The aqueous product refinery, like the coal liquid refinery, was retained. The product slate was simplified and the main chemicals presently produced from FT synthesis are waxes, wax derivatives, paraffins, and oxygenates. The fuel gas and slack wax (Waksol A and B) are sold onto the energy market.

5.3.2
Sasol Synfuels Facility

The Sasol Synfuel facility in Secunda, South Africa, is the second oldest operational FT facility and was previously known as Sasol 2 and 3. It is still a Fe-HTFT-based CTL plant, although capacity expansion using natural gas instead of coal is likely in future. The gas loop (Figure 5.3) includes cryogenic separation and makes provision for recycle processing of methane. Due to the selection of the coal gasification, coal liquids are coproduced with FT syncrude. The Sasol Syncrude facility has seen many changes since its commissioning, especially in the refinery (Figure 5.4) [8]. The most significant changes in the facility were the following:

a) Conversion from circulating fluidized bed to fixed fluidized bed HTFT technology.
b) Development and construction of many different processes for the extraction of linear α-olefins (n-1-alkenes) from syncrude. At present, the main olefins that are extracted as chemicals are ethylene, propylene, 1-pentene, 1-hexene,

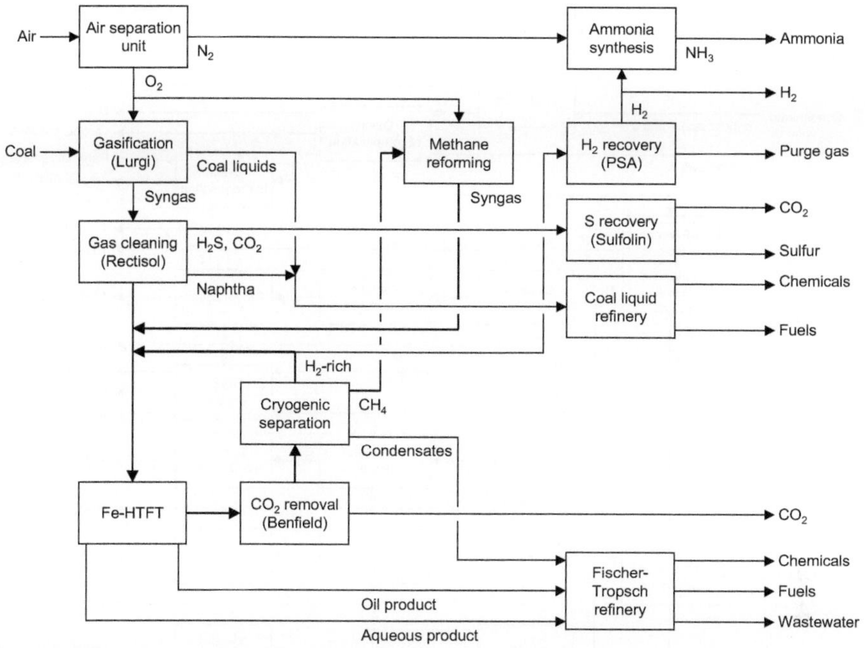

Figure 5.3 Sasol Synfuels coal-to-liquids facility; the gas loop is shown with thicker lines.

1-heptene, 1-octene, and 1-dodecene/1-tridecene. Some are sold as linear α-olefins and some are used as intermediates for downstream chemical production.

c) Reorganization of the refining from on-specification fuels as main product, to a tightly integrated facility with combined fuels and chemicals production. Chemical production approaches almost half of the capacity after the implementation of project Turbo [47], which diverted a significant fraction of the light naphtha to a new high-temperature fluid catalytic cracking unit.

CTL conversion is more complex than the GTL conversion. Coal preparation and coal gasification contribute significantly to the overall capital cost and complexity of the facility. Due to the high ash content of the local coal, Lurgi moving bed dry ash gasifiers are employed for syngas generation. Coal liquids are coproduced, which require a more complicated design for gas cleaning. However, this does supply the facility with valuable materials for chemicals and fuel production. Syngas cleaning is performed with a Rectisol process (cold methanol wash) and the design includes downstream H_2S removal.

At present, only fixed fluidized bed Fe-HTFT synthesis reactors are employed, using a fused iron catalyst produced on-site. The FT gas loop includes a second CO_2 removal unit (Benfield carbonate wash) and cryogenic separation. The presence of two CO_2 removal units is not because HTFT produces more CO_2 than LTFT, but because the gas that is cryogenically separated must be CO_2-free. Although cryogenic separation is expensive, it is necessary for the recovery and

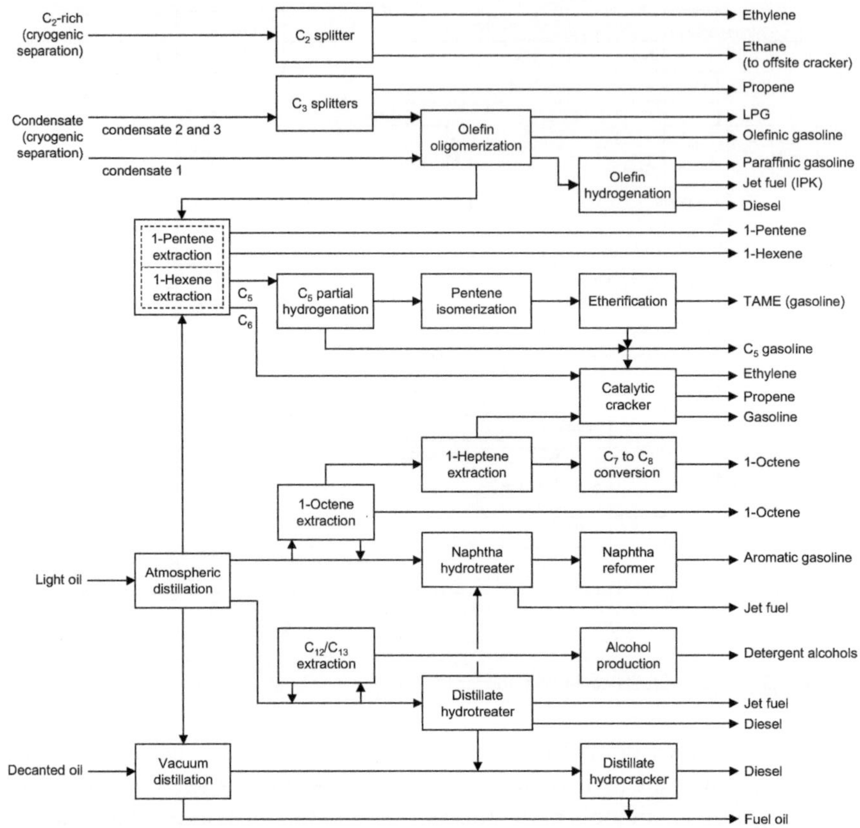

Figure 5.4 Current Sasol Synfuels FT oil refinery (aqueous and tar refineries not shown).

separation of ethylene from the HTFT syncrude. It has the added benefit of producing methane as a separate product that is recycled via reforming to produce syngas. The inclusion of cryogenic separation reduces the carbon footprint of the FT gas loop [48], as it provides a pathway for recovering the methane from the gasifier and from the HTFT synthesis. Methane is the most hydrogen-rich product in the CTL facility and is the lowest CO_2-footprint source of H_2.

The refinery at the Sasol Synfuels site is complex and resembles a crude oil refinery with integrated petrochemical production (Figure 5.4). However, the refinery is only a small part of the overall CTL complex. The original design was a departure from previous FT refineries as it was modeled on a crude oil refinery design. However, this made it less efficient than the Hydrocol HTFT refinery, which was of a simpler design [49].

The Sasol Synfuels became more complex over time, due to the addition of a number of integrated chemical production units. The original design produced chemicals only from the FT aqueous product and cryogenic separation. Motor gasoline, synthetic jet fuel, and diesel fuel are still produced, but in lower volumes as much material is extracted for chemicals production. Light olefins (ethylene and propylene) and linear α-olefins are major products as well as oxygenate chemicals, including detergent alcohols, light alcohols (ethanol, propanol, and butanol), and ketones (acetone and butanone). Coal liquids are also refined and phenol and cresols are extracted from the gas liquor [8].

5.3.3
Shell Middle Distillate Synthesis (SMDS) Facilities

The Bintulu facility in Malaysia was the first industrial application of the Shell middle distillate synthesis process (Figure 5.5) [8]. It is a gas-to-liquids facility, which employs the Shell gasification process (SGP) for the production of syngas from methane, with steam methane reforming to adjust the $H_2:CO$ ratio of the syngas for Co-based LTFT synthesis.

The SMDS process employs multitubular reactors operating at above 30 atm and 200–230 °C. The catalyst is robust and stable, and it has been reported that the Co-LTFT catalyst achieved over 8 years of operation without regeneration [20]. Synthesis is aimed at high wax production. The syngas passes through the fixed bed reactors, where it is converted into heavy hydrocarbons and water. In order to increase the overall syngas conversion, the design employs reactors in series. After separation of wax and condensates by simple cooling of the reaction stream, the unconverted syngas is partially recycled back to the LTFT reactors and partially released from the recycle loop as tail gas. In the SDMS process, the tail gas is sent directly to a steam reformer without recovering the light syncrude products. Methane, C_2–C_4 hydrocarbons, and some of the C_5 and heavier hydrocarbons are converted into a hydrogen-rich syngas, which is mixed with the it from the SGP. The SGP is a noncatalytic partial oxidation process and it produces a syngas with $H_2:CO$ ratio of 1.7. In this way, the combined reformer product meets the requirement of the Co-LTFT synthesis.

The syncrude has two refining pathways. The first pathway is hydrocracking to produce fuel blending materials and intermediates, namely, naphtha, kerosene,

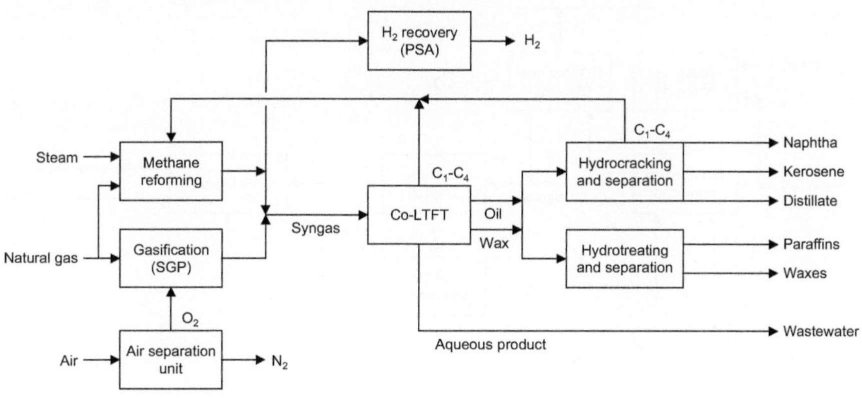

Figure 5.5 Shell middle distillate synthesis (SMDS) process, Bintulu facility.

and distillate. These products are not further refined on-site to produce transportation fuels, but are used as feed and/or blending materials during crude oil refining. The second pathway is hydrotreating to produce chemicals, mainly paraffinic solvents and waxes. Refining of the LTFT syncrude by either pathway can also produce a waxy product that is suitable for further refining into lubricant base oils.

The Pearl GTL facility differs from the Bintulu facility in size and by the inclusion of natural gas liquid co-processing in the refinery design. The facility consists of two Co-LTFT-based synthetic trains, each producing about 3 000 000 t/a (70 000 bbl/day) [20]. The natural gas liquids resemble the light oil from Co-LTFT synthesis and corefining these products provides additional economy of scale. Lubricant base oil production is included in the design of the Pearl GTL facility.

5.3.4 PetroSA GTL Facility

The PetroSA GTL facility, formerly known as Mossgas, generates syngas by methane reforming of natural gas (Figure 5.6) [8]. Like the SMDS process, it employs two different types of reformers to balance the $H_2:CO$ ratio, but the configuration is different. Part of the natural gas is steam reformed. The product from steam reforming is then combined with the remainder of the natural gas and tail gas recycle. In the original design, the syncrude was produced by three circulating fluidized bed Fe-HTFT reactors, but in 2005 a new 42 000 t/a (1000 bbl/day) Co-LTFT reactor was commissioned running in parallel with the Fe-HTFT reactors. This is at present the only operating industrial HTFT–LTFT facility, since Sasol 1 was converted from HTFT–LTFT into an LTFT only facility.

The tail gas from HTFT synthesis is not subject to cryogenic distillation and only the C_3 and heavier hydrocarbons are recovered from the syncrude. The C_1–C_2 hydrocarbons are recycled to the autothermal reformer. The aqueous product is

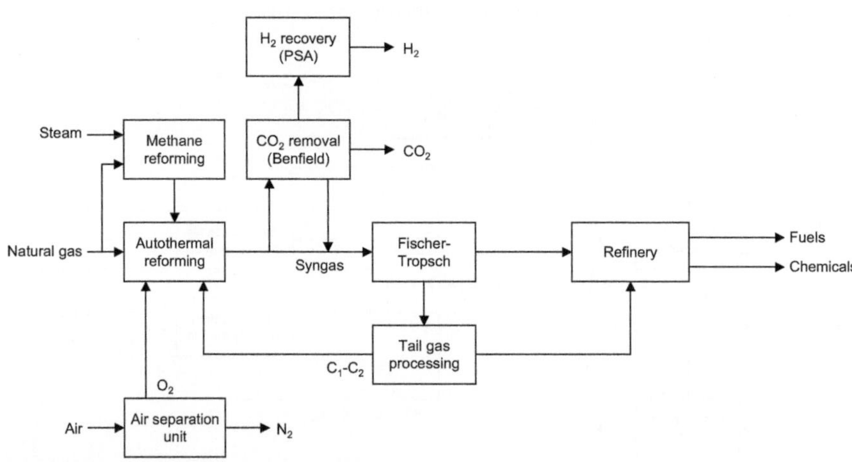

Figure 5.6 PetroSA gas-to-liquids facility.

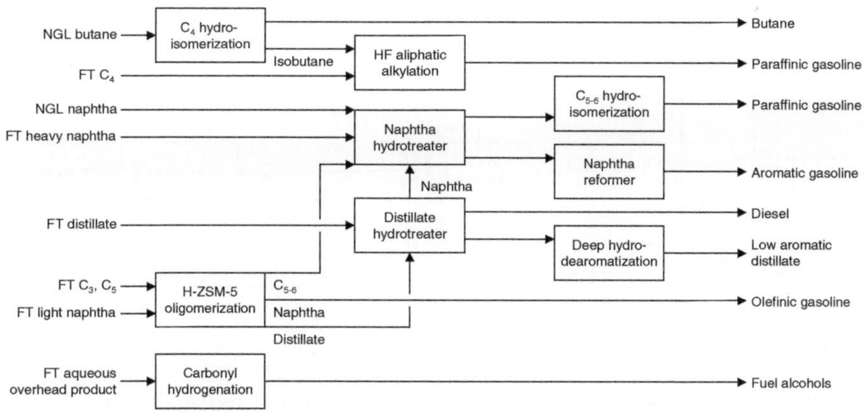

Figure 5.7 PetroSA gas-to-liquids refinery after atmospheric distillation of FT syncrude, natural gas liquid (NGL), and FT aqueous product.

treated separately and the alcohols are recovered after partial hydrogenation of the carbonyls. The wastewater is anaerobically digested to product methane, which is recycled to the process. The syncrude is co-refined with natural gas liquids, as in the SMDS Pearl GTL facility. The natural gas liquids and the Co-LTFT syncrude are quite similar, and there was no need to change refinery design to accommodate the Co-LTFT syncrude. The FT oil refinery of the PetroSA facility was designed to cope with a unique blend of HTFT syncrude and natural gas liquids in mind (Figure 5.7) [8, 49], and is much simpler than the Sasol Synfuels oil refinery. The main products are on-specification transportation fuels for the South African market, as well as some low aromatic distillate and alcohols.

5.3.5
Oryx and Escravos GTL Facilities

The Oryx GTL facility represents the first industrial application of Sasol's Co-LTFT slurry-phase distillate process. Unlike the Fe-LTFT slurry bed at the Sasol 1 facility, a number of technical problems arose in the initial Co-LTFT design (Section 5.2.3). The Escravos GTL facility is a carbon copy of the Oryx GTL facility and likely includes the changes to deal with the technical problems arising from the FT catalyst attrition. An overview of the general process flow is given in Figure 5.8 [8].

Most of the syngas employed for FT synthesis is produced by autothermal reforming of natural gas. A steam reformer is used to produce hydrogen for refinery use, as well as to adjust the H_2 : CO ratio of the syngas. The syncrude from the Co-LTFT slurry bubble column is recovered as liquid wax and lighter gaseous products. The wax must be treated to remove metals from FT catalyst leaching and attrition, before it is combined with oil as feed to a hydrocracker. Unlike the SMDS process, the hydrocracker technology was not specifically developed for FT wax and it requires sulfur addition to keep the catalyst active and is operated at more severe conditions. The products are liquid petroleum gas, naphtha, and distillate. The naphtha and

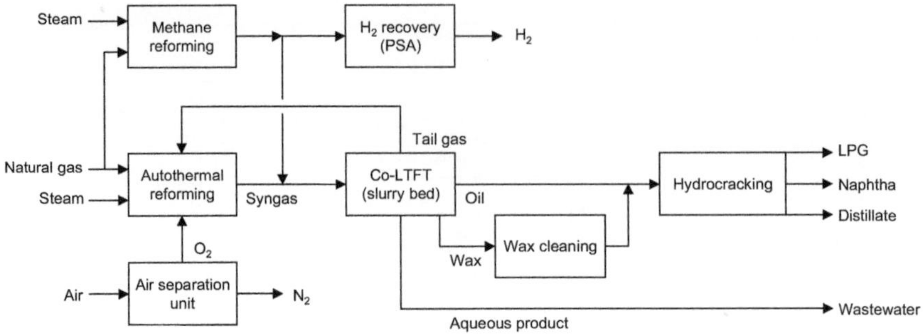

Figure 5.8 Oryx and Escravos gas-to-liquids facilities.

distillate are intermediate products that can be used as cracker feed or blending material. The tail gas from Co-LTFT synthesis is employed in the recycle and as fuel gas. The light alcohols are recovered from the aqueous product and incinerated. The remainder of the aqueous product is wastewater that is biologically degraded.

The Oryx and Escravos GTL facilities are the least complex of all the current industrial FT facilities and produce lower value products.

5.4 Perspectives on Industrial Developments

5.4.1 Further Investment in Industrial FT Facilities

FT-based development is still being debated and issues such as energy security, economic uncertainty, environmental impact, and political expediency make it difficult to predict what the future holds. It is clear that security of conventional crude oil supply will come to the fore as global energy demand keeps on increasing. FT-based facilities have the advantage of increasing supply security by converting a variety of raw materials, including coal, natural gas, waste, and biomass into liquid fuels and petrochemicals. Furthermore, the industrial facilities now in operation are economically sound. Despite that, investment in FT-based facilities has been slow, which can be traced to some of the following problems and misunderstandings:

a) There is considerable complexity associated with clean syngas production and the FT gas loop (syngas conditioning, FT synthesis, and syncrude separation and recovery). The product is a multiphase syncrude, which must still be refined to produce useful products. With many units in series, the risk of problems increases and it is higher than that encountered in conventional crude oil production and refining.
b) Confidence in FT technology was dented by some early problems in industrial implementation. Many of the industrial facilities faced start-up problems that constrained throughput or failed to deliver the anticipated performance in some respects.

c) The capital cost associated with FT-based facilities is considerable (Chapter 7), and though the economics may be favorable, the investment risk is significant. It may be noted that even in the more conventional business of crude oil refining, few new refineries are being constructed. For example, in the United States, no new crude oil refineries have been built since the 1970s! The huge capital cost of a large FT facility could however be overcome by investing in smaller projects giving similar returns, but with greater distribution of risk, which is not possible in a single project.
d) The water and CO_2 footprints of FT-based facilities are high (Chapter 15). Even though these may be comparable to those of some other energy conversion technologies, such as coal-fired power generation, they are more than those for crude oil production.
e) The absolute cost and the cost differential between crude oil and the feed materials employed for FT facilities fluctuate widely over time, making it difficult to accurately forecast the economic viability of an FT facility and also making it a rather more uncertain investment decision.

5.4.2
Technology Lessons from Industrial Practice

Despite the limited number of industrial FT facilities, it is quite clear that there are divergent views on which FT technology to select and what to do with the products. There is no roadmap or a decision tree that can be followed. Each FT-based facility must be designed in accordance with local conditions, the available feed, and product requirements. Some general observations are nevertheless possible.

a) All industrial FT facilities make use of syngas production technologies that require air separation units to limit the volume of difficult to remove inert material (e.g., N_2 and Ar) in the syngas. In order to exploit economy of scale, the size of an industrial FT-based facility should be large enough to justify at least one world-scale air separation unit.
b) The least technology risk for FT synthesis is associated with the operation of multitubular fixed bed reactors.
c) The FT technology should be selected based on the products desired. Different FT technologies improve the efficiency for different products.
d) The processes of syngas generation and FT synthesis are inherently complex and capital intensive. Simplification of syncrude refining only erodes the product value associated with the facility.
e) There seems a natural tendency for FT facilities to shift toward increasing chemical production over time only if the chemical market demands are positive. This is logical considering the cost of syncrude preparation and the possibly added value offered by chemicals over transportation fuels, blending materials, and intermediates.
f) FT syncrude is very different from crude oil and requires a different approach to refining. Imposing a crude oil refining approach on syncrude leads to inefficiency and some technologies may not work at all.

g) The diversity of products produced from a typical FT-based facility may be greater than that produced from a typical crude oil refinery.
h) Both in scale and in complexity, FT-based facilities must be compared with petrochemical complexes and not with crude oil refineries. A million ton per year facility (~22 000 bbl/day) may be considered world scale.

5.4.3
Future of Small-Scale Industrial Facilities

In order to apply FT technology to biomass (BTL) conversion, the capacity of each BTL plant must be relatively modest. This is due to the logistics and cost associated with biomass transportation. It may also be desirable to employ smaller sized facilities for waste-to-liquids (WTL) conversion to enable decentralized solid waste processing. Likewise, there may be local sources of stranded natural gas that can be exploited for gas-to-liquids conversion, but only if the capacity of the facility is commensurate with the size of the gas source. Consequently, there is a clear need for the development of small-scale industrial facilities, typically with capacities less than 100 000 t/a product (<2500 bbl/day).

The risk associated with investment in FT-based facilities can be reduced by reducing the capacity, which may enable investment by smaller players in the market. This is a clear benefit. There is also less risk that markets will react negatively to a small-scale facility, because it represents only a small increment in global production capacity. However, engineers will be quick to point out that reducing the capacity goes against the principle of "economy of scale" (Box 5.1). In order to maximize profitability, the capacity of process facilities has been increasing over time. Unfortunately, this trend has also stifled all but incremental innovation, because the risk associated with new technology also depends on the size of the implementation.

Box 5.1 Economies of Scale

The cost of equipment is scaled based on capacity. This relationship is not linear, but follows a power law called the "six-tenths factor rule." The cost ratio ($C_{A/B}$), which is the ratio of the cost of A to the cost of B, is related to the capacity ratio ($X_{A/B}$), which is the capacity of A over the capacity of B, through the relationship: $C_{A/B} \sim X_{A/B}^{0.6}$. In practice, the exponent varies depending on the nature of the equipment, unit, or facility. However, in the present context, the implication is that a larger facility is cheaper per unit capacity. For example, when doubling the capacity, the capital cost increases only by half. This is also called "economy of scale." However, some units, even in large-scale facilities, cannot benefit from economy of scale. Equipment is not infinitely scalable and the maximum viable capacity for a single unit is limited. Sometimes this limit is reached rather quickly. For example, the maximum syngas capacity of a Lurgi Mark V moving bed dry ash coal gasifier is sufficient only for around 75 000 t/a (1500 bbl/day) Fischer–Tropsch production.

Although at a first glance the prognosis for small-scale industrial FT facilities does not seem good, closer consideration shows that this is a limited view, as the design requirements are different for small-scale facilities and there are also significant advantages. Small-scale facilities represent a radical departure from the design paradigm of large-scale facilities and can exploit opportunities other than economy of scale. The lessons learnt in large-scale industrial facilities should not be ignored, but economically viable and technically efficient small-scale FT-based conversion plants can be run. Some pointers are as follows:

a) All large-scale industrial FT facilities that employ autothermal reforming or gasification technology for syngas generation include air separation units however, this is not desirable for smaller facilities. Although reformers and gasifiers can be operated with air, instead of O_2, the inert gases (N_2 and Ar) increase the size of all downstream equipment and reduce efficiency. Steam reforming, or steam gasification, where combustion heat is externally applied can avoid the introduction of air into the process. It is anticipated that these types of syngas generation will become more prevalent in small-scale facilities. Such technology has the further advantage of producing a high $H_2:CO$ syngas.
b) Syngas cleaning in GTL can be as efficient as for large facilities, but is likely to be less efficient for small BTL and WTL facilities. Some level of FT catalyst deactivation due to syngas contaminants is therefore inevitable.
c) The volume fraction of light gases produced by high-temperature FT technology makes such technology more difficult to implement efficiently on a small scale. It is therefore likely that small-scale industrial facilities will benefit from LTFT and MTFT technologies where synthesis has a higher chain growth probability (α-value).
d) The optimum α-values for small-scale facilities that are not designed to produce wax as final product are likely to be smaller than those found in large-scale industrial LTFT operation.
e) Despite the fact that it is inconvenient to use fixed bed FT reactors for iron catalysts, Iron-based low-temperature FT catalysts have a number of advantages over cobalt-based catalysts for small-scale applications. They are less sensitive to contaminants in syngas, are active for the water-gas shift reaction, and can process syngas over a wide range of $H_2:CO$ ratios. Besides, Fe-LTFT catalysts have lower methane selectivity, produce a more olefin- and oxygenate-rich product making refining of the lighter products easier, deactivate in a way that makes refining easier, are much cheaper, and, last but not least, Fe-LTFT catalysts are easier and less complex to produce on-site.
f) Fixed bed reactors are preferred for small-scale facilities, as they are more robust and less complex to operate. Fixed bed reactors are beneficial when syngas quality is variable since the top of the FT catalyst bed acts as a guard bed to protect the catalyst.
g) It is unlikely that small-scale facilities will be designed to produce on-specification transportation fuels or petrochemicals, even though such products have a higher value. The refinery design will be geared toward a limited product slate producing liquid intermediates or blending materials.

h) Transportability of equipment is important, and remote locations may preclude the use of large items.

i) The presence of multiple smaller units in parallel is an advantage in a small-scale facility, because it may increase the overall robustness of the facility. The reason is simple: when one of the multiple units in parallel goes off-line, the facility can remain in production, albeit at reduced capacity. However, when a single large unit goes off-line, the whole facility goes off-line and production drops to zero.

References

1 Gillingham, J. (1985) *Industry and Politics in the Third Reich*, Columbia University Press, New York.

2 Stranges, A.N. (2007) *Stud. Surf. Sci. Catal.*, **163**, 1–27.

3 King, D.L. and De Klerk, A. (2011) *ACS Symp. Ser.*, **1084**, 1–24.

4 Steynberg, A.P. (2004) *Stud. Surf. Sci. Catal.*, **152**, 1–63.

5 Stranges, A.N. (2003) Germany's synthetic fuel industry 1927–45, in *Proceedings of the AIChE 3rd Topical Conference on Natural Gas Utilization* (eds C.H. Chiu, R.D. Srivastava, and R. Mallison), New Orleans, pp. 635–646.

6 Schulz, H. (1999) *Appl. Catal. A*, **186**, 3–12.

7 Davis, B.H. (2003) An overview of Fischer–Tropsch synthesis at the U.S. Bureau of mines. Presented at the AIChE 2003 Spring National Meeting, New Orleans, LA, March 30–April 3.

8 De Klerk, A. (2011) *Fischer–Tropsch Refining*, Wiley-VCH Verlag GmbH, Weinheim.

9 Kölbel, H. (1959) Die Fischer–Tropsch synthese, in *Winnacker–Küchler Chemische Technologie 3, Organische Technologie I*, Karl Hauser Verlag, Munich, pp. 439–520.

10 Kölbel, H., Ackermann, P., and Engelhardt, F. (1956) *Erdöl Kohle*, **9** (153), 225–303.

11 Kölbel, H. and Ackermann, P. (1951) Hydrogenation of carbon monoxide in the liquid phase. Proceedings of the 3rd World Petroleum Congress, The Hague, 1951, Section IV pp. 2–12.

12 Kölbel, H. and Ackermann, P. (1956) *Chem. Ing. Tech.*, **28**, 381–388.

13 Schanke, D., Wagner, M., and Taylor, P. (2009) Scale-up and demonstration of Fischer–Tropsch technology, in *Proceedings of 1st Annual Gas Processing Symposium* (eds H. Alfadala, G.V. Rex Reklaitis, and M.M. El-Halwagi), Elsevier, Doha, Qatar, pp. 370–377.

14 Rytter, E., Schanke, D., Eri, S., Wigum, H., Skagseth, T.H., and Bergene, E. (2007) *Stud. Surf. Sci. Catal.*, **163**, 327–336.

15 Rytter, E., Skagseth, T.H., Eri, S., and Sjastad, A.O. (2010) *Ind. Eng. Chem. Res.*, **49**, 4140–4148.

16 Forbes, A. (2007) *Petrol. Econ.*, **74** (7), 30.

17 *Petrol Econ.* (2008) **75** (6), 36–38.

18 Köhler, E., Schmidt, F., Wernicke, H.J., De Pontes, M., and Roberts, H.L. (1995) *Hydrocarb. Technol. Int.*, 37–40.

19 Sie, S.T., Senden, M.M.G., and van Wechem, H.M.H. (1991) *Catal. Today*, **8**, 371–394.

20 Overtoom, R., Fabricius, N., and Leemhouts, W. (2009) Shell GTL, from bench scale to world scale, advances in gas processing, in *Proceedings of the 1st Annual Gas Processing Symposium* (eds H. Alfadala, G.V. Rex reklaitis, and M.M. El-Halwagi), Elsevier, Doha, pp. 378–386.

21 Cao, D.B, Zhang, F.Q., Li, Y.W., and Jiao, H.J. (2004) *J. Phys. Chem. B*, **108**, 9094–9104.

22 Chen, Y.H., Cao, D.B., Yang, J., Li, Y.W., Wang, J.G., and Jiao, H.J. (2004) *Chem. Phys. Lett.*, **400**, 35–41.

23 Cao, D.B., Zhang, F.Q., Li, Y.W., Wang, J.G., and Jiao, H. (2005) *J. Phys. Chem. B*, **109**, 833–844.

24 Huang, D.M., Cao, D.B., Li, Y.W., and Jiao, H. (2006) *J. Phys. Chem. B*, **110**, 13920–13925.

25 Liao, X.Y., Cao, D.B., Wang, S.G., Ma, Z.Y., Li, Y.W., Wang, J.G., and Jiao, H. (2007) *J. Mol. Catal. A*, **269**, 169–178.

26 Ma, Z.Y., Huo, C.F., Liao, X.Y., Li, Y.W., Wang, J.G., and Jiao, H. (2007) *J. Phys. Chem. C*, **111**, 4305–4314.

27 Zhang, C.H., Yang, Y., Teng, B.T., and Li, T.Z. (2006) *J. Catal.*, **237**, 405–415.

28 Teng, B.T., Chang, J., Zhang, C.H., Cao, D.B., Yang, J., Liu, Y., Guo, X.H., Xiang, H.W., and Li, Y.W. (2006) *Appl. Catal. A*, **301**, 39–50.

29 Teng, B.T., Zhang, C.H., Yang, J., Cao, D. B., Chang, J., Xiang, H.W., and Li, Y.W. (2005) *Fuel*, **84**, 791–800.

30 Teng, B.T., Chang, J., Yang, J., Wang, G., Zhang, C.H., Xu, Y.Y., Xiang, H.W., and Li, Y.W. (2005) *Fuel*, **84**, 917–926.

31 Yang, Y., Xiang, H.W., Tian, L., Wang, H., Zhang, C.H., Tao, Z.C., Xu, Y.Y., Zhong, B., and Li, Y.W. (2005) *Appl. Catal. A*, **284**, 105–122.

32 Yang, Y., Xiang, H.W., Xu, Y.Y., Bai, L., and Li, Y.W. (2004) *Appl. Catal. A*, **266**, 181–194.

33 Yang, Y., Xiang, H.W., Zhang, R., Zhong, B., and Li, Y.W. (2005) *Catal. Today*, **106**, 170–175.

34 Hao, X., Dong, G., Yang, Y., Xu, Y.Y., and Li, Y.W. (2007) *Chem. Eng. Technol.*, **30**, 1157–1165.

35 Li, Y.W., Xu, J., and Yang, Y. (2010) *Biomass to Biofuels: Strategy for Global Industries* (eds A.A. Vertès, N. Qureshi, H. P. Blaschek, and H. Yukawa), John Wiley & Sons, Inc., New York, pp. 123–140.

36 Perego, C, Bortolo, R., and Zennaro, R. (2009) *Catal. Today*, **142**, 9–16.

37 ENI (2007) World Oil & Gas Review.

38 Steynberg, A.P. (2004) *Stud. Surf. Sci. Catal.*, **152**, 64–195.

39 Collins, J.P., Font Freide, J.J.H.M., and Nay, B. (2006) *J. Nat. Gas Chem.*, **15**, 1–10.

40 Font Freide, J.J.H.M., Collins, J.P., Nay, B., and Sharp, C. (2007) *Stud. Surf. Sci. Catal.*, **163**, 37–44.

41 Waddacor, M. (2005) *Fundamentals of Gas to Liquids*, 2nd edn, Petroleum Economist, London, pp. 41–44.

42 Agee, M.A. (2012) Excerpted from the Wall Street Transcript CEO/Company Interview, Syntroleum Corporation.

43 Schubert, P.F., Bayens, C.A., and Weick, L. (2001) *Oil Gas J.*, **99** (11), 69–73.

44 Lerou, J.J., Tonkovich, A.L., Silva, L., Perry, S., and McDaniel, J. (2010) *Chem. Eng. Sci.*, **65**, 380–385.

45 Collings, J. (2002) *Mind over Matter. The Sasol Story: A Half-Century of Technological Innovation*, Sasol, Johannesburg.

46 Aman, A. (2009) *Chem. Eng. News*, **87** (49), 23.

47 (2005) *Chemical Technology* (South Africa), October, 13–16.

48 De Klerk, A. (2011) *ACS Symp. Ser.*, **1084**, 215–235.

49 De Klerk, A. (2009) *Advances in Fischer–Tropsch Synthesis, Catalysts, and Catalysis* (eds B.H. Davis and M.L. Occelli), Taylor & Francis, Boca Raton, pp. 331–364.

6
Other Industrially Important Syngas Reactions

Peter M. Maitlis

Synopsis

There are a number of industrially important syngas reactions other than the Fischer–Tropsch synthesis (FT-S). They include CO hydrogenation to methanol, which is the major route to methanol and produces over 30 million tons annually. The methanol synthesis reactions and those that lead, via additional steps, to dimethyl ether, ethanol, and even acetic acid, formates, and carbonates are reviewed. Several mechanisms have been proposed; the reasons for the sharp differences are considered.

6.1
Survey of CO Hydrogenation Reactions

It would be wrong to give the impression that the Fischer–Tropsch synthesis is the only industrially important CO hydrogenation reaction based on syngas: indeed, by volume it is not even the biggest process, as this distinction belongs to the production of methanol (Equation 6.1), with over 30 million tons being manufactured each year.

$$CO + 2H_2 \rightleftarrows CH_3OH \quad (\Delta H_{298\,K} - 91\,kJ/mol \quad \Delta G_{298\,K} + 25\,kJ/mol) \qquad (6.1)$$

In addition to methanol and the FT-S, there are many other industrially important catalytic processes for which syngas provides the basic feedstock. These include the manufacture of dimethyl ether, dialkyl carbonates, formic acid and formates, long-chain aldehydes and alcohols (via hydroformylation), and, critically, hydrogen gas production. The production of methane and water from syngas is greatly favored thermodynamically; although this is not an industrially useful process, the reverse reaction, the reforming of methane (as natural gas) is the most usual way of making syngas today.

In methanol synthesis, as in the FT-S, no further substrate is used and the reaction just involves contacting syngas with a heterogeneous catalyst. In contrast to the linear hydrocarbons produced by the FT-S using an iron, cobalt, ruthenium,

Greener Fischer-Tropsch Processes for Fuels and Feedstocks, First Edition. Edited by Peter M. Maitlis and Arno de Klerk.

or rhodium catalyst, usually on an "inert" oxide support, the catalyst for methanol synthesis is composed of copper and zinc oxide, usually supported on alumina.

In order to put our discussions of the FT-S into perspective, we devote this chapter to a summary of other reactions in which syngas is used as a building block for the manufacture of some key chemicals. Most of these processes incorporate a metal-catalyzed step that involves both CO and hydrogen. Hydrogen itself is manufactured largely from syngas via the metal-catalyzed water-gas shift reaction (WGSR) (Equation 6.2) (also see Section 2.3) that links CO, water, hydrogen, and CO_2. Thus, reactions mainly involving hydrogen or carbon monoxide also have their roots in syngas; some of these are listed in Sections 6.5 and 6.8.

$$CO + H_2O \rightleftarrows CO_2 + H_2 \quad (\Delta H_{298\,K} - 41\,kJ/mol \quad \Delta G_{298\,K} - 28\,kJ/mol) \tag{6.2}$$

Many of these reactions take place only over heterogeneous catalysts, but some are best conducted in solution under homogeneous conditions. The latter include the hydroformylations of alkenes to aldehydes and alcohols, the acetic acid syntheses, and the formation of alkanoates. The WGSR can be conducted either homogeneously or heterogeneously, though the latter is practiced in conjunction with the very large-scale methanol and the FT synthesis plants.

The most important organic compounds that are produced via metal-catalyzed reactions from syngas are illustrated in Figure 6.1. The Haber–Bosch ammonia

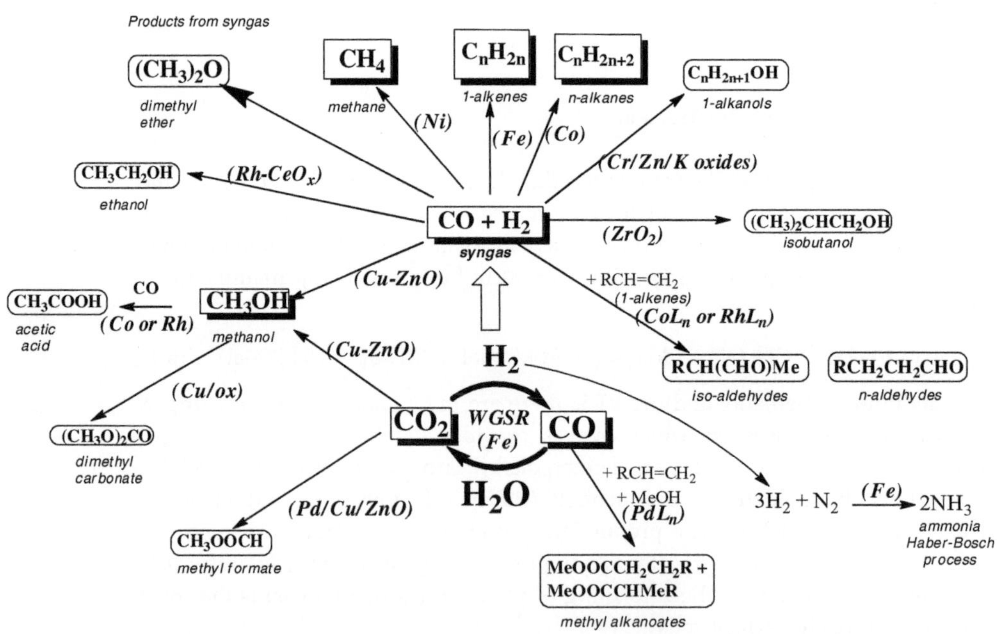

Figure 6.1 Chart showing organic compounds that are produced via metal-catalyzed reactions from syngas. The major products and the water-gas shift reaction (linking $CO_2 + H_2$ and $CO +$ water) are in bold and ringed; metals that play important roles in the individual reactions are italicized and are in brackets.

synthesis is also included as a representative hydrogenation; although it does not involve CO directly, the hydrogen is usually generated from syngas via the WGSR.

There are also other syngas reactions, which require substrate molecules in addition to CO and hydrogen. In hydroformylation (Equation 6.3) (also see Section 6.7), this is an alkene and the immediate products are linear and branched chain aldehydes with one more carbon than the original alkene.

$$RCH{=}CH_2 + CO + H_2 \rightarrow RCH_2CH_2CHO + RCH(CHO)CH_3 \qquad (6.3)$$

Although hydroformylations can be carried out on internal alkenes, most industrially significant reactions involve 1-alkenes.

There are a number of variations on this theme in which an olefin and water or an alcohol are combined with CO (Section 6.6).

A noteworthy feature of many of these processes is that they have been improved to give the products by methods that are environmentally more favorable than the older energy demanding ways. Higher selectivity using better catalysts reduces the need for expensive distillations. A good example is the synthesis of acetic (ethanoic) acid, which used to be made by fractional distillation of the complex oxygenate mixtures obtained on oxidizing (burning) alkanes. It is now made under relatively mild conditions (30–60 atm, 150–200 °C) by reacting methanol and CO over a rhodium plus iodide or on an iridium–ruthenium plus iodide catalyst.

6.2
Syngas to Methanol

6.2.1
Introduction

Even though the methanol synthesis reaction (Equation 6.1) appears simple, this is deceptive and it has taken much time and effort to develop a good process and to elucidate the details. The main problem is that water and hydrocarbon formation (especially of methane) are thermodynamically favored as in the FT-S, and thus the selectivity of the syngas reaction must be completely changed to give methanol. It is a tribute to the engineers and chemists that more than 99% selectivity to methanol is now routinely achieved.

The first industrial methanol "high-pressure" process, operating at 340 atm and 320–380 °C, was initiated by BASF in the 1920s following the original discovery of the methanol synthesis by Mittasch and Pier. This was the dominant technology for over 45 years. The zinc oxide/chromia (ZnO/Cr_2O_3) catalysts used in the early methanol plants were chosen as they were relatively poison-resistant and able to cope with a syngas feedstock that contained considerable chlorine and sulfur impurities, as it was made from low-grade German coal (lignite). This syngas was barely selective enough for practical operation, and forcing conditions close to those that encouraged hydrocarbon formation were necessary.

As the thermodynamic equilibrium (Equation 6.1) is most favorable for methanol synthesis at high pressures and low temperatures, considerable efforts were made to improve the process. In the early 1960s, a much cleaner syngas became available from the steam reforming of hydrocarbons: naphtha and natural gas. This allowed ICI to develop a new copper-based catalyst, comprising Cu–ZnO on alumina, which was active at 50–100 atm and 240–260 °C. The enhanced reactivity saved energy as it allowed milder operating conditions, and this became known as the "low-pressure" process, when it was introduced in 1972. Today, this represents the main route to make methanol. The engineering is critical, especially the mode by which the heat of reaction is removed [1].

It was found that ZnO was a perfect dispersant for the copper. In addition to a structural role in stabilizing the copper catalysts, ZnO is also able to limit the effects of poisons as it removes H_2S from gas streams with the formation of zinc sulfide.

To retain the long-term activity of Cu catalysts, it was found empirically that gas-phase sulfur concentrations needed to be kept below 1 ppm and preferably below 0.1 ppm. In fact, the copper catalysts have some sulfur tolerance and it was found that if a feedstock containing an average of 2% sulfur was employed, the methanol synthesis activity of a $Cu/ZnO/Al_2O_3$ catalyst was approximately 80% of the unpoisoned activity.

Transition metals such as iron also had deleterious effects on methanol synthesis as they can increase the hydrogenation activity and promote the dissociation of CO and CO_2, leading to formation of methane and long-chain paraffins and/or waxes by Fisher–Tropsch reactions. Methane formation was always a problem with the original high-pressure methanol synthesis catalysts, and it is thought this was mainly caused by iron impurities. Iron is sometimes deposited on synthesis catalysts during use by the decomposition of gaseous iron pentacarbonyl, $Fe(CO)_5$ formed from rust that may be present in the makeup gas system.

6.2.2 Synthesis Reaction

It is now agreed that the key player in methanol synthesis is carbon dioxide and that the overall reaction occurring on the surface of Cu/CuO catalysts is predominantly Equation 6.4 rather than Equation 6.1:

$$CO_2 + 3H_2 \rightarrow CH_3OH + H_2O \qquad (6.4)$$

In other words, methanol is formed from carbon dioxide. Catalysts made by a controlled precipitation procedure were used in the first low-pressure synthesis plants. They had lives of more than 3 years, and produced methanol of a higher purity than the older high-pressure process. Continued development resulted in catalysts suitable for operation at 100 atm, which is around the optimum operating pressure for high-capacity plants producing more than 360 000 t/a of methanol. However, the rate of catalyst deactivation was higher at the higher pressure.

ZnO/Al_2O_3 is not a sufficiently refractory support at 100 atm; however, a more refractory support is provided by introducing some of the zinc components as zinc spinel [$ZnAl_2O_4$]. The effect of addition of even more refractory oxides, such as MgO, probably contributes further to the stability of the mixed oxide support rather than stabilizing the copper crystallites directly. The optimization of the multi-component catalysts for both maximum initial activity and maximum stability has not been easy.

6.2.3
Mechanism

There have been many attempts to understand the mechanism of the methanol synthesis over the Cu/ZnO catalyst. EXAFS under realistic conditions combined with time-resolved X-ray diffraction studies and HRTEM (high-resolution transmission electron microscopy) showed that copper metal is the active phase. Surface intermediates such as a symmetric carbonate $\{CO_3\}$ of copper and a formate $\{HCO_2\}$ have been spectroscopically detected. A surprising and very significant discovery was that although CO is the normal feedstock, carbon dioxide was actually the immediate precursor to methanol. This led to a *static kinetic* model for the reaction that comprised some 13 elementary reaction steps, including the formation of carbon dioxide and surface formates $\{HCOO\}$ in going from CO to CH_3OH [Box 6.1] [2].

Although the "static" model fits much of the experimental data, a better agreement was obtained when allowance was made for changes in the inlet gas compositions, that is, the reduction potential of the gas phase. This has led the Danish

Box 6.1 Representation of steps postulated to occur in the "static" model for methanol synthesis

The following equations represent the main steps postulated to occur in the synthesis of methanol over a Cu–ZnO–alumina catalyst. Surface species are only partially defined at best and are therefore written within curly brackets. The active precursor is carbon dioxide (*not* CO) and spectroscopically detected intermediates include surface carbonate $\{HCO_3\}$ and formate $\{HCOO\}$ [2, 3]. It is suggested that carbon monoxide acts mainly to remove surface oxygen atoms.

$$\{H_2O\} \rightleftarrows \{OH\} + \{H\}$$
$$2\{OH\} \rightleftarrows \{H_2O\} + \{O\}$$
$$CO \rightleftarrows \{CO\}$$
$$\{CO\} + \{O\} \rightleftarrows \{CO_2\}$$
$$\{CO_2\} \rightleftarrows CO_2$$
$$\{CO_2\} + \{H\} \rightleftarrows \{HCOO\}$$
$$\{HCOO\} + \{H\} \rightleftarrows \{H_2COO\}$$
$$\{H_2COO\} + \{H\} \rightleftarrows \{H_3CO\} + \{O\}$$
$$\{H_3CO\} + \{H\} \rightleftarrows \{CH_3OH\}$$
$$\{CH_3OH\} \rightleftarrows CH_3OH \uparrow$$

group to develop a *dynamic microkinetic model* in which the state of the catalyst during the reaction is determined by the changing composition of the gas phase at that point with time. It was presumed that the observed changes in particle morphology are caused by a change in the contact surface free energy between the Cu particles and the support, which is postulated to arise if the concentration of oxygen vacancies at the Zn–O–Cu interfaces vary with changes in the reduction potential of the gas phase.

6.2.4
Catalyst Deactivation

Copper catalysts are susceptible to thermal sintering due to the migration of copper atoms on the surface. This is the main cause of deactivation, which is markedly accelerated by the presence of even traces of chloride. Care must therefore be taken to eliminate halides from copper catalysts during manufacture, and from reactants during use. Modern copper catalysts contain oxides to minimize thermal sintering. Typical formulations contain Cr_2O_3 and/or Al_2O_3, in addition to CuO and ZnO, the thermal stability of which is significantly higher than that of the early catalysts. When they are operated under well-controlled conditions, neither poisoning nor coking is normally a significant cause of deactivation; but operating temperatures must be restricted, usually to below 300 °C. Minimizing the side reactions that occur at higher temperatures requires rapid heat removal either by introducing cold gas or by using multiple tube reactors.

The form of the catalyst, its morphology, and its method of preparation are important but they vary widely. The work carried out at ICI has been summarized [4, 5].

As disposal of spent catalyst has been a common practice in industrial methanol synthesis, ways of reusing the catalyst have been evaluated for both economic and environmental reasons. Thus, for example, Rahimpour [6] found that the spent catalyst retains considerable activity and can be reused if mixed with fresh catalyst.

The finding that carbon dioxide is the precursor to methanol also suggests a useful, convenient, and environmentally benign way of using some of the excess CO_2 that is generated on earth. However, this challenge does not yet seem to have been taken up by any industry. The reason is presumably that more hydrogen is needed to make methanol from CO_2 than from CO. And, of course, hydrogen today still comes largely from syngas! Once a convenient method for generating hydrogen *without* involving syngas and fossil fuels is developed in a hydrogen economy (say by sunlight-induced catalytic water cleavage), the methanol will be a very admirable feedstock.

6.2.5
Uses of Methanol

The largest use of methanol by far is in making other chemicals: some 40% is converted into formaldehyde, and from there into plastics, plywood, paints, explosives, and permanent press textiles. Smaller percentages (~10%) are used to make acetic acid and acetic anhydride, dimethyl ether, and dimethyl carbonate (DMC). A methanol to gasoline process was developed by Mobil and a plant was built in New Zealand; however, this did not prove to be viable and has since been closed. Methanol has been directly used in internal combustion engines, especially as a high-energy fuel for motor racing, and was also employed to make the gasoline additive methyl *tert*-butyl ether (MTBE). More recently, methanol has been used in the transesterification of triglycerides to make a biodiesel fuel.

6.3 Syngas to Dimethyl Ether (DME)

Today, DME is primarily produced from methanol by dehydration in the presence of, for instance, a silica–alumina catalyst. It can also be produced directly from syngas using a dual catalyst system that permits both methanol synthesis and dehydration in the same process unit, without methanol isolation and purification, a procedure that promises efficiency advantages and cost benefits. Thus, for example, a highly efficient synthesis of DME from syngas over a catalyst comprising CuO–ZnO–Al_2O_3 and HZSM-5 zeolite modified with antimony oxide has been reported [7]. Approximately 50 000 t/a of DME were manufactured in Western Europe, but DME is now being marketed as a "multiuse, multisource low-carbon fuel" and major production facilities are being planned around the world. Although usually derived from hydrocarbons, DME can also be made using organic waste or biomass.

6.3.1 DME Uses

Currently, the largest use of DME is as substitute for propane as a fuel, especially in China. Two other important applications are as a replacement for chlorofluorocarbons as a propellant in aerosol canisters and as a precursor to dimethyl sulfate by reaction with sulfur trioxide. It can also be used as a feedstock for acetic acid synthesis and as a refrigerant. As DME has a very low boiling point (−23 °C), this can limit its use; however, this property facilitates its removal from reaction mixtures.

6.4 Syngas to Ethanol

6.4.1 Introduction

Ethanol is very useful industrially as a solvent and also as a feedstock. Currently, it is produced by hydration of ethene (from fossil fuels) catalyzed by phosphoric acid or by fermentation of biomass-derived sugars from sugarcane (Brazil) or corn (the United States). Sugars containing six carbons are readily fermented, whereas 5-carbon sugars and lignins, which are also present in the biomass, largely remain behind. However, there are problems with both these routes and a good new industrial ethanol synthesis would be welcome.

6.4.2
Direct Processes

Although a path involving gasification of biomass to syngas ($CO + H_2$) followed by a catalytic conversion of syngas into ethanol has been extensively explored, no commercial process currently exists. The literature has been well reviewed by Subramani and Gangwal [8]. Modification of the catalysts and conditions for the FT-S and for methanol synthesis can give rise to higher oxygenates. Thus, the FT reaction can be partially diverted to give significant proportions of ethanol, for example, by the use of rhodium metal plus a rare earth oxide as catalyst. The known homogeneously catalyzed processes are unattractive commercially as they need expensive catalysts, high operating pressures, and tedious workup procedures to separate and recycle the catalyst. The heterogeneous catalytic processes for converting syngas into ethanol suffer from low yield and poor selectivity. It has been suggested that this is due to the slow kinetics of the initial C—C bond formation and fast chain growth of the C_2 intermediate.

The Institut Francais du Petrole/Idemitsu process based on a copper–cobalt alloy catalyst made ethanol in a 950 t/a pilot plant near Tokyo. The process used steam reforming of natural gas followed by multiple synthesis reactors to give mixed linear C_1–C_7 alcohols suitable for blending. The purity of the alcohol phase was very good.

Snamprogetti, Enichem, and Haldor Topsoe (SEHT) used a modified methanol synthesis catalyst (in a 400 t/day plant that operated between 1982 and 1987) in a series of fixed bed adiabatic reactors operated in the temperature range of 260–420 °C and pressures as high as 180–260 atm to give mixed alcohols. The crude mixture containing 20% water was purified using three distillation columns; the first column removed methanol and ethanol, the second removed water, while the third recovered C_{3+} alcohols by an azeotropic distillation using cyclohexane. The final water content of the mixed alcohol product was below 0.1%; it was blended at 5 vol% in gasoline and marketed successfully as a premium fuel.

In contrast, the Lurgi–Octamix process used a low-pressure, low-temperature modified methanol synthesis catalyst, reported to contain 25–40 wt% CuO, 10–18 wt% Al_2O_3, 30–45 wt% ZnO, and 3–18 wt% promoter oxides. Typical operating conditions used were ≈350 °C and 100 atm. The process gave a 21–28% CO conversion, a 66–79% selectivity to alcohol products, and 17–25% selectivity to CO_2. The selectivity to methanol was 41–58%, but that to ethanol was only 1–9%.

An analysis of the possible mechanism of ethanol formation based on the results of ethanol oxidation and of decomposition studies on a single-crystal Rh(110) surface concluded that the transformation of syngas into ethanol occurred via surface acetate [9]. However, a DFT calculation of ethanol synthesis concluded that the CO was hydrogenated to ethanol via surface formyls on Rh(111) [10].

6.5 Syngas to Acetic Acid

Acetic (ethanoic) acid is a major commodity chemical, with a current world production capacity of about 9 Mt/a. It has long been a mainstay of the organic chemicals industry, both as a solvent and as a feedstock. The older processes produced a range of carboxylic acids by alkane (naphtha) oxidation and the acetic acid then needed to be isolated and purified by successive distillations. These energy-demanding routes were superseded when syngas became readily available as it allowed the manufacture of acetic acid in high purity from C_1 feedstocks (CO and methanol) and thus in two steps from CO and hydrogen. This was an early success story of homogeneous catalysis and one of the triumphs of modern coordination chemistry [11]. The overall reaction is given by Equation 6.5:

$$CO + MeOH \rightarrow MeCOOH \tag{6.5}$$

6.5.1 Acetic Acid Processes

The first process, commercialized by BASF in the 1960s, used a cobalt and iodide catalyst, but needed very stringent conditions (~700 atm and ~250 °C), and had a selectivity of about 90% to acetic acid. This was quickly superseded by the Monsanto (later Monsanto-BP) process. Although this was based on a rhodium (plus iodide) catalyst that was about 100× more expensive than the cobalt catalyst on a molar basis, the much milder reaction conditions (30–60 atm and 150–200 °C) that were needed as well as a higher selectivity allowed great savings in capital costs that more than made up for the difference in the price of the catalyst.

In the Monsanto process, the reaction involves carbonylation of a homogeneous *solution* of the feedstock in the presence of a rhodium catalyst precursor (e.g., the rhodium β-diketonate, $[Rh(OCMeCHCMeO)_3]$) [11, 12]. The reactants are methanol plus iodomethane (methyl iodide (MeI), a convenient source of iodide) and the main product is acetic acid. Thus, under reaction conditions, the "solvent" is largely methyl acetate.

The Rh/I^- catalyst gave >99% selectivity to acetic acid that was purified from water, the main contaminant, and most new plants commissioned in the 1970s and 1980s were based on the Monsanto technology. As this is a well-defined example of a homogeneously catalyzed process of industrial importance, it is shown in some detail in Box 6.2.

More recently, BP Chemicals introduced the Cativa™ process, based on an iridium/iodide catalyst, promoted by ruthenium [13]. This used conditions similar to that of the Monsanto process; however, since the Cativa process operated at low

Box 6.2 Steps occurring in the Rh/I⁻ homogeneously catalyzed carbonylation of methanol to acetic acid (Monsanto-BP process)

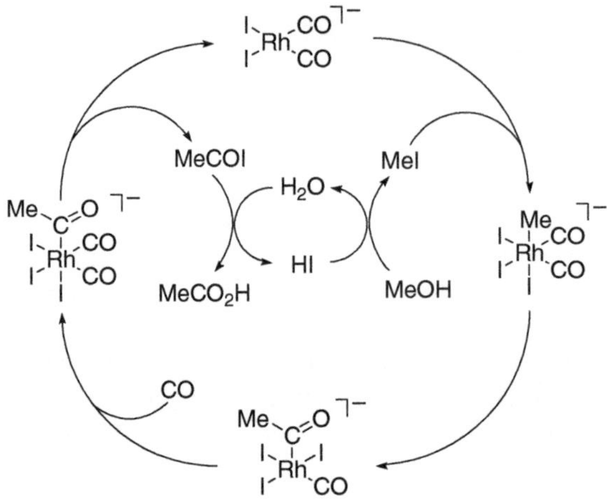

The mechanism comprises two cycles: (i) the "rhodium cycle" that involves reactions of organometallic complexes, and (ii) the "iodide cycle" that involves organic reactions of the iodide cocatalyst. Initially, methanol (or methyl acetate) reacts with the hydrogen iodide promoter to give iodomethane. The iodomethane undergoes an oxidative addition reaction with the square planar Rh(I) complex *cis*-$[Rh(CO)_2I_2]^-$ to give a product –$[Rh(Me)(CO)_2I_3]^-$, where both methyl and iodide ligands bind to the rhodium center. This is the slowest reaction in the cycle, that is, the rate-determining step. The product is an octahedral rhodium(III) complex in which the methyl ligand is coordinated *cis* to two CO ligands. This complex is highly reactive (and only present in small concentration) due to a rapid migratory insertion reaction that brings together the methyl group and a CO to give an acetyl (C(O)Me) ligand. This key step in the cycle results in a new C—C bond being formed between ligands derived from the two C_1 feedstocks. The migratory insertion opens a vacant coordination site on rhodium, which can take up another molecule of CO to give an acetyl dicarbonyl complex. The final step in the organometallic cycle is a reductive elimination reaction that releases acetyl iodide and regenerates the active Rh(I) complex. Hydrolysis of acetyl iodide gives the acetic acid product and regenerates the cocatalyst HI to complete the iodide cycle [11, 12].

water levels, considerable energy savings were possible as less water had to be removed to make the desired anhydrous acid. This is a remarkable story in that all three Group 9 metals are active and there has been a progressive improvement in the iodide-promoted process from Co via Rh to Ir (with Ru) [14].

The major units of a Monsanto-BP methanol carbonylation plant are illustrated in Figure 6.2 [15]. The MeOH and CO feedstocks are continuously fed to the reactor vessel. In the initial product separation step, the reaction mixture is passed

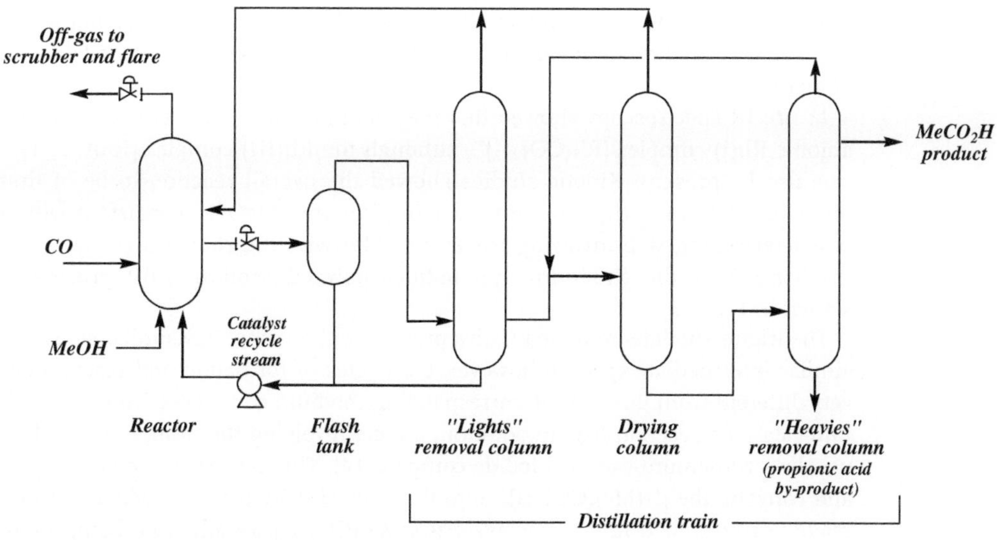

Figure 6.2 Schematic representation of an acetic acid plant using the Monsanto-BP technology for the carbonylation of methanol. Adapted from Ref. [15].

from the reactor into a "flash tank" where the pressure is reduced to induce vaporization of most of the volatiles. The catalyst remains dissolved in the liquid phase and is recycled back to the reactor vessel. The product stream from the flash tank is directed into a distillation train that removes iodomethane, water, and small amounts of heavier by-products (e.g., propionic acid) from the acetic acid. Since acetic acid is produced, the methanol feedstock is largely esterified into methyl acetate, which acts as solvent for the reaction. For the rhodium catalysts, quite a high concentration of water (about 10 M) is required to maintain high rates and prevent deactivation by precipitation of the catalyst.

However, to make the desired anhydrous acetic acid, the water must then be separated from the acetic acid product by distillation, which incurs substantial costs. Another problem arises from the high water levels as they increase the rate of the WGSR (also catalyzed by the rhodium/iodide system) leading to loss of significant amounts of CO as CO_2.

One point that should be mentioned relates to the carbonylation of methyl acetate to acetic anhydride (Equation 6.6):

$$\mathrm{MeOCOMe} + \mathrm{CO} \rightarrow (\mathrm{MeOCO})_2\mathrm{O} \qquad (6.6)$$

While virtually all carbonylation processes were until quite recently based on syngas from natural gas or naphtha, as the Eastman acetic anhydride plant was sited on a major coalfield in the United States, the CO was generated from coal.

6.5.2
Mechanisms

The mechanisms of the various methanol carbonylation reactions, as well as those for the closely related methyl acetate carbonylation to acetic anhydride, have been extensively studied, and are now well understood. The reactions occurring in the original Monsanto rhodium- and iodide-catalyzed processes serve as a model for all of them.

In situ IR spectroscopy showed that the rhodium catalyst exists mainly as the anionic Rh(I) complex $[Rh(CO)_2I_2]^-$, although the Rh(III) complex $[Rh(CO)_2I_4]^-$ can also be present. Kinetic studies showed the overall reaction to be of first order in [MeI] and [Rh] and zero-order in [CO] and [MeOH], consistent with a rate-determining step involving reaction of MeI with the rhodium catalyst. The mechanism of the rhodium- and iodide-catalyzed Monsanto-BP process is shown in Box 6.3.

The iridium/iodide-promoted Cativa process contains basically similar steps and similar intermediate species: however, their rates of formation and reaction are very different from those of the corresponding rhodium complexes. Thus, the iridium-catalyzed reaction has, in addition, a cycle involving the promoter, which is usually a ruthenium carbonyl iodide complex [14]. This acts as an iodide acceptor and converts the $[Ir(Me)(CO)_2I_3]^-$ into the neutral $[IrMe(CO)_2I_2]$, which is more readily carbonylated into the acetyl $[Ir(COMe)(CO)_2I_2]$. In addition, being more kinetically inert, the anionic iridium complexes are less susceptible to reduction and deactivation via the WGSR.

6.5.3 Catalyst Deactivation

Most attempts to improve upon the original Monsanto technology have involved strategies to allow operation at lower water concentration. Hoechst-Celanese developed a modified catalytic system using an iodide salt (e.g., LiI) to stabilize the rhodium catalyst at relatively low water concentration [16]. Their acid optimization (AO) technology was used to increase production at a plant in Clear Lake, TX, from its original capacity of 270 kt/a to 1200 kt/a in 2001. The LiI-promoted low-water process is operated at higher methyl acetate concentration, which helps to stabilize the Rh(I) catalyst by moderating the equilibrium HI concentration. A lower HI concentration results in a decreased tendency for the active Rh(I) form of the catalyst $[Rh(CO)_2I_2]^-$ to be oxidized to the Rh(III) form, $[Rh(CO)_2I_4]^-$, which then readily loses CO to deposit an *inactive* Rh-I material.

Another approach, developed by Chiyoda/UOP, uses a rhodium catalyst heterogenized on a polymeric cation exchange resin. This takes advantage of the fact that the rhodium-catalyzed carbonylation involves anionic complexes. Since the heterogenized catalyst is essentially confined to the reactor, problems with precipitation in the product separation and catalyst recycle stages are minimized. The removal of solubility constraints allows operation at lower water concentration with increased catalyst loading, and rates comparable to the homogeneous reaction can be achieved. The lower water concentration is also claimed to result in reduced by-product formation. Mechanistic studies have indicated that the catalytic cycle for the supported catalyst is essentially identical to the homogeneous process [17].

6.6 Higher Hydrocarbons and Higher Oxygenates

Most short-chain olefins such as propene and the butenes are made by cracking, in other words, breaking down larger hydrocarbons. Several variants including steam and catalytic cracking are practiced. However, some higher hydrocarbons and also higher oxygenates can be made from syngas, as mentioned in the following sections.

6.6.1 Isobutene and Isobutanol

There has been considerable commercial interest in making isobutene (2-methylpropene) and isobutanol (2-methylpropanol) as they are intermediates in the synthesis of the gasoline additives (antiknock agents and octane enhancers), methyl–*tert*-butyl ether (MTBE, $MeOCMe_3$) and ethyl-*tert*-butyl ether (ETBE, $EtOCMe_3$). MTBE was initially popular but environmental concerns have led to a decline in its use and a corresponding increase in the use of ETBE, which offers equal or greater air quality benefits than ethanol, while being technically less difficult than MTBE. Unlike ethanol, ETBE does not induce evaporation of gasoline, which is one of the causes of smog, and does not absorb moisture from the atmosphere. MTBE and ETBE are made by etherification of isobutanol with methyl or ethyl alcohol or by addition of MeOH or EtOH to isobutene.

The demand for tertiary alkyl ethers as gasoline additives has attracted attention to alternative pathways for their production. One possibility is by the synthesis of isobutanol–methanol mixtures via CO hydrogenation, however, the processes disclosed so far do not sound attractive. For example, the isobutyl oil synthesis that was operated (by BASF with a total capacity up to 390 000 t/a during the Second World War) strongly resembled the early high-pressure methanol synthesis in catalyst ($ZnO/Cr_2O_3/K_2O$) and reactor choice, the main difference being the more extreme reaction conditions 420–460 °C and 325 atm. Even then under optimized conditions, the catalyst lifetime was only 90 days and a typical product mix consisted of water (about 25%), methanol (50%), isobutanol (11–13%), and smaller amounts of higher alcohols [18].

Another group of publications has reported higher alkene and *iso*butanol (2-methylpropanol) syntheses from syngas; in each case the significance of the use of zirconia was emphasized. Thus, Erkey *et al.* reported selective formation of isobutene from syngas over zirconia in a laboratory slurry reactor [19]. Experiments were conducted to determine the effects of space velocity and CO/H_2 ratio on CO conversion and hydrocarbon product distribution. Maruya *et al.* have also reported the use of ZrO_2 in syngas catalysts for isobutene [20]. Isobutanol can conveniently be made by hydrogenation of isobutanal, which is made by hydroformylation of propene (Section 6.7). The more usual hydroformylation product is *n*-butanal, but formation of the branched chain aldehyde can be facilitated by using special ligands on either Co or Rh catalysts.

6.7
Hydroformylation

The hydroformylation (or "oxo") reaction was discovered in 1938 by Roelen who was working on the formation of oxygenates as by-products of the FT reaction over cobalt catalysts. It soon became clear that the aldehydes and alcohols found were the products of secondary reactions undergone by the 1-alkenes formed in the FT reaction. In this new cobalt-catalyzed reaction, H and CHO were added to an olefin: hence, the name, *hydroformylation*. It was later found that the true catalyst was not cobalt metal but a substituted cobalt carbonyl, such as the hydride, $CoH(CO)_4$, and that the reaction occurred preferentially in solution. The process was rapidly developed and commercialized by BASF.

1-Alkenes are hydroformylated to give both linear (normal, *n*-) and branched chain (iso, *i*-) aldehydes (Equation 6.7). The best-known application is the reaction with propene to give *n*-butanal, which is then hydrogenated to *n*-butanol. As the hydroformylation can add CO either to the terminal or to the internal carbon of the double bond, two isomeric butanals are formed. The *i*-butanal formed can be hydrogenated to *iso*-butanol (2-methylpropanol), also a commercially useful chemical (see Section 6.6.1).

$$MeCH{=}CH_2 + CO + H_2 \rightarrow MeCH_2CH_2CHO + MeCH(CHO)CH_3 \qquad (6.7)$$

Hydroformylation is generally carried out in homogeneous solution under moderate conditions of pressure and temperature, but using a metal catalyst solubilized and activated by special ligands. In the search for high selectivity, various ligand and metal combinations have been tested. The original "unmodified" cobalt carbonyl-hydride catalyst is in an equilibrium, where a vacant site is created by CO dissociation at which the alkene can coordinate (Equation 6.8):

$$CoH(CO)_4 \rightleftarrows CoH(CO)_3 + CO \qquad (6.8)$$

The basic cobalt process was widely used up to the 1970s when Shell introduced the trialkylphosphine-modified catalysts, the active constituents of which are CoH $(CO)_3(PR_3)$, where R is typically a bicyclic phosphine such as a 9-substituted-9-phosphabicyclononane. Since the aldehydes are easily hydrogenated to alcohols, which are the most frequently required end products, the reactions are often run to make the alcohol. A concise summary has been given [21].

Nobel Laureate Sir Geoffrey Wlkinson working with Johnson Matthey and Davy Powergas later discovered that a rhodium triphenylphosphine complex was a much better catalyst for many hydroformylation reactions as it required milder conditions and gave a higher selectivity. Triphenylphosphine is a bulky ligand that helps to give the desirable high *n*-/*i*-ratio in the butanal product by facilitating the addition of CO to the terminal carbon of the double bond in the intermediate steps (Box 6.3).

One important variant of the Rh/PPh_3 catalysis is the two-phase catalyst system developed by Kuntz at Rhone-Poulenc in 1981, but using a sulfonated triphenylphosphine ligand, $P(C_6H_4\text{-}m\text{-}SO_3Na)_3$ (known as TPPTS), to generate the water-soluble catalyst $RhH(CO)[TPPTS]_3$. Since the catalyst has a very high (9-) formal

Box 6.3 Simplified mechanism of the rhodium–triphenylphosphine-catalyzed hydroformylation of propene to give *n*- and *i*-butanal

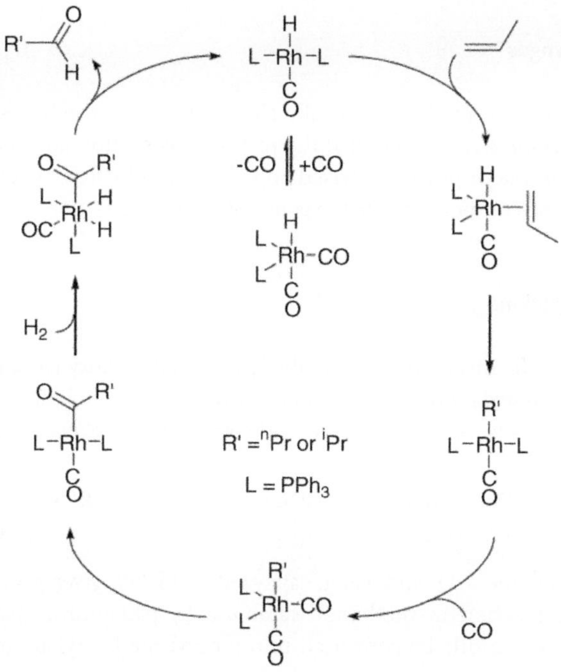

The starting complex is a rhodium bis(triphenylphosphine)-(dicarbonyl) hydride, $[RhL_2(CO)_2H]$. In the first step, one CO is replaced by the alkene. Internal migratory insertion of the alkene into the Rh—H bond leads to the formation of two compounds, containing the rhodium *n*-alkyl $Rh(CH_2CH_2Me)$ and the rhodium *i*-alkyl $Rh(CHMe_2)$ ligands. These intermediates then undergo carbonylation to give the corresponding rhodium-acyls, $Rh(COCH_2CH_2Me)$ and $Rh(COCHMe_2)$, which are then cleaved by an internal H-transfer to give the free aldehydes and regenerate the starting rhodiumhydride, as shown.

anionic charge, it is insoluble in all but the most polar solvents. The resultant two-phase catalyst system has the advantage over the completely homogeneous system in that the organic product butanal is essentially all in the organic phase and can easily be separated. Similarly, recovery of the catalyst is straightforward as it all stays in the aqueous phase. This helps the economics of the process by making it easier to isolate the desired product and to recycle the expensive catalyst plus ligand. An excess of the phosphine ligand is again needed for good *n*-/*i*- selectivities, but lower concentrations are required because the TPPTS phosphine dissociation equilibrium in water is shifted toward the Rh-TPPTS coordinated complexes. High *n*-/*i*- regioselectivities of 16–18: 1 for making butanal from propene can be obtained using this water-soluble catalyst, but rates are slower than those with conventional Rh/PPh_3 catalysts.

6.8
Other Reactions Based on Syngas

Since both CO and hydrogen are made industrially from syngas via the WGSR (Equation 6.2) (also see Section 2.3), to complete the picture this section summarizes very briefly a number of major processes based on *either* CO *or* hydrogen. The WGSR is used to enhance one component at the expense of the other.

6.8.1
Hydroxy and Alkoxy Carbonylations

Reactions related to the hydroformylations include the hydroxy and alkoxy carbonylations, involving alkene carbonylation in the presence of water or alcohols (Equation 6.9). No extra hydrogen gas is used as the H needed comes from the alcohol or water:

$$R^1CH{=}CH_2 + CO + R^2OH \rightarrow R^1CH_2CH_2COOR^2 + R^1CH(COOR^2)CH_3 \quad (6.9)$$

For example, reaction of ethene, CO, and water, catalyzed by nickel, gives propionic acid, while the same reaction in methanol, catalyzed by palladium, gives methyl propionate, which can readily be converted into methyl methacrylate, for the manufacture of strong transparent polymers. Butadiene can be converted into dimethyl adipate (for nylon manufacture) by a cobalt-catalyzed dimethoxy carbonylation (Equation 6.10):

$$CH_2{=}CHCH{=}CH_2 + CO + MeOH \rightarrow MeO_2C(CH_2)_4CO_2Me \quad (6.10)$$

These processes have been reviewed [14].

6.8.2
Methyl Formate

Industrial methyl formate is usually produced by the combination of methanol and carbon monoxide in the presence of a strong base, such as sodium methoxide (Equation 6.11):

$$CH_3OH + CO \rightarrow HCOOCH_3 \quad (6.11)$$

This process, practiced commercially by BASF among other companies gives 96% selectivity toward methyl formate, although it may suffer from catalyst sensitivity to water, which can be present in the carbon monoxide feedstock derived from synthesis gas. Very dry carbon monoxide is, therefore, an essential requirement. Methyl formate is used primarily to manufacture formamide, dimethylformamide, formic acid, and also some agrichemicals and pharmaceuticals.

Since there is much interest in using carbon dioxide as a feedstock, a process for making methyl formate based on carbon dioxide fixation by surface coupling over a Pd/Cu/ZnO nanocatalyst should be noted. This claims that the catalyst is capable of activating gaseous CO_2 to methyl formate in high yield (>20%) with excellent selectivity (>96%) [22].

6.8.3
Dimethyl Carbonate (DMC)

In the last few years, the use of DMC in the chemical industry has grown, due to its chemical properties and its nontoxicity; thus, it allows the production of aromatic polycarbonates without involving phosgene. The excellent physical properties of organic carbonates have given rise to new industrial syntheses of dimethyl carbonate. These include the copper chloride-catalyzed one-step liquid-phase process, by EniChem [23], and the palladium-catalyzed two-step gas-phase process, via methyl nitrite, by UBE [21]. Two further technologies are likely to be developed: the gas-phase direct methanol oxycarbonylation and the alkylenecarbonate transesterification process. Large-scale applications of DMC as solvent and as oxygenate in reformulated fuels offer very promising areas for the future.

6.8.4
Ether Gasoline Additives

As methyl *tert*-butyl ether is no longer the preferred gasoline additive for environmental reasons, interest is now centered on the less-volatile ethyl *tert*-butyl ether (ETBE) and methyl-*tert*-amyl ether (TAME) [24].

6.8.5
Hydrogenation

This is a very vital aspect of syngas chemistry as large amounts of hydrogen are used for a myriad of processes. The major process is the Haber–Bosch ammonia synthesis (Equation 6.12), which contributes substantially to the 130 billion tons of ammonia manufactured each year.

$$N_2 + 3H_2 \rightarrow 2NH_3 \tag{6.12}$$

Other hydrogenation reactions such as hydrotreating in the petroleum industry and the conversion of animal and vegetable fats into margarine (now known as "spread") for human consumption also use substantial amounts. More than 90% of the hydrogen used comes from syngas via the WGSR.

Note Added to Text Grabow and Mavrikakis have carried out DFT calculations for methanol synthesis and the WGSR on Cu/ZnO/alumina [25]. They found that both

CO and CO_2 were involved in MeOH formation, but that under typical industrial conditions, the CO_2 path was responsible for two-thirds and the CO path for one-third of methanol formation.

References

1 Twigg, M.V. and Spencer, M.S. (2003) *Top. Catal.*, **22**, 191.
2 Molenbroek, A.M., Helveg, S., Topsoe, H., and Clausen, B.S. (2009) *Top. Catal.*, **52**, 1303–1311.
3 Ovesen, C.V., Clausen, B.S., Schiùtz, J., Stoltze, P., Topsùe, H., and Nùrskov, J.K. (1997) *J. Catal.*, **168**, 133–142.
4 Chinchen, G.C., Denny, P.J., Jennings., J. R., Spencer, M.S., and Waugh, K.C. (1988) *Appl. Catal.*, **36**, 1–65.
5 Waugh, K.C. (1992) *Catal. Today*, **15**, 51–75.
6 Rahimpour, M.R., Moghtaderi, B., Jahanmiri, A., and Rezaie, N. (2005) *Chem. Eng. Technol.*, **28**, 226–234.
7 Mao, D.S., Xia, J.C., Zhang, B., and Lu, G. Z. (2010) *Energy Convers. Manage.*, **51**, 1134–1139.
8 Subramani, V. and Gangwal, S.K. (2008) *Energy Fuel.*, **22**, 814–839.
9 Bowker, M. (1992) *Catal. Today*, **15**, 77.
10 Choi, Y.-M. and Liu, P. (2009) *J. Am. Chem. Soc.*, **131**, 13054.
11 Dekleva, T.W. and Forster, D.F. (1986) *Adv. Catal.*, **34**, 81.
12 Maitlis, P.M., Adams., H., Sunley, G.J., and Howard, M.J. (1996) *J. Chem. Soc., Dalton Trans.*, 2187–2196.
13 Sunley, G.J. and Watson, D.J. (2000) *Catal. Today*, **58**, 293–307.
14 Haynes, A., Maitlis, P.M., Morris, G.E., Sunley, G.J., Adams, H., Badger, P.W., Bowers, C.M., Cook, D.B., Elliott, P.I.P., Ghaffar, T., Green, H., Griffin, T.R., Payne, M., Pearson, J.M., Taylor, M.J., Vickers, P.W., and Watt, R.J. (2004) *J. Am. Chem. Soc.*, **126** (9), 2847–2861.
15 Maitlis, P.M. and Haynes, A. (2006) Chapter 4, in *Metal-Catalysis in Industrial Organic Processes* (eds G.P. Chiusoli and P.M. Maitlis), RSC Publishing, Cambridge.
16 Yoneda, N., Kusano, S., Yasui, M., Pujado, P., and Wilcher, S. (2001) *Appl. Catal. A – Gen.*, **221**, 253–265.
17 Haynes, A., Maitlis, P.M., Quyoum, R., Pulling, C., Adams, H., Spey, S.E., and Strange, R.W. (2002) *J. Chem. Soc., Dalton Trans.*, 2565.
18 Verkerk, K.A.N., Jaeger, B., Finkeldei, C.-H., and Keim, W.I. (1999) *Appl. Catal. A – Gen.*, **186**, 407–431.
19 Erkey, C., Wang, J.H., Postula, W., Feng, Z.T., Philip, CV., Akgerman, A., and Anthony, R.G. (1995) *Ind. Eng. Chem. Res.*, **34**, 1021–1026.
20 Maruya, K.-I., Komiya, T., and Yashima, M. (2000) *Stud. Surf. Sci. Catal.*, **130**, 3693–3698.
21 Maitlis, P.M. and Haynes, A. (2006) *Metal-Catalysis in Industrial Organic Processes* (eds G.P. Chiusoli and P.M. Maitlis), RSC Publishing, Cambridge, p. 134.
22 Yu, K.M.K., Yeung, C.M.Y., and Tsang, S. C. (2007) *J. Am. Chem. Soc.*, **129**, 6360.
23 Delledonne, D., Rivetti, F., and Romano, U. (2001) *Appl. Catal. A –Gen.*, **221**, 241–251.
24 Ancillotti, F. and Fattore, V. (1998) *Fuel Process. Technol.*, **57**, 163–194.
25 Grabow, L.C. and Mavrikakis, M (2011) *ACS Catal.* **1**, 365.

7
Fischer–Tropsch Process Economics

Roberto Zennaro

Synopsis

The economics of the Fischer–Tropsch reaction as a source of liquid fuel and as a chemicals feedstock is reviewed, and methods of obtaining useful numbers for capital costs, operating costs, and revenues are outlined. Sensitivity analyses give further important information. In addition to the obvious factors that must be considered in building a new facility such as the type of feedstock (gas, coal, biomass, etc.), the reactor design, and the location of the plant, a number of other aspects need to be taken into account. An adequate supply of water is needed and facilities for recovering, cleaning, and reusing water are necessary. The economics of a conversion process can be completely changed by outside circumstances; thus, the widespread availability of shale gas (and other forms of natural gas) is making GTL the currently favored technology. Although coal reserves are huge, coal-to-liquids (CTL) conversions based on coal are less favored due to the high cleanup costs. Although biomass may in time become important as a reusable resource for liquid fuels and energy, a variety of problems that need to be overcome will make this questionable in the near future.

7.1
Introduction and Background

GTL via Fischer–Tropsch is an industry that underwent dramatic changes in ambitious projects after 2006. In February 2007, ExxonMobil canceled its "Palm GTL project" in Qatar of 7.7 million t/a (154 000 bpd) capacity (that would have been larger than the Shell Pearl project), in favor of the Barzan domestic gas supply project costing about \$10.3 billion. In April, the Algerian Ministry of Energy & Mines stopped the international bidding round for the Tinrhert integrated GTL project in which several international oil and gas companies had participated, including Eni and the Statoil-PetroSA consortium. This was probably a consequence of the difficulties that Sonatrach, the Algerian national oil & gas company,

Greener Fischer-Tropsch Processes for Fuels and Feedstocks, First Edition. Edited by Peter M. Maitlis and Arno de Klerk.

had in managing a large integrated project with two large LNG initiatives (the rebuilt Skikida and the Gassi Touil projects) going on in parallel.

A more immediate concern to the GTL industry was Sasol's announcement in May 2007, a year after inaugurating the Oryx plant, regarding problems with the startup of the first GTL train, due to technical difficulties in the fine control of the FT catalyst [1]. This announcement also had implications for the NPC-Chevron Escravos project, under construction in Nigeria, that was based on the same Sasol technology. However, there was a good outcome as Sasol implemented new downstream facilities, and today the Oryx plant is on stream at its planned capacity.

Since 2007, the most recent oil crisis and a dramatic cost escalation (of over 50%) have jeopardized plans by oil and gas companies to invest in capital-intensive initiatives, such as LNG or GTL. Limited operating experience was also one of the main reasons for the dramatic cost escalation experienced by the Shell Pearl GTL project. When originally announced in 2003, its EPC (engineering, procurement, and construction) cost was estimated at around $5 billion, but in the actual figure turned out to be $19 billion, nearly four times the original estimate.

In late 2011, the combination of new factors such as the balance in supply and demand of natural gas and middle distillates and the high oil price have renewed interest in GTL initiatives. These factors combine to increase the profitability of GTL initiatives by increasing the spread between the cost of the feedstock and the price of the GTL products, which is linked to the oil price.

We consider the outlook for the natural gas and GTL product market by analyzing the investment and operating costs that influence the attractiveness of the various routes for gas monetization.

7.2
Market Outlook (Natural Gas)

Estimates of global natural gas reserves have continuously increased, from 153 Tscm in 2000 to over 193 Tscm in 2010. The gas demand has been far more robust than the demand for oil that increased only 9% in the same period. This was due to strongly rising consumption of gas in emerging and developing countries (+13.8%), as well as of growth in industrialized countries (+3.9%), particularly in the last few years. The environmental benefits of gas in the industrialized nations compared to other fossil fuels foreshadow further growth in the natural gas market, particularly in the power generation sector.

Following the so-called "unconventional gas revolution" (primarily in shale gas), the United States became the leading world producer with 583 Bcm in 2009, overtaking Russia, the historical leader in this area. The US growth in gas production is one of the most important trends in the natural gas market, showing an increase of approximately 95 Bcm (+19%) over the 2005–2010 period.

Although shale gas was originally largely a North American phenomenon, developments have also happened fast in Europe. Detailed studies have been carried out across the continent, and exploration wells have already been drilled in Germany, Sweden, Austria, Hungary, France, the United Kingdom, and Poland. Poland is likely to be the first European country to start production from its large deposits of shale gas (5.3 Tcm) with 90 exploration licenses awarded in 2010–2012 to companies including Chevron, ExxonMobil, and Eni. The first shale gas production test well was successfully drilled horizontally and hydraulically fractured (*hydrofracking*) in July 2011 in Lieben, 90 km from Gdansk. If these sources can be effectively tapped and exploited, it could change the European energy scene, by reducing dependence on Russian imports, and will have an impact on the gas price similar to that observed in North America. Shale gas developments in China, South America, and North Africa are also progressing rapidly. This is particularly so in China which is forecast to become the largest shale gas producer by 2030 due to its huge identified gas resources (35 Tscm).

However, the full potential of the recoverable shale gas resources in these countries is not yet clear, since there is increasing concern about the impact of shale field developments on the environment, and in particular on drinking water. Water is an increasingly scarce resource in many parts of the world and if drinking water is deemed to be contaminated by incorrect disposal of the flowback fracking fluids, this will increase opposition and also the development costs since special water reuse schemes will need to be implemented. About 10 000 barrels of water per stage are required during a fracking job, so a typical well of 6000 ft (about 1830 m) of horizontal length with 15 stages implies a water volume close to 150 000 barrels.

In addition to these unconventional gas discoveries, conventional gas production has also risen rapidly, for example, in Qatar, where the agreed reserves were doubled between 2005 and 2010 (Figure 7.1).

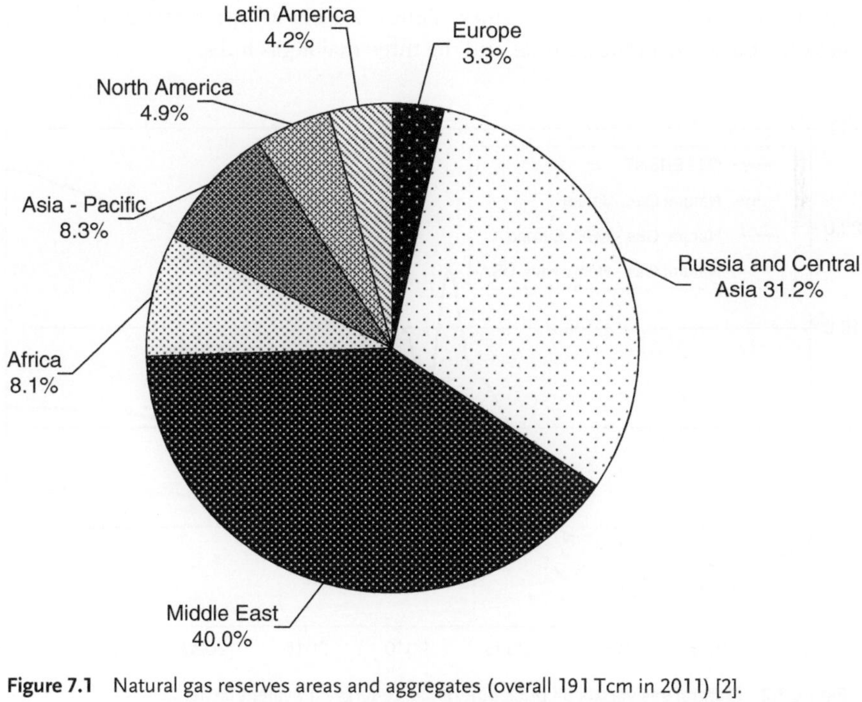

Figure 7.1 Natural gas reserves areas and aggregates (overall 191 Tcm in 2011) [2].

Unconventional gas developments are now changing the LNG industry and in the long term may even diminish the need for imports of LNG, given the increased endogenous production of natural gas. For example, the LNG industry in North America has been heavily affected by shale gas developments and many previously projected LNG terminals have lost their viability. As a result, the gap between natural gas prices and crude oil prices, on a per energy equivalent basis, has been expanding since 2003, and the current Henry Hub natural gas price is currently ca. 20% of the WTI oil price. This compares to a factor of 1.0 in 2003. The natural gas price in the United States seems set to remain much cheaper than the oil price, by a factor of 0.3 in the 2010s and by a factor of 0.4 by 2020 [3].

The price differential between natural gas and crude oil in Europe and Asia-Pacific has also increased; however, here the gap is considerably smaller and remains not far from the historical average. This is related to the fact that while natural gas prices in these regions are historically linked to oil prices, the LNG flooding this market, especially that diverted from North America to Europe, contributes to keeping the natural gas spot prices relatively low. However, the Asian LNG market has seen significant improvement in LNG demand in 2009, in particular in South Korea and Japan, where spot prices of LNG have increased. Following the Fukushima nuclear disaster, the natural gas prices have risen by nearly a third due to increased Japanese demand as an alternative source of power.

According to IEA forecasts, the ratio of natural gas prices to crude oil prices on an energy equivalent basis will remain relatively low throughout the next decades. Such expectations of sustained lower prices of natural gas in comparison to oil lead one to expect a gradual switch from oil to natural gas, especially for some transportation uses and/or as GTL feedstock. Figure 7.2 shows the forecast differential between the Brent crude oil price and the three major gas hubs.

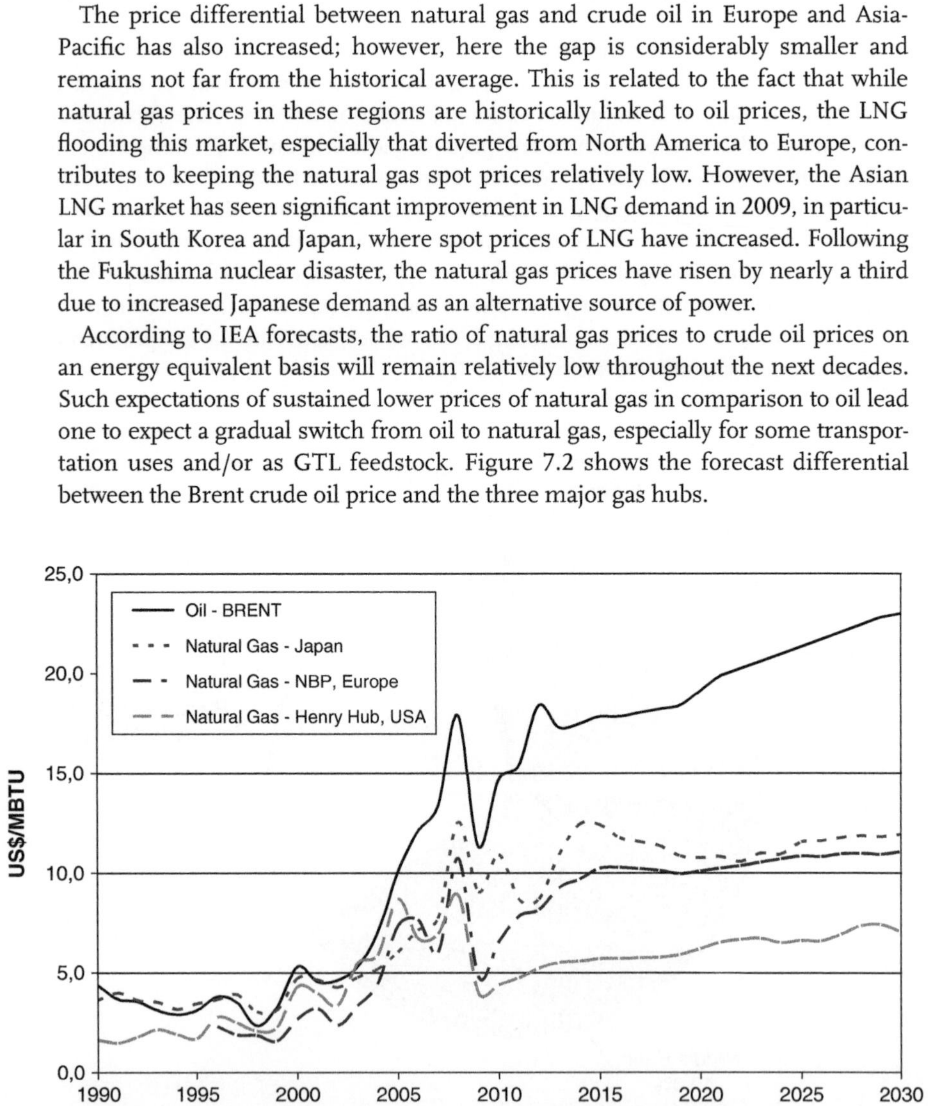

Figure 7.2 Natural gas versus oil price differential (energy equivalent basis).

The huge availability of natural gas, especially in Eastern Europe, the Middle East, and North America, thanks also to the unconventional gas developments, means that we must reconsider the major drivers expected to influence future developments in supply and demand in the oil markets in order to fully evaluate the GTL process and its products.

The growth of future global oil demand (about 1% per year between 2009 and 2035) is expected to be driven by trends in transportation fuels and in petrochemical feedstocks. The latter include naphtha demand for ethylene and aromatics production.

Middle distillates demand will be the major factor, with incremental requirements strongly dependent on automotive diesel fuel and jet kerosene (predicted to grow by 2.20% and 2.16%, respectively, per year to 2035). Increasing commercial activity in developing countries, combined with the increase in the proportion of diesel cars for private motoring, for example, in Western Europe, is expected to support diesel demand. Increased air travel and cargo movements will support jet fuel demand. While growing in importance, the contribution of biofuels is expected to remain relatively limited and to be mainly as biodiesel, which represents 5–6% of total middle distillates on an incremental basis.

Developments in petrochemical demand of major relevance are also expected, resulting in growing naphtha demand for ethylene and aromatics production. In this context, total naphtha demand is expected to grow at an average of 2.9% per year between 2009 and 2035, resulting in incremental demands above those of gasoline.

These development patterns will have major implications on supplies from world refineries, particularly within the context of potential overinvestment in the near future. During the early 2000s, rapid growth in oil demand driven by developing countries, including China and India, has tightened global refining capacities. But the rapid growth of new refining capacity spurred on by large investments has coincided with the recent economic crisis and a major drop in oil demand. As a consequence, a dramatic drop in operating rates has recently occurred, sharply affecting margins in most key refining centers. Within this context, a major rationalization is expected to take place, particularly in the more mature OECD markets (North America, Europe, and Japan), eventually prompting closure of some existing refining capacity in OECD countries during 2010–2015.

Illustrating the potential supply/demand balance on a product-by-product basis, Figure 7.3 points not only to a period of oversupply in the short term but also to the possibility of an oversupply of the lighter products (gasoline, kerosene, and gas oil) for an extended period, possibly up to 2018, and this is concomitant with a potentially tighter fuel oil balance. The key message is that not only is there a risk if too much refining capacity is added in the short to medium term, but also that an excess of conversion capacity is planned by refiners on a global basis.

The latter trend is reflected in the design of new refineries with "zero fuel oil" output, while existing refineries are implementing incremental conversion capacity. This is also in line with efforts to reduce the output of fuel oil, the lowest value product (and which is also facing reduced market demand), and it also reveals a trend toward overinvestment in conversion capacity globally. This situation is

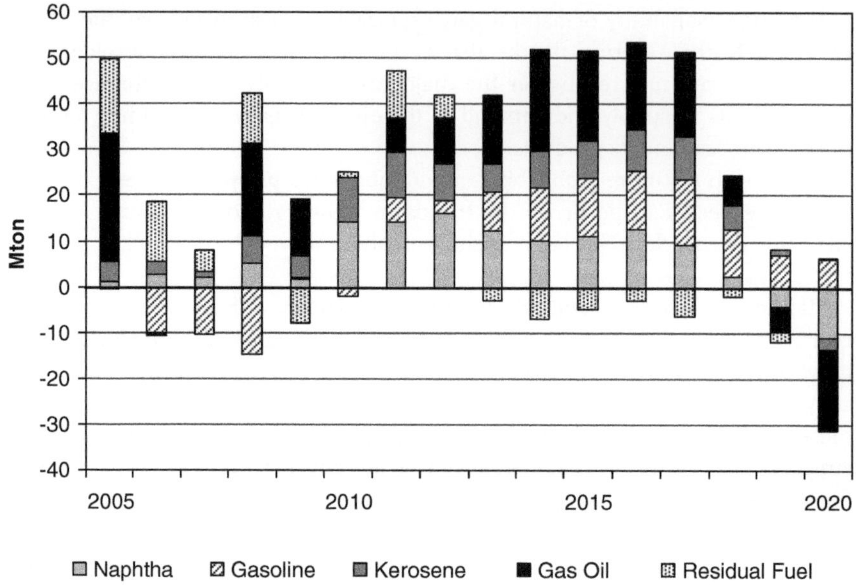

Figure 7.3 Global product demand/supply balances.

expected to seriously affect refining economics over the next few years, further supporting the need for rationalization.

It may be noted that gas oil and kerosene supplies are rapidly expected to shift into deficit after 2015, since a persistent gasoline surplus is forecast. This is another indication of the need for structural rationalization of refining capacity in Europe, North America, and Japan.

Post 2018, a global deficit of combined gas oil, kerosene, and naphtha (all GTL products) is predicted. By 2020, the worldwide demand could require an additional 20 million t/a of gas oil and kerosene and another 11 million t/a of naphtha. Should rationalization take place at North American and European refineries, these numbers could be significantly higher.

The other important element of the global oil markets is the regional development of refining capacity based on two major waves of additions. The first wave (2009–2011) relates to new capacity in India (Asia and Pacific) and China (North East Asia), followed by middle distillates (gas oil and kerosene) oriented investments in North America and Europe. The second wave is expected in 2013–2014. More capacity is expected in Russia and Latin America as well as major capacity additions in the Middle East. In line with demand, most of this capacity is targeting middle distillates, including hydrocracking capacity.

In general, new GTL units are seeking to maximize diesel production, while the older units target gasoline/olefins (South Africa) or other specialty products such as paraffins/wax in Bintulu, Malaysia. Other streams such as naphtha and kerosene with high added value give specialty products, including LPG, paraffins, base lube,

Table 7.1 Present and planned GTL capacity (kbpd).

	2000	2009	2010	2015	2020	2025	2030	2035
Mosselbay (PetroSA), South Africa	24.0	24.0	24.0	24.0	24.0	24.0	24.0	24.0
Bintulu (Shell), Malaysia	14.7	14.7	14.7	14.7	14.7	14.7	14.7	14.7
Sasolburg (Sasol I), South Africa	—	8.0	8.0	8.0	8.0	8.0	8.0	8.0
Oryx I, Ras Laffan (QP, Sasol), Qatar	—	34.0	34.0	37.4	37.4	37.4	37.4	37.4
Pearl (Shell), Qatar	—	—	—	140.0	140.0	140.0	140.0	140.0
Escavros (Chevron, NNP), Nigeria	—	—	—	34.0	34.0	34.0	34.0	34.0
Secunda (Sasol II/III) – Phase I, South Africa	—	—	—	11.0	11.0	11.0	11.0	11.0
Uzbekistan (Sasol/Uzbekneftegaz/ Petronas)	—	—	—	—	38.0	38.0	38.0	38.0
Montney Shale Play, Canada (Sasol)	—	—	—	—	90.0	90.0	90.0	90.0
Total	38.7	80.7	80.7	269.1	397.1	397.1	397.1	397.1

and linear alcohols. GTL-quality diesel (negligible sulfur and high cetane number) is an ideal blending component in new-generation diesel fuel.

Other key products such as GTL kerosene may be used as substitute or blender for petroleum-based jet fuel or kerosene. Negligible sulfur content is at a premium, but density may be an issue for jet fuel. GTL naphtha is an excellent ethylene feedstock, with better yields than conventional material ex-refinery. Co-produced water may be an added value in arid locations.

The largest GTL developments are expected to take place in Qatar. The scenario assumes the gradual ramping up of the first new-generation plant (Sasol/Chevron/QP, 1.7 million t/a (34 000 bpd), started in 2007, and expected to have a minor expansion), followed by two large 3.5 million t/a (70 000 bpd each) units in subsequent years (Shell/QP).

Table 7.1 updates present and predicted GTL capacity, the latter based on major project announcements. The forecast assumes only few large plants, such as the one in Escravos, Nigeria, owned by NNPC (Nigerian National Petroleum Corp.) and Chevron and based on Sasol SBCR technology. The plant, which had long been planned, faced a number of difficulties, but is now being put on stream. The second large plant was announced in Uzbekistan. This again involves Sasol SBCR technology, which is also being used to expand the capacity of the Secunda GTL plant, fed with natural gas from Mozambique. Finally, the first shale gas-based GTL initiative in Canada (Montney shale play formation) was announced in March 2011 by Sasol. The engineering company Forster Wheeler was awarded a contract in June 2011 for the completion of a technical feasibility study for a GTL unit up to 4.8 million t/a (96 000 bpd).

By 2015, some 9 million t/a (about 180 000 bpd) of GTL capacity is expected on stream in the Middle East, with this number not changing in the longer term.

Based on the global refinery scenario and the size of the potential reference market, the impact of GTL products on the global oil product market is expected to be minimal.

However, there would appear to be new opportunities for GTL initiatives since the GTL product mix fits well into other global developments. From 2015 onward, about 9 million t/a (180 000 bpd) GTL capacity is expected to be available on stream in the Middle East over the longer term.

There also seem to be new opportunities for GTL initiatives since the GTL product mix fits well with global developments from a demand standpoint, especially in providing high-quality middle distillates for the transportation sector. This potential advantage, however, will be highly sensitive to the actual capacity rationalization that is likely to take place in the OECD countries, in addition to possible investments in developing countries. However, these are extremely difficult to predict.

From a capacity viewpoint, any eventual world-scale GTL complex potentially starting after 2015 is unlikely to have any considerable impact on global gas oil, naphtha, and kerosene balances.

CTL industrial developments based on FT processes have been announced in China and India, as well as some expansion of the large older capacities in South Africa (4 million t/a, 80 kbpd). Due to the boom in unconventional shale gas developments, some announced coal-based CTL initiatives in the United States have recently been replaced by GTL. The overall picture described above will change only marginally, even allowing an impact from CTL. The impact of BTL biofuels, given the level of initiatives at research or pilot projects, is likely to be negligible.

The main critical factors influencing the viability of GTL initiatives are the feedstock, namely, the natural gas production costs, the oil product prices, and the initial investment costs. These will benefit from large economies of scale and the leverage from the learning curve. The latter has not yet been observed due to the limited experience so far available. Integration with the infrastructures available on the selected site and the process and utility optimizations are of utmost importance to reduce the investment cost. Finally, the outstanding qualities of GTL fuels (diesel, jet kerosene), the GTL naphtha as feedstocks for petrochemical use, and specialties such as lubricant bases, linear alcohols, and olefins can boost the projected profitability when the market assigns a premium value to these products.

7.3 Capital Cost

We here present an analysis of the total investment needed for the facilities and the preoperating costs of a typical GTL unit of about 850 000 t/a (17 000 bpd) of GTL products, based on a single FT module with slurry bubble column reactor technology.

Table 7.2 EPC cost estimate breakdown.

EPC cost package	Million US$ (2011)
Engineering services	133
Supply of equipments	498
Supply of bulk materials	369
Transportations	47
Construction	830
Supervision to erection	127
Precommissioning, commissioning, test run[a)]	35
Total investment cost	2039

a) Commercial acceptance test run.

This cost estimate has been developed from a front-end engineering design (FEED) deployed by Eni for a North African location where GTL initiatives can have a potential application [4]. It is representative of the work required by the EPC contractors to execute the engineering, procurement and construction of the GTL facilities. In particular, the cost figures have been based on values for similar equipment and unitary cost data for bulk material through feedback from purchasing departments that are periodically updated with current published cost databanks [5–7].

The GTL plant capital cost estimate can be broken down into the following seven discrete packages (Boxes 7.1–7.7). For each package, the facilities cost estimate is shown in Table 7.2; this adds up to a total investment cost around US$2 billion. Boxes 7.1–7.7

Box 7.1 Engineering services

The engineering services are estimated in terms of work-hours against each activity and/or deliverable under the project scope and determined on the basis of the proposed project resource assignments.

This cost package does not include expenses for the contractor's personnel site accommodation (living, lodging, and local transportation) and site security costs.

- Front-end engineering design
- Detailed engineering
- Management
- Project control and administration
- Quality assurance and HSE
- Procurement

Box 7.2 Supply of equipment

The costs of equipment covered in this package are estimated using in-house estimating software, order values for similar equipment, or using data obtained from vendors. (The cost of capital spares, when not provided by vendors, has been estimated as a percentage of the cost of the equipment.)

- Air coolers
- Blowers
- Boilers
- Capital spare parts
- Catalysts (first charge)
- Chemicals (first charge)
- Columns
- Compressors
- Exchangers
- Filters
- Firefighting
- Heaters
- Laboratory equipment
- Packages
- Pumps
- Reactors
- Tanks
- Turbines
- Vessels

Box 7.3 Supply of bulk materials

Cost of bulk materials is determined on the basis of in-house unit price lists (periodically updated from vendor-specific quotations or vendors unit price).

- Civil materials
- DCS (distributed control system), ESD (emergency shut down), FGS (fire gas system), BMS (burner management system), and MMS (machine monitoring system)
- Electrical equipment and materials
- Instrument equipment and materials
- Insulation materials
- Paints
- Piping, valves, and ladders
- Steel structures
- Telecommunications

Box 7.4 Transportation on-site CIF (cost, insurance, and freight)

Costs of transportation ex-work to the site that are covered under this heading are evaluated as a percentage of the cost of the equipment and bulk materials (customs duties are excluded).

- Transportation of materials offshore
- Ocean freight
- Transportation of materials onshore

Box 7.5 Construction

The cost of construction is based on the materials to be installed from subcontractor quotations, in response to specific inquiry packages.

This cost heading also includes the expenses for the provision of temporary construction facilities as well as the costs for the provision of all utilities needed to run the facilities such as site offices and furniture, supplies, cabling, warehouse, stockyard, and means of transportation to/from site for the contractor's personnel, vendors, and licensors, including the rent of vehicles with relevant insurance.

- Buildings (including necessary workshops, offices, canteens)
- Civil and concrete
- Heavy lift
- Mechanical erections (equipment, piping, structure, and tanks)
- Electrical/instrument/telecommunication erections
- Painting and insulation
- Work carried out at daily rates
- Assistance for commissioning
- Field running costs and temporary construction facilities

Box 7.6 Supervision of erection

This package includes costs for personnel such as Camp Manager Assistant, Camp Supervisors, Camp Maintenance Supervisor and Assistance, General Service Assistant, Document and Reproduction Staff, Secretary, Clerk, General Service Assistant, CAD Operator to support Field Engineering Staff, First Aid Nurse, Document Controller, Drivers, Administration Officer, Security Guards, Local Procurement Officer, Personnel Officer, Labor Relation Officer, and so on, which are determined on the basis of the proposed plan of project resource assignment.

- Expert EPC contractor personnel
- International personnel
- Local personnel
- Vendors Personnel

Box 7.7 Precommissioning, commissioning, and test run

This package includes costs for the provision of all services, materials, and tools required for precommissioning, commissioning, startup, and test run of the facilities. They are determined in terms of estimated resources.

- Expert EPC contractor personnel
- International personnel
- Vendor personnel
- Licensor personnel
- Training
- Tools, chemicals, and consumables for precommissioning and commissioning

The cost breakdown based on the main GTL process, the utilities, and the off-site units is given in Figure 7.4. The Syngas unit is the most expensive, covering about 26% of the EPC cost if the Air Separation unit is included but excluding part of the investment for the hydrogen unit necessary for hydrodesulfurization. The Fischer–Tropsch synthesis unit costs about 24% of the total investment. Significant are the expenditures for utilities and off-site work that cover up to 36% of the overall plant investment. A typical list of the utilities and off-site is presented in Table 7.3.

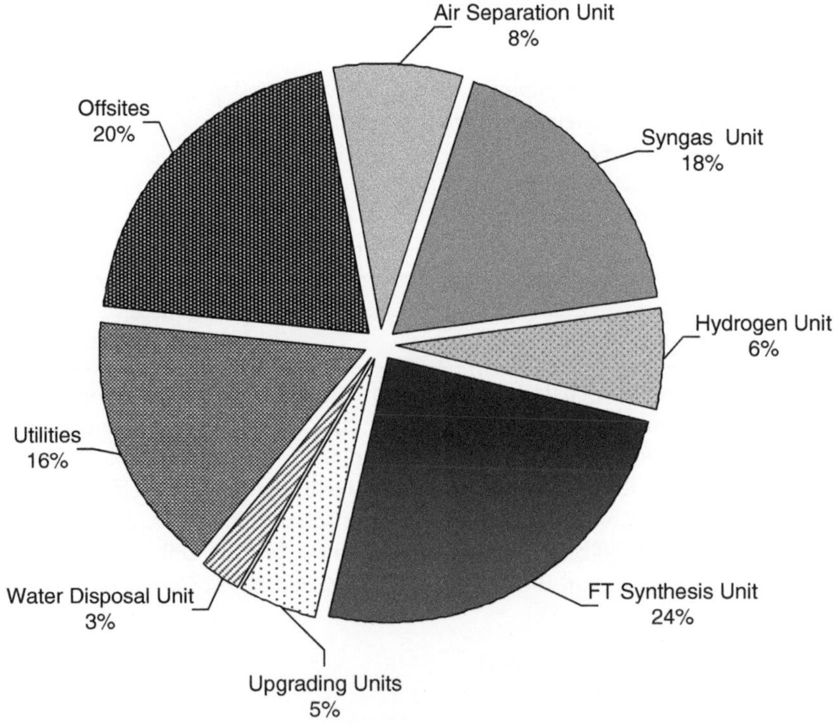

Figure 7.4 EPC cost breakdown for a 17 000 bpd GTL plant (process, utility, and off-site).

Table 7.3 Utilities and off-sites main list for a typical GTL plant.

Utilities	Off-sites
Cooling water tower	Fuel gas unit
Desalination water	Flare vent and blow down
Plant air and nitrogen	Product storage tanks
Demineralization water	Intermediate storage tanks
Steam and condensate	Laboratories
Freshwater	Interconnecting
Aerobic treatment	Buildings
Service water	
Firefighting water	
Power generation	

The cost partition between the material and equipments cost, and its transport to, installation at the site is almost the same and covers 86% of the overall EPC investment. The remainder is for services such as the detail engineering, supervision of erection, and the expenses associated with commissioning and testing the plant as far as the commercial acceptance and startup of the operations (Figure 7.5).

The overall (total investment) cost is normally completed by including in the facilities costs the EPC contractor's profit (typically assumed at 10% of the facilities

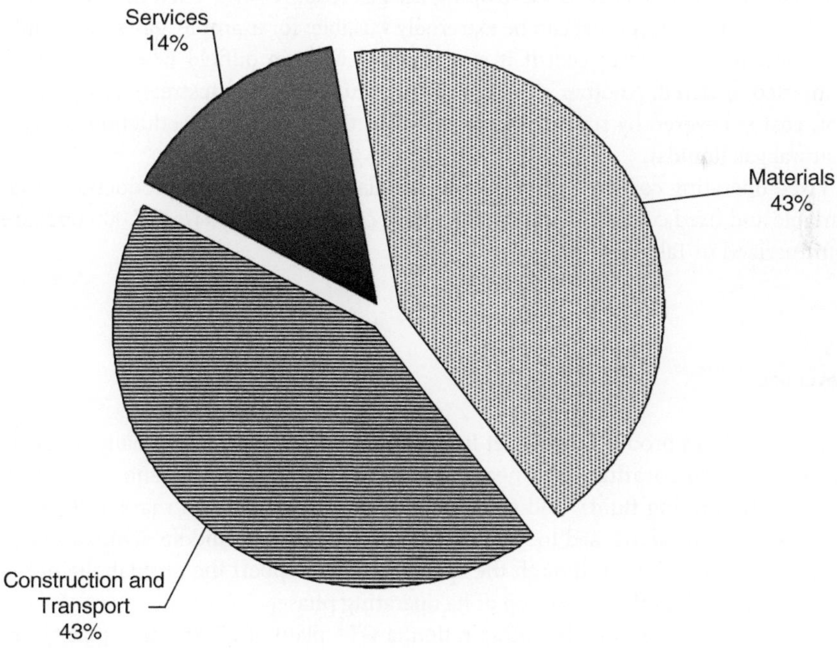

Figure 7.5 EPC cost breakdown for a 17 000 bpd GTL plant (materials, construction, and transport services).

cost), project's owner costs, prestart and startup costs, staff training, customs duties, withholding tax, financial charge, all risk insurances and funally project owner's contingency.

If an overall investment cost estimate is necessary for economic studies of the project profitability, a detailed evaluation of these latter items will be required. However, such extra costs could differ substantially from company to company and will depend on the project characteristics. Notwithstanding, they must be considered as part of the investment as they are not negligible and could be as high as 20–25% of the total. Since there are many parameters that could influence this estimate, we compare only the EPC costs here, following similar approaches adopted in the literature.

7.4 Operating Costs

The typical variable operating cost for a GTL plant is given by the ratio of the expenses to the percentage of capacity utilization. The expenses comprise the feedstock (the natural gas) and the utility costs, which include the catalyst and chemicals consumption figures. The natural gas cost depends on the consumption, which is related to the energy efficiency reached by the GTL process, the utility configuration, and the gas purchasing price, which is in turn related to the upstream production costs of developing the gas resource and delivering the gas to the GTL plant. The gas cost can be extremely variable: for example, gas access could be at low (or even zero) cost if it is associated with an oilfield where the gas is reinjected or flared. Another situation could occur when the upstream gas production cost is covered by revenues accrued from the associated production of NGL (natural gas liquids).

Fixed operating costs are the expenses unrelated to the rate of production. Both variable and fixed costs for a typical GTL plant of about 850 000 t/a (17 000 bpd) are summarized in Table 7.4.

7.5 Revenues

A GTL plant can produce diesel, jet fuel, and naphtha as well as specialty products such as normal paraffins for linear alkyl sulfonates or alkylbenzene production (LAS, LAB), drilling fluids, and waxes. Given the size of the fuel market, the GTL plant is normally addressed to maximize the production of middle distillates such as diesel or jet fuel, even though the specialties can support the profitability of the project especially at the beginning of its operating phase.

Depending on the utilities configuration, a GTL plant of 17 000 bpd capacity can produce about 20 MW of electrical power for export in addition to that produced for internal use. This can increase the profitability of the project; of course, a distribution grid for delivering such electrical power must be available.

Table 7.4 Typical GTL operating.

Equipment operating cost	Maintenance (preventive and corrective)	Maintenance specialist Spare parts Civil works
	Chemicals	Catalyst treatment and disposal Catalyst and chemicals
	Operating insurance	
	Others	ICT Catering cost Site running cost connected with field operation
General and administration cost	Direct personnel	Production and utilities personnel Ordinary maintenance and laboratory personnel Administration and management personnel HSE personnel
	General and administration	Training Headquarter
Service cost	Logistic	Industrial transport services Civil transport services

The overall operating expenditure (OPEX) is estimated at US$8.7/bbl with the breakdown given in Figure 7.6.

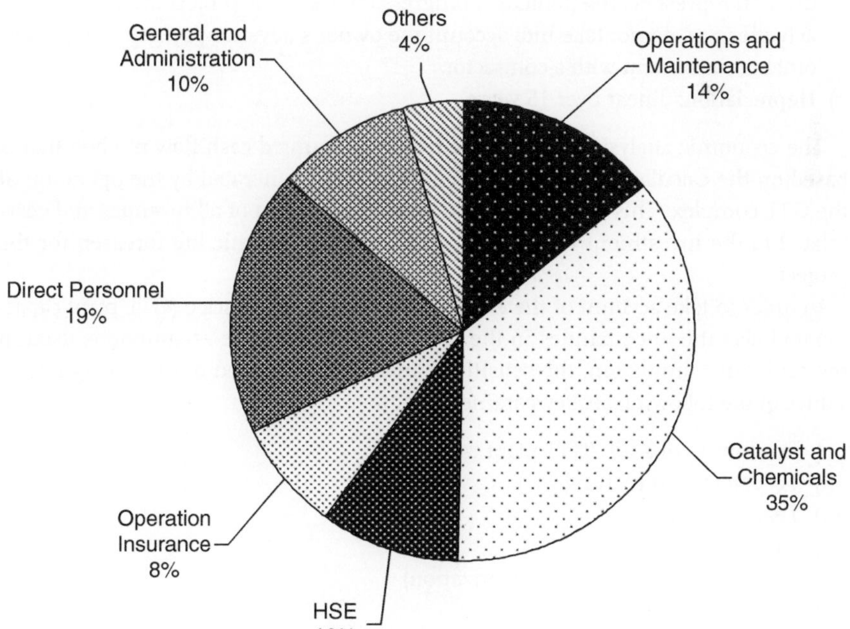

Figure 7.6 Typical GTL operating costs distribution.

7.6
Economics and Sensitivity Analysis

The standard way to evaluate the project profitability is by an economic analysis, but this is often susceptible to criticism unless a precise definition of the project basis can be provided. Unfortunately, an appropriate project definition requires very specific information about the GTL initiatives, such as the right location, the site infrastructures available, the fiscal regime, the product marketability, the project owners costs, and so on.

We here carry out an economic simulation with the aim of identifying the key technoeconomic and market parameters that influence the profitability of a GTL project. For each parameter, a sensitivity analysis is then presented in order to provide a quantitative estimate of the chosen profitability index.

The following bases have been adopted for the economic analysis:

a) **Currency:** US$.
b) **Project economic life:** 25 years of operations with 50% production in the first year and 100% from the second production year on.
c) **Inflation rate (on reference currency):** 2% per year throughout the project economic life.
d) **Income taxation:** 30% of net income as corporation tax, paid in the year in which the income is calculated.
e) **Investment cost disbursement:** The construction of a GTL complex is considered to last 52 months with a capital disbursement breakdown of 5%, 20%, 35%, 30%, and 10% each year over the assumed EPC period 2011–2015. This division represents the (tentative) progress curves of the project implementation activities and do not take into account the owner's advance payment and/or any other commitment with a contractor.
f) **Depreciation:** linear over 15 years.

The economic analysis was assessed by the discounted cash flow method that is based on the calculation of the annual net cash flow generated by the operation of the GTL complex. This calculation requires the projection of all revenues and costs related to the investment initiative throughout the economic life foreseen for the project.

In order to take account of the effect of the economies of scale (GTL plant capacity) and also the uncertainties in the cost estimate and price assumptions used in the economic simulation, the sensitivity analysis was carried out by changing the values of the following six key parameters:

i) Economies of scale (GTL plant capacity)
ii) Feedstock cost (natural gas)
iii) Learning curve
iv) Tax regime
v) Product price (GTL diesel valorization)
vi) Oil price scenario

Table 7.5 Process units configuration at different GTL plant capacity.

GTL plant	Base case	Base x2	Base x3	Base x4	Base x5
Capacity (bpd)	17 000	34 000	51 000	68 000	85 000
Gas treatment	1	1	1	2	2
Air separation unit (ASU)	1	2	3	4	5
Syngas preparation (ATR)	1	2	3	3	4
Fischer–Tropsch	1	2	3	4	5
Hydrogen plant	1	1	1	1	1
Product upgrading	1	1	1	1	1
Water treatment	1	1	1	1	1

7.6.1 Sensitivity to GTL Plant Capacity (Economy of Scale Effects)

The search for economies of scale is normally considered when we evaluate a new industry such as the GTL. Large plant capacity for GTL is usually considered when associated with large gas reservoirs (80–150 Bscm). This is similar to the LNG technology that has experienced a constant increase in the single liquefaction train capacity to sustain its profitability, and dates from the first trains built in the 1970s with capacities in the range of 0.7–1.6 million t/a to the 7.8 million t/a train capacity recently built in Qatar (AP-X Air Products' process).

This sensitivity calculation is for five different plant capacities obtained by increasing the number of the FT reactors and the associated process units. Table 7.5 details the process unit configurations for each case.

The total capital costs have been recalculated based on the estimate for the single train GTL plant of 17 000 bpd discussed in Sections 7.1–7.3.

The exponential factoring method [8] has been applied to both single bulk materials and equipment to estimate the EPC cost of the scaled-up GTL plants.

Figure 7.7 shows the calculated facilities cost (in US$ per bbl of product) for the five plant capacities considered. The specific investment costs tend to level off at US$83 000 bpd for the highest capacity considered in this study.

Operating costs have been recalculated according to the same division between variable and fixed costs used in Table 7.4. The result is an OPEX decrease, while increasing the plant capacity to about US$5.0/bbl for the highest capacity considered in the study.

7.6.2 Sensitivity to Feedstock Costs

For a GTL initiative, the natural gas cost depends on the consumption, which is associated with the energy efficiencies reached by the GTL process and utility configuration and the gas purchasing price, linked in turn to the upstream production cost to develop the gas resource and deliver it to the GTL plant.

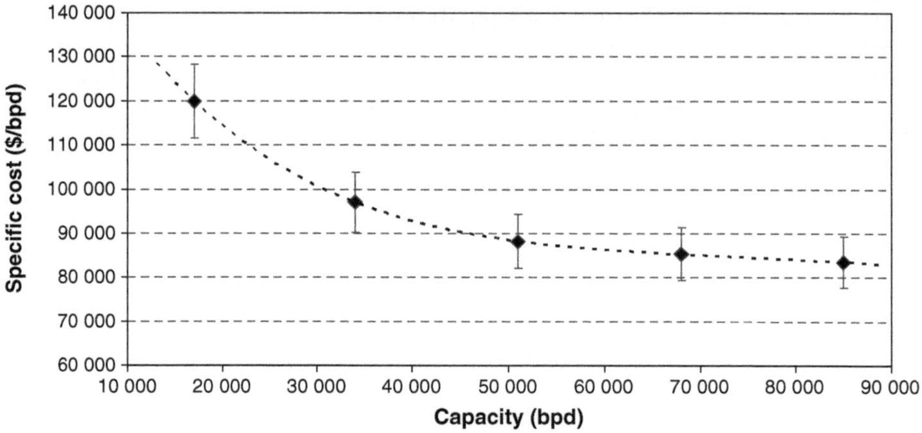

Figure 7.7 Specific investment cost for a GTL complex.

The sensitivity analysis considers an initial value of $2.5/MBtu, typical upstream production costs for an offshore gas development (including subsea, offshore platforms, flowlines, and export pipeline to shore and onshore gas receiving facilities), which then reduce this value to $1.0/million Btu, a situation with low-cost gas from an oilfield where the gas is reinjected or flared. The "stranded gas" situation, gas reserves that cannot be developed for economic reasons and/or limited access to the market, may also be associated with gas production costs in this range. This will depend on the characteristics of the gas reserves (e.g., high CO_2 concentration, deepwater offshore, long distance to the GTL plant site), which can result in very high gas production costs that will diminish the attractiveness of a GTL project.

7.6.3
Sensitivity to GTL Project Cost (Learning Curve Effect)

This sensitivity study has been performed to assess the impact of a learning curve on plant construction costs. For the plant capacity of 68 000 bpd, the study has indicated a decrease of 10% for the total investment cost. This figure considers an increase in experience and optimization of GTL Plant construction as, presently, there are limited worldwide analogues and slurry bubble column reactor development is at an early stage.

7.6.4
Sensitivity to Tax Regime

The tax regime is a very important factor in the economics of a GTL plant. In this sensitivity analysis, an average taxation of 30% is considered. A 5-year tax holiday and exemption from import taxes on all relevant equipment and services for the GTL project has been introduced to simulate possible government support.

7.6.5 Sensitivity to GTL Diesel Valorization

The GTL diesel product has the potential advantage of benefiting from a premium price compared to the petroleum-based products. This is because its excellent properties meet environmental regulations aimed at reducing emissions for light- and heavy-duty diesel vehicles.

The low availability of GTL fuels still makes it premature to speak about a real GTL product market, and this limits the possibility of setting a level for the premium. Nevertheless, the future existence of a premium is confirmed by marketing operations on top-tier fuel products, which, improved by GTL diesel blending, enable extra profit margins as well as reinforcing or enhancing market shares. Several years before the startup of the Pearl GTL plant in Qatar, Shell had already promoted its V-Power Diesel™, based on the addition of a small percentage of GTL diesel.

Apart from the premium, the sale price of the GTL products is estimated by reference to the crude oil price. As a first approximation, the price of naphtha, diesel, or jet fuel can be calculated on the basis of a linear relationship with the price of crude oil. For instance, the ultralow sulfur diesel (ULSD) (5 ppm S) price is approximately 1.2 times the Brent oil price. This correlation was confirmed over the period 2005–2011, which was also characterized by large crude oil price fluctuations. This sensitivity analysis was carried out to assess the effect of quality premium GTL diesel relative to prevailing ULSD quotations (average 10% higher).

7.6.6 Sensitivity to Crude Oil Price Scenario

Based on the sale price of GTL products relative to the crude oil price, it is normal to refer the economic evaluation to a crude oil price scenario. This is of course subject to the oil price fluctuation as well as to the uncertainties in the forecast scenarios.

In this sensitivity analysis, a crude oil price variation of US$90–100/bbl has been considered. For comparison over this period, the annual WTI and Brent average prices are US$93/bbl and US$110/bbl, respectively.

7.6.7 Effects of Key Parameters on GTL Plant Profitability

The effects of the selected parameters on project profitability are measured by comparing the net present value (NPV), representing the difference of all cash inflow and outflow accruing throughout the entire project life, discounted at a predetermined interest rate (discount rate). This is one of the discounted cash flow techniques used in comparative appraisal of investment proposals. The discount rate should be at least equal either to the interest rate paid by the borrower to the financing institutions or to the "opportunity cost of capital" that represents the possible return that the investor would get, employing the same amount of capital in an alternative initiative with similar enterprise risk.

For the purpose of this study, the discount rate is assumed equal to a weighted average cost of capital (WACC) of 9%, assuming the 75% of total investment cost of the initiative borrowed at a rate of 7%/year and a cost of equity equals to 15%/year.

The relative effect of each key parameter on the project profitability, expressed as NPV discounted @9%, is proposed in Figure 7.8 as incremental variation from the base.

It is evident that the influence of the economies of scale, typical for initiatives based on capital intensive technologies (e.g. LNG), plays a major role in the economics of the GTL initiative. This is mainly due to the reduction in the specific investment as already presented in the previous paragraph. For sake of information, a 68.000 bpd plant capacity would lead to internal rate of return (IRR) after tax of the initiative double in respect to the discounting rate used for the NPV calculation.

Another key parameter impacting the project' profitability is the natural gas cost. That leads to seriously consider the economic viability of a GTL initiative in remote inland areas where cheap gas is available and its delivery is not prohibiting in terms of cost and inter-countries permitting procedure.

Such consideration is even more valid considering an higher oil price scenario. The above Figure 7.8 shows that an increase of oil scenario from 90 to 100 USD/bbl would lead to an increase of NPV@9% of about 2 Billion US$ over a 25 years

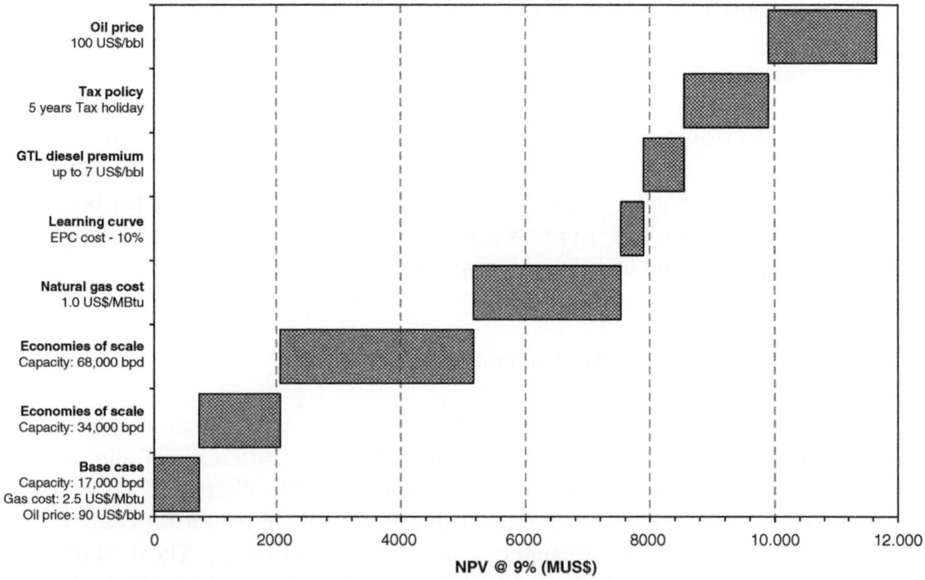

Figure 7.8 Key parameters effect on GTL project profitability (NPV @ 9%).

economic life of the project. It is worth to highlight that associated gas cost is reasonable independent from any increase of GTL's products price.

The above sensitivity analyses show also that, due to the significant annual operating margin, EBITDA (earn before interest, taxes, depreciation, amortization) that a GTL initiative is able to generate, a 5 years tax holiday period (that may be granted by a government to attract investors) can be considered an important feature that very much impact the project profitability.

It can be concluded that according to the basis of this study and targeting the simultaneous occurrence of all the benefits herein considered, a 68.000 bpd GTL initiative can experience an IRR after tax three times higher than the here considered discount rate.

References

1 Forbes, A. (2011) http://www.petroleum-economist.com/Article/2732396/Reality-check.html (accessed December 14).

2 ENI (2011) World Oil & Gas Review, 10th edition.

3 ODS Petrodata (2010) Market Survey System on LNG Regasification Terminals Market Segment Analysis, ODS Petrodata Version 16, September.

4 Iovane, M. and Rossini, S. (2010) Eni, development of gas-to-liquid technology. *Hydrocarb. World*, **5** (1), 7–10.

5 CRU – Commodities Research Unit. UK http://www.crugroup.com (2011).

6 OECD (2011) Main Economic Indicators, Organization for Economic Co-operation and Development, Volume 2011/1.

7 Industria Meccanica, ANIMA http://www.industriameccanica.it (2011).

8 Westley, R.E. (1997) *The Engineer's Cost Handbook*, Marcel Dekker, New York.

Part Three
Fundamental Aspects

8
Preparation of Iron FT Catalysts

Burtron H. Davis

Synopsis

The main industrial Fischer–Tropsch (FT) catalysts are based on either iron or cobalt. Iron is effective and cheap, albeit with reported poor activity and selectivity that need to be enhanced. Methods of improving the iron catalyst for both the low-temperature Fischer–Tropsch (LTFT) and high-temperature Fischer–Tropsch (HTFT) syntheses are described. The precipitation of an oxo/hydroxo iron compound from aqueous solution still appears to be the most important route for the preparation of an industrial iron catalyst. Although it is very difficult to replace the existing route, there is hope that an approach such as that developed by Süd-Chemie involving only the reaction of iron with oxygen can decrease the environmental problems associated with the preparation and activation of the iron catalyst.

8.1
Introduction

Even though many metals have been involved in FT syntheses (FT-S), the most commonly used industrial FT catalysts are based on iron as it is effective and cheap. However, both the activity and selectivity of iron are poor and thus they need to be enhanced. This is usually done by changing the physical format of the catalyst particles or by incorporating special additives. There are many recipes that have been used to improve iron FT catalysts, and this chapter offers a summary of the more important methods of doing so. The preparation of iron FT catalysts offers many options. First, there are two working temperature ranges, known as the low-temperature Fischer–Tropsch and high-temperature Fischer–Tropsch regimes, which, in general, utilize different catalyst formulations. To attempt to control both the catalyst life and the selectivity, promoters are added to the iron base and this adds complexity to the final catalyst formulation. These promoters can be classified in general into those that exert a chemical influence and those that impact the physical properties of the catalyst.

Greener Fischer-Tropsch Processes for Fuels and Feedstocks, First Edition. Edited by Peter M. Maitlis and Arno de Klerk.

The chemistry of iron is very complex and 13 iron oxides, oxyhydroxides, and hydroxides have been identified to date [1, 2]. All these consist of Fe, O, and/or OH^-, and so the different compounds differ in composition, iron oxidation state, and crystal structure. While all these compounds can be made in the laboratory, equipment and cost must be considered when the preparation is adapted to the industrial scale that is needed for the FT synthesis.

8.2 High-Temperature Fischer–Tropsch (HTFT) Catalysts

Sasol reports an operating temperature of about 350 °C using iron catalysts for the HTFT synthesis [3]. At this temperature iron oxides have a low surface area, so there is little, if any, advantage of starting with a high surface area precursor. Thus, the early Sasol operations utilized the iron mill scale from a nearby steel plant as the feedstock for their catalyst and it appears that they continue to do so even today. To this catalyst they add at least an alkali promoter at levels that are not divulged. Sasol has about 50 years of operating experience with this catalyst that currently is being used to produce about 5 million t/a (100 kb/day) of products. "The choice of this catalyst was largely based on the mechanical integrity required to survive the abrasive environment inside the circulating reactor bed" [4].

PetroSA operates a 1.8 million t/a (36 kb/day) plant in South Africa that utilizes circulating fluid bed reactors with iron catalysts. These catalysts are apparently based on an iron ore that is imported from Brazil. It has been reported that this iron catalyst offers " . . . limited service life and needs to be replaced periodically. It is also indicated that scaling up the PetroSA [iron catalyst] technology is cost-prohibitive" [5]. In spite of these shortcomings, PetroSA still uses the iron catalyst.

A significant amount of the early literature was based on fused iron catalysts. Many of the mechanistic studies by Emmett and coworkers as well as those at the US Bureau of Mines in the 1950–1960s were conducted with this type of catalyst. Emmett and coworkers utilized fused iron catalysts prepared at the Fixed Nitrogen Laboratory that were prepared initially for ammonia synthesis. This procedure uses electrical heating with carbon electrodes immersed in the metallic iron and the promoter components, which form a molten or semimolten mass that becomes rather uniform in composition. After cooling, the mass is crushed to produce the desired particle size. It is understood that the United Catalyst C-73 synthetic ammonia catalyst that has been used for many Fischer–Tropsch synthesis studies, especially by Satterfield and coworkers, was prepared using this technique. However, Satterfield *et al.* used the C-73 catalyst under low temperature conditions as a baseline catalyst since it is mechanically stable and resistant to process upsets [6]. Davis and coworkers found that although the C-73 catalyst was very stable for FT synthesis, it had only about 1/10th the activity of a high surface area precipitated catalyst for LTFT synthesis.

It is reported that the circulating fluid bed catalysts for Sasol were of the fused iron type [7]. Here, alkali and other promoters were added to a melt of magnetite at temperatures above 1400 °C. The molten mixture was cast into ingots, cooled, and then crushed and milled to produce the desired particle sizes. Phase separation apparently occurs upon cooling so that the promoters are not homogeneously distributed. Although these disadvantages limit the flexibility one has in designing and preparing these fused catalysts, they have been used at Sasol for about 90% of their production in South Africa.

A precipitation method has been described for the preparation of a HTFT catalyst [8, 9]. Here, the iron may initially be in solution or as a suspension of a noncalcined iron containing solid, together with a promoter P1 selected from a number of elements, including B, Ge, N, P, As, S, Se, and Te. The solvent is then removed and a second promoter P2 that is an alkali or an alkali earth is added to the solid after it has been calcined. It has also been reported that addition of low levels of chromium oxide to the iron catalyst increases the production of oxygenates and branched hydrocarbons. Although such catalysts offer advantages over the fused catalyst, it does not appear that they are able to replace it in commercial operations.

Because of its low surface area, the fused catalyst is a challenge to activate. Most procedures involve a lengthy activation at high temperatures in a hydrogen flow that reduces the catalyst essentially completely to metallic iron. The metallic iron catalyst may then be treated at a lower temperature to form carbide or directly subjected to the syngas at the synthesis temperature. The surface area of the catalyst, after exposure to the synthesis gas, will initially be low (usually well below 10 m^2/g). Because of the large crystal size, the chemical composition of the catalyst changes in the surface and near-surface layers rather quickly, but the interior of the particle may remain metallic iron for a long time. Figure 8.1 represents the catalyst

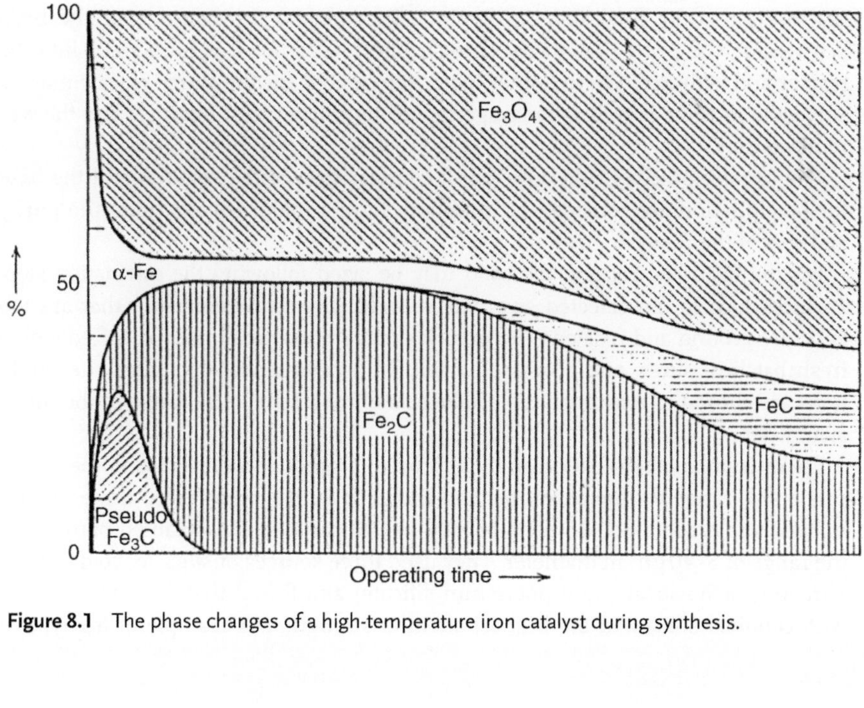

Figure 8.1 The phase changes of a high-temperature iron catalyst during synthesis.

changes. Although this has been taken to be a general representation of the changes that an iron catalyst undergoes during FT-S, this is not the case. While this diagram is representative of the changes for the HTFT, it does not describe the iron catalyst changes when used in the LTFT synthesis.

8.3
Low-Temperature Catalysts

The LTFT synthesis is today carried out at the commercial scale in two types of reactor: the fixed bed and the slurry phase reactors. The physical demands of the two reactors are sufficiently different, that is, at least some differences in the catalysts are required. Thus, although a common precipitated precursor may be used in both reactors, the final catalyst will require the physical properties that are dictated by the reactor type. Since the precipitated catalyst is the most widely used today for the LTFT, it will be described first.

Dry [10] reports that the surface area and the pore volume are the major factors in determining the properties of the iron FT synthesis catalyst. These physical properties are rather easy to determine for the fresh catalyst, but may be very difficult for the activated and used catalysts, as much of the iron will then be present as a carbide phase. In addition, wax will be present in the used catalyst and the removal of the wax without altering the catalyst is a very demanding task. Thus, the properties identified by Dry may not be easily determined for the activated or used catalyst.

The catalyst preparation may be the same for the fixed bed and the slurry reactors up to the final shaping for introduction into the reactor. For the fixed bed reactor, the pellet is not subject to strong forces once it is introduced into the reactor and activated. On the other hand, the catalyst for the slurry reactor may be subject to erosion as well as breakage caused by contact with internals and the wax removal unit.

The variables for both the iron salt for preparation of the solution and the base for precipitation are considerable and both have a major impact upon the purity and cost of the catalyst.

For the fixed bed reactor, the solid may be sized following the calcination step. Particles having the selected size will be used directly, while those that are too large are ground and resized. The particles that are too small may be mixed with a fresh batch of slurry and processed again. In a commercial operation, the conditions are chosen to maximize the production of the desired size range in the initial sizing step.

Spray drying is the most common approach to produce a catalyst with the appropriate size range for the slurry reactor. Bukur *et al.* [11] spray dried three catalyst materials that contained 16 parts of SiO_2 to 100 parts Fe and obtained particles in the range of 5–40 μm in diameter. They used three sources of silica (colloidal silica, tetraethyl-orthosilicate, and potassium silicate) and found that the one prepared with colloidal silica had the highest attrition strength. An example of the type of

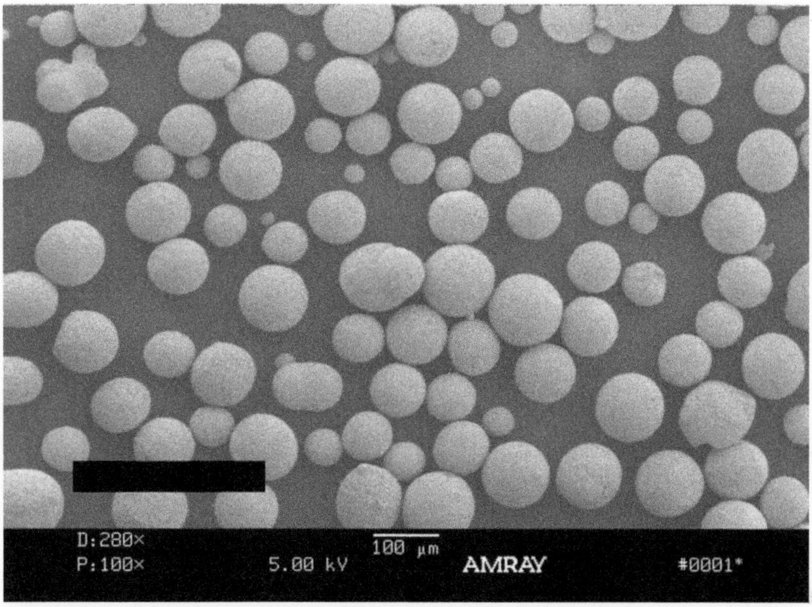

Figure 8.2 Examples of particles of iron oxides formed by spray drying.

spheres that can be produced is illustrated in Figure 8.2 [12]. Depending upon the spray drying technique, the formulation of the precipitate, and the additive used, particles can be prepared that will have a very narrow size range. An important advantage of spray drying is that the water-soluble promoters such as the alkali and copper may be added to the slurry prior to spray drying and these will then be incorporated into the dried spherical particle when the water is evaporated.

8.4 Individual Steps

8.4.1 Oxidation of Fe^{2+}

A cheap source of iron for the catalyst is the ferrous sulfate obtained as a by-product from the steel industry. Even though the raw material is cheap, a significant cost will be encountered in the extensive washing that is needed to reduce the sulfate levels to a low level where no catalyst poisoning will result when the catalyst is activated. At least two pathways are available for the use of this material. One method is to oxidize the iron by bubbling air or oxygen through the solution. In one example, the pH of the ferrous sulfate was adjusted by the addition of sodium hydroxide to yield a Fe^{2++}: OH^- ratio of 0.583 producing a suspension with a pH of 8.2 [13]. Oxygen was bubbled through the solution and samples of the solid were removed at

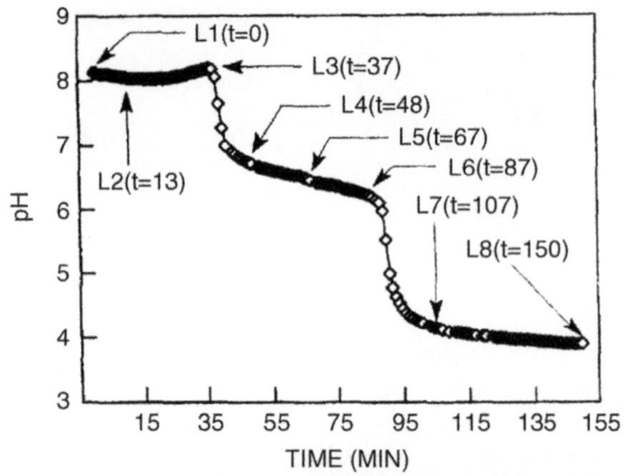

Figure 8.3 The variation in pH of the suspension with time. The samples L1 through L8 were withdrawn at several stages in this curve. The times the samples were taken have been mentioned within parenthesis after the sample number.

the times indicated in Figure 8.3 and examined by TEM. Initially, the pH remained essentially constant during the conversion of Fe^{2+} to Fe^{3+}, but when about 30% of the Fe^{2+} had been oxidized, there was a rapid drop in pH from about 9.2 to about 6.2. As more oxygen was passed through the solution, there was a small, continuous decrease in pH to about 6.2 and a sudden decrease in pH occurred to reach a final pH of slightly below 4.0 as the remaining Fe^{2+} was oxidized.

Samples were withdrawn during the oxidation at the points labeled in Figure 8.3. The morphology of the sample prior to oxidation was a mixture of dense, elongated particles and shapeless thin regions, and the appearance of the particles at point L2 at constant pH 8.2 was very similar (Figure 8.4). The L3 sample contained large thin hexagonal crystals that contained many small holes and is commonly referred to as Green Rust II. Many of the particles in sample L4 taken after the abrupt pH change resembled those of L3, but dense Fe_3O_4 hexagonal particles were also present. Continued oxidation led to more morphological changes, with some of the particles in sample L5 being needle-like γ-FeOOH, while those in sample L8 were all essentially the needle-like γ-FeOOH. The oxidation of Fe^{2+} to Fe^{3+} is shown in Figure 8.4. In summary, the oxidation of iron(II) sulfate involves complex solution chemistry and produces a variety of particulate shapes, while the presence of sulfur as sulfate in the final material presents challenges for catalyst activation.

In the FT-S, the lower carbon number products exit the reactor together with the unconverted syngas, water, and CO_2 in the gas phase. However, the higher carbon number hydrocarbons (wax) are not volatile and must be removed as a liquid. The wax is usually removed through a filter and this requires that the catalyst particles remain large enough not to plug the filter or to exit through the filter with the wax.

One of the problems encountered in using iron catalysts is the separation of the wax product from the catalyst–wax slurry. If large high surface area FeOOH particles (Figure 8.4) can be converted into the active carbide phase, this should make

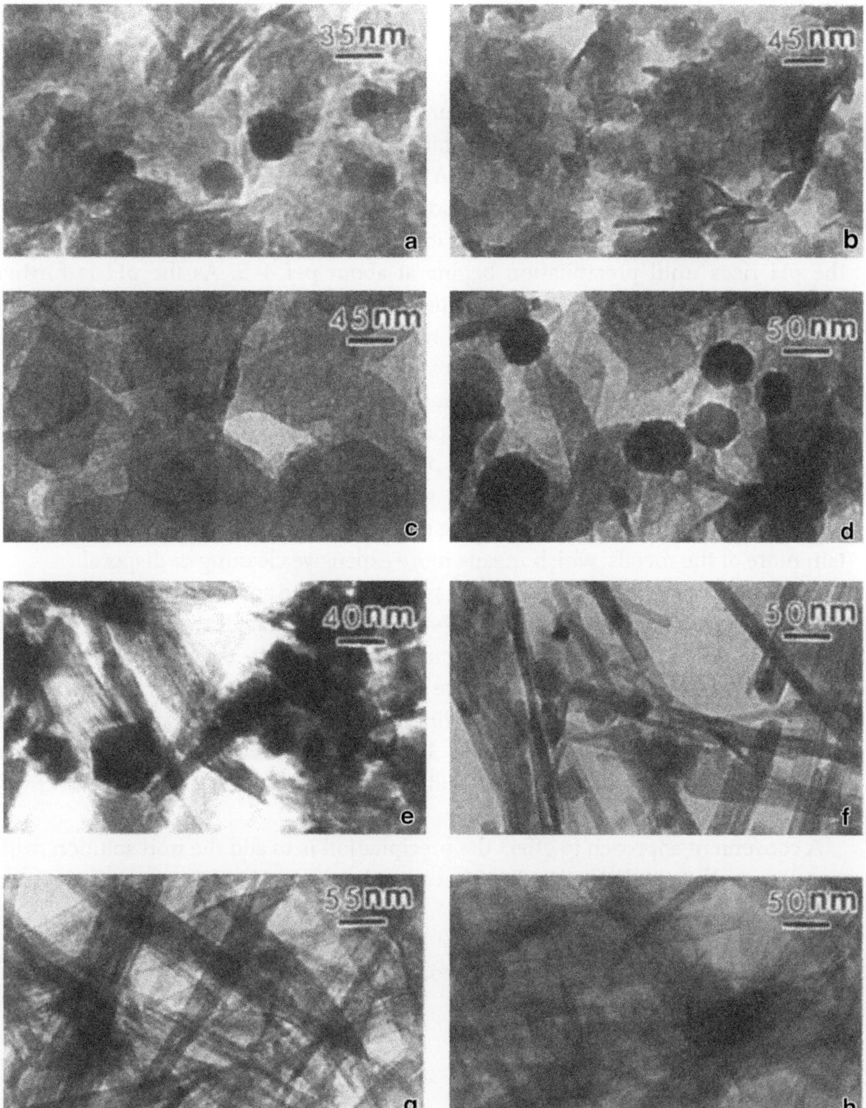

Figure 8.4 A schematic representation of a pattern of changes in shape/structure morphology from samples L1–L8.

separation much easier. To this end, a sample of needle-shaped FeOOH was prepared and converted into the carbide phase with a minimum of agitation [14]. Even under these mild conditions, during conversion into the carbide, the structure changed from needle to approximately spherical shaped particles whose diameters were about the width of the needle-shaped rods. It does not appear that a Fe oxyhydroxide particle retains its original shape as it transforms into the carbide phase.

8.4.2
Precipitation of Fe^{3+}

To avoid poisoning the catalyst by sulfur during activation, an iron nitrate solution is frequently utilized for catalyst preparation even though it is more expensive. The iron in the aqueous nitrate solution is in the oxidation state Fe^{+3} (or Fe(III)).

Since the iron nitrate solution is acidic, the solution with the concentration of iron normally used for catalyst preparation will be at a pH ≤ 4. As base is added, the pH rises until precipitation begins at about pH 4–5. As the pH is further increased, the solubility remains essentially constant until the pH reaches about 10 after which the solubility increases as negatively charged oxy-iron complexes are formed. The solubility versus pH curves for iron, silicate, and aluminum are illustrated in Figure 8.5. It is evident that for the preparation of an iron catalyst containing an alumina or silica promoter, the precipitation will restrict operations to a pH range of about 5–10. Operating at a pH much above or below this range will enhance the Ostwald ripening effect, which will decrease the particle size and thereby decrease the surface area of the resulting catalyst. The filtrate will also contain more of the metals, which means more expensive cleaning or disposal.

Many of the early iron catalysts were prepared by the addition of sodium carbonate to a hot Fe^{3+} solution to a pH of 7–8. This precipitated a very poorly ordered hematite (Fe_2O_3) that generally had a surface area of 200–300 m^2/g. The resulting slurry was added to a potassium silicate solution, where the silicate maintained or even increased the surface area of the final material. The mixed oxide that resulted was then calcined by heating in air at 300 °C. Varying the preparation procedure can influence the properties of the resulting solid and the activity and/or selectivity of the final catalyst.

A convenient approach to effect the precipitation is to add the iron solution, with or without the metal ion that is to become the physical promoter, and the base

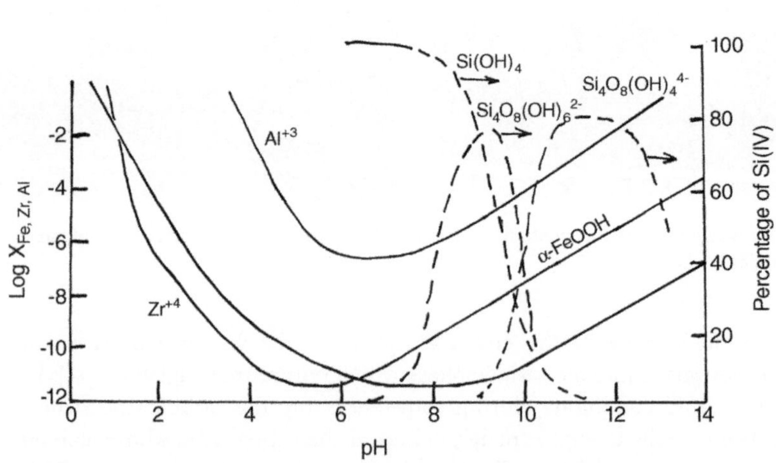

Figure 8.5 Solubility versus pH for iron, silica, and aluminum species in aqueous solution.

separately at room temperature into a continuously stirred tank reactor (CSTR) where the precipitation occurs [15]. The average residence time in the reactor depends upon the flow rates into the reactor and the size of the reactor, and is easy to adjust. The slurry leaving the CSTR can be directed to a wash tank containing clean water. After calcining at 350 °C, the samples that were obtained using this method had surface areas in the range of 250–300 m^2/g.

Kölbel and Ralek [16] carried out the precipitation with hot solutions and vigorous stirring in as short a time as possible and indicated that brief boiling of the slurry facilitated the filtration. The pH should be at 7.0–7.3 when the precipitation is complete. These authors described a laboratory reactor, illustrated in Figure 8.6, which should be scalable to a commercial operation. The preheated metal salt solution and the precipitation solution enter the bottom of the reactor tangentially and flow up the reactor column. The precipitation pH and temperature are measured at the top of the reactor. The precipitate then passes from the top of the reactor and is transferred directly to a filtration unit.

The German workers in the 1930–1950 period usually added a sodium carbonate solution to the acidic iron solution. Although this was an effective way to change and to control the pH at which the precipitation occurs due to the buffering by the bicarbonate or carbonate ion, the resulting precipitate contained high levels of sodium that could only be significantly reduced by many washings. At that time they did not have to be concerned with the environmental regulations that must now be adhered to. As the disposal of the catalyst washing solution can be a major

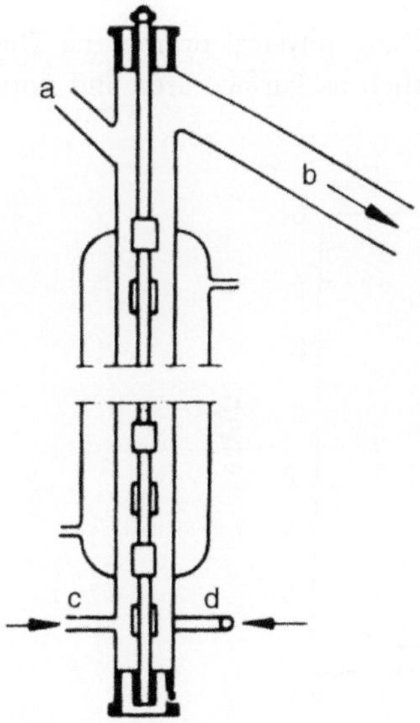

Figure 8.6 Construction of the zone reactor for the continuous precipitation of catalysts. a: pH and temperature measuring site. b: overflow for removing the precipitate. c: inlet for metal salt solution. d: inlet for precipitant.

problem and greatly increase the cost of catalyst production, these problems led some to utilize aqueous ammonia solutions for precipitation; the washings could then be used as a nitrogen fertilizer on nearby farms. However, this approach is also problematic because of restrictions on trace metals that may be present in the ground. Even so, the use of ammonia for precipitation is a reasonable approach in comparison to the use of sodium carbonate and has an additional advantage in that the ammonium nitrate that is formed can be removed from the catalyst by thermal decomposition during the calcination step, thereby minimizing the number of times that washing is required. However, this approach requires capturing the nitrogen compounds that are produced.

An exotherm occurs when the solid obtained from the precipitation and washing steps is heated to a given temperature, as illustrated for the unpromoted sample in Figure 8.7 [17]. There appears to be a general trend in that the smaller the ionic radius of the promoter, the higher the temperature where the exotherm occurs (Figure 8.8). The exotherm is due to the crystallization of the solid and the loss of a major portion of the surface area. However, while these exotherms are of scientific interest, they can be avoided as long as one restricts the calcination temperature to ≤350 °C.

Iron oxide is one of those materials that lose surface very rapidly as the calcination temperature is increased. As shown in Figure 8.9, there is a nearly linear decline in surface area as the solid is calcined in the temperature range of 100–400 °C followed by a gradual, slower decline above 400 °C [18]. Depending on the starting surface area of the iron oxide sample, the activation by carbiding may increase the surface area of a low surface area material or decrease the surface area of a high surface area material.

The iron catalyst may be modified by chemical and physical promoters. The physical promoters primarily impact properties such as surface area and pore

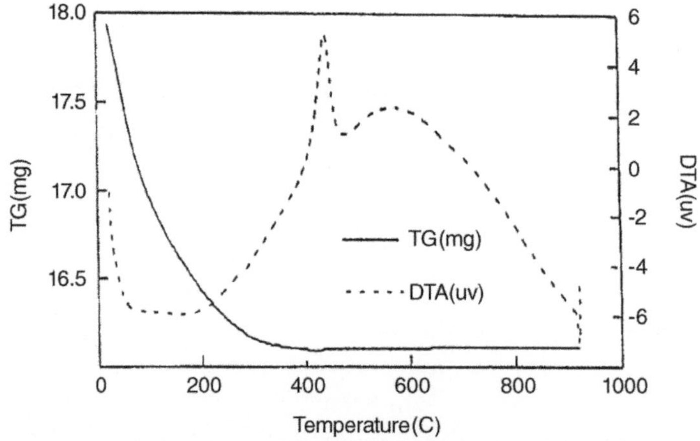

Figure 8.7 Thermal gravimetric (TG) and differential thermal analysis (DTA) curves for the unpromoted iron oxide sample.

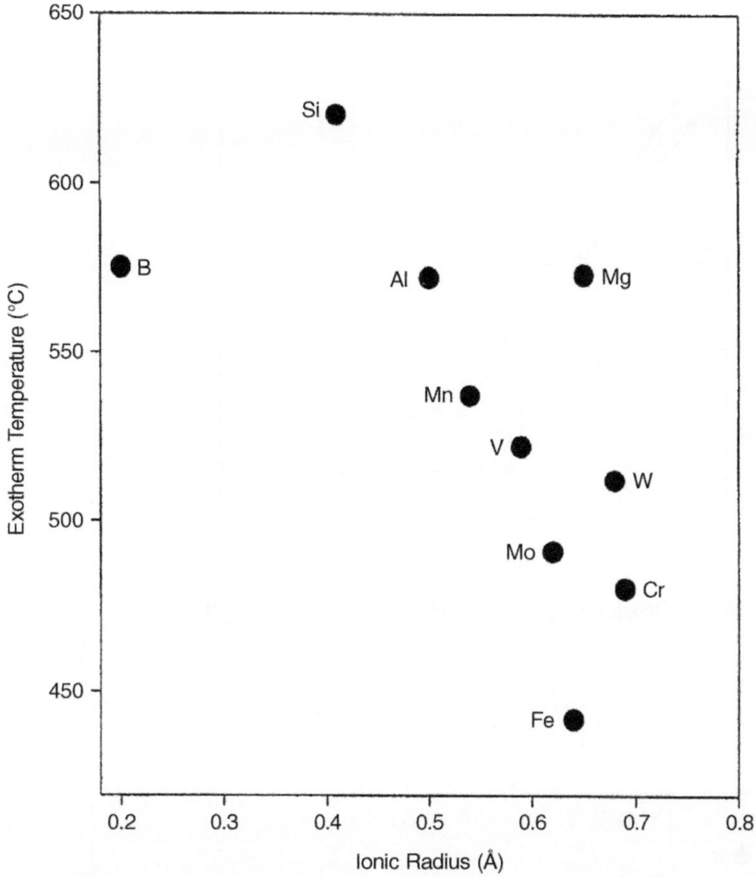

Figure 8.8 Exotherm temperatures for unpromoted and promoted (6% metal oxide) iron oxides.

structure but do not alter the composition of the products, while the chemical promoters (usually alkali or alkaline earth metals) impact product selectivity as well as physical properties.

The physical promoter will have a significant effect on the surface area of the calcined material. This is illustrated in Figure 8.10 [19], which shows that those metals with an ionic radius smaller than that of Fe^{3+} increase the surface area, whereas those metals with a radius larger than that of Fe^{3+} lead to a lower surface area than that for the unpromoted iron material. There is a similar, but less well-defined, trend in the total pore volume of the calcined material. In spite of the differences in the surface area and the pore volume, the type of nitrogen adsorption isotherm is similar. For all catalysts, a type IV adsorption isotherm is obtained and the hysteresis loop is a type H2 (IUPAC classification) [20].

The surface areas reported in Figure 8.10 were measured at a constant 6 wt% loading of the metal oxide; however, the surface area may depend upon the at.% loading. This is illustrated in Figure 8.11, which shows that a surface area gain of

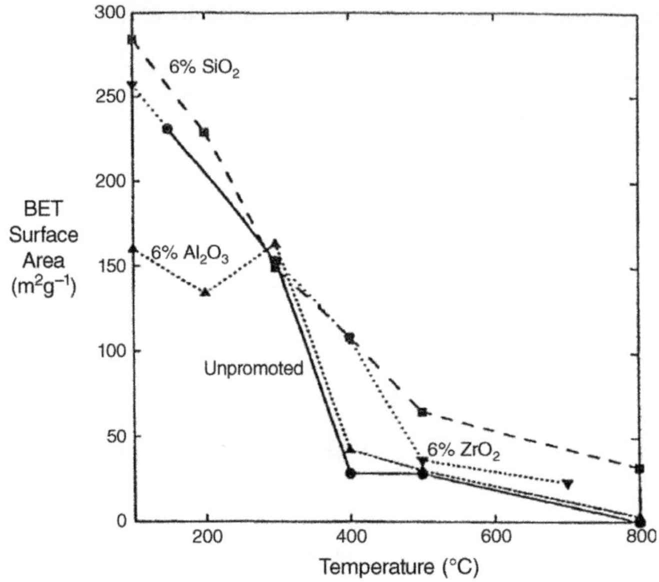

Figure 8.9 Effects of heating temperature on BET surface areas of FeOOH catalysts.

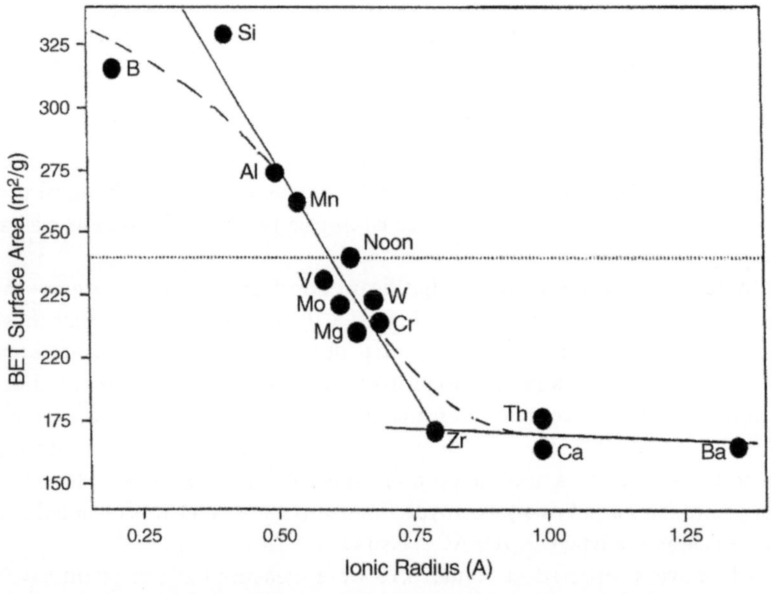

Figure 8.10 The dependence of the surface area of the catalyst precursor upon the ionic size of the metal promoter (6 wt%) (dotted line indicates the value of the unpromoted iron oxide).

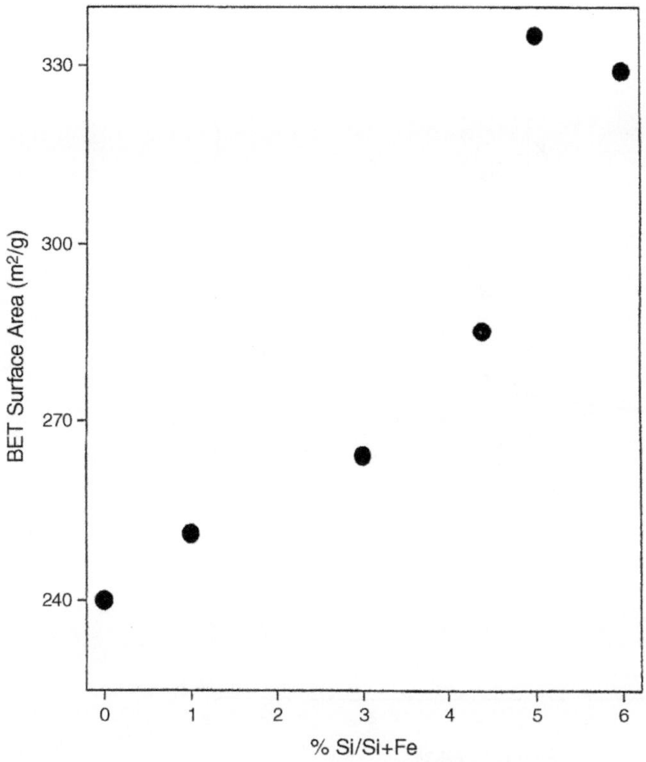

Figure 8.11 Variation of the surface area of the catalyst precursor with increasing silica content (at.%).

about 90 m^2/g is obtained by the addition of only about 1.5 wt% silica. This is a much higher surface area than could be obtained if the silica was present as a separate phase of SiO_2. An increase in silica content above about 6 at.% does not cause a further increase in the surface area. Thus, silica must be enhancing the surface area by at least partly being incorporated into the iron oxide or by forming a surface coating that inhibits sintering. A similar but weaker effect is obtained for the aluminum promoter.

The addition of alkali to an aluminum-promoted catalyst is illustrated in Figure 8.12. The addition of either K or Ca as the alkali promoter leads to a lowering of the surface area as the alkali metal loading is increased [19]. The addition of both K and Ca (K/Ca curve) leads to an even lower surface area than that the addition of either promoter alone could lead to. The results for the silica-containing iron sample were similar to those shown for the aluminum-promoted sample. Thus, it appears that the alkali promoter has a small negative impact on the surface area of the catalyst precursor.

It is not a simple task to define the impact of the promoter on the activity of the finished catalyst. As shown in Figure 8.13, the initial activity of four catalysts (unpromoted iron, K-promoted iron, Si-promoted iron, and K- and

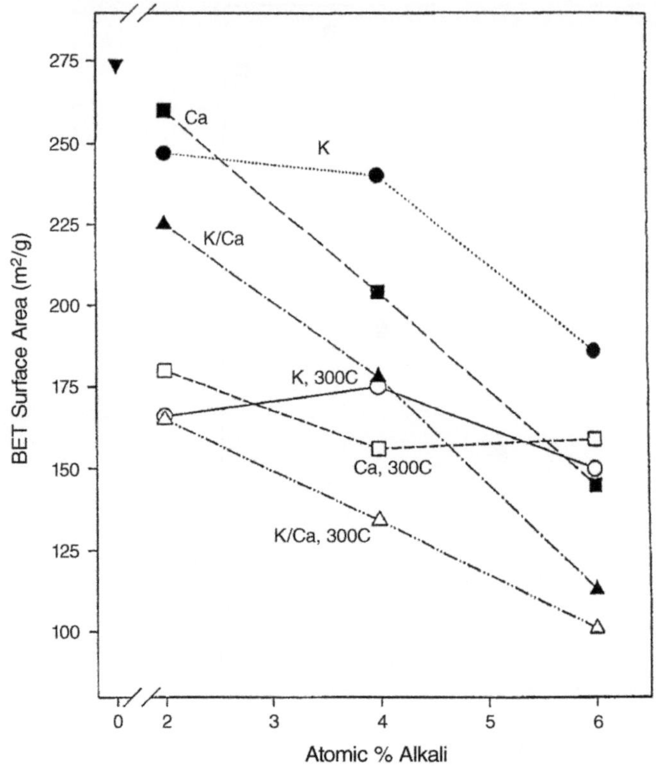

Figure 8.12 BET surface area of an alumina-promoted iron oxide (Al/(Al + Fe) = 0.06) with increasing amounts of K^+, Ca^{2+}, or ($K^+ + Ca^{2+}$) following drying at 100 °C or calcination at 300 °C.

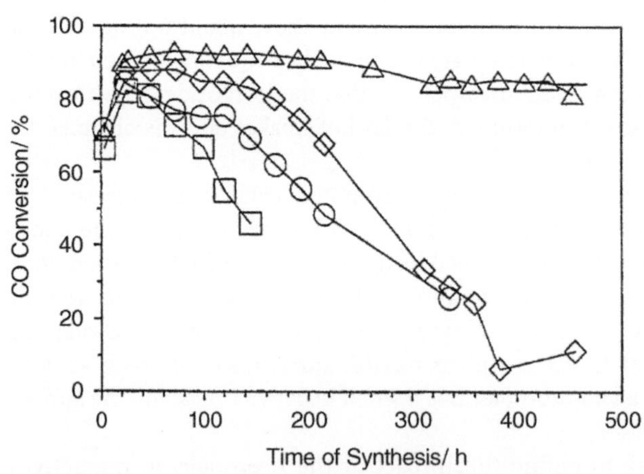

Figure 8.13 CO conversion with time on stream for catalyst containing: ○, only iron; □, iron + 0.72 K; ◇, iron + 3.6 Si; Δ, iron + 3.6 Si and 0.71 K.

Si-promoted iron) varies only slightly at their initial highest conversion. These catalysts were activated at 270 °C in CO and, as described below, should have exhibited the maximum activity of each catalyst. Thus, it is difficult to discern a significant impact of the promoter on the initial maximum activity of the four catalysts. However, it is apparent that the promoter does have a major impact, when one compares the lifetime activity. The activity of the unpromoted iron catalyst is initially high, but decreases during 400 h to approach a very low conversion. The addition of the K promoter does not improve the activity of the catalyst, and probably decreases the lifetime activity relative to the unpromoted iron catalyst. The silicate promoter causes the lifetime activity of the catalyst to be only slightly, if any, better than the unpromoted material. However, the presence of both K and silicate results in a synergistic effect and the catalyst is much more stable than the unpromoted iron catalyst. In fact, the activity decline of the silicate and the K-promoted catalyst is less than 1% CO conversion per week, as shown in Figure 8.14. These results clearly show the need to include both a physical and a chemical promoter in the catalyst.

Attempts were made to relate the retention of the activity to a particular phase of iron, but to date that has not been successful. For the low-temperature synthesis, it has been shown that Fe_3O_4 is not a catalyst for either the WGSR or the FT synthesis (CAER studies (unpublished)). For the unpromoted iron carbide catalyst that is shown in Figure 8.13, it was greater than 90% iron carbides following activation, but was converted rather quickly to the oxide Fe_3O_4 when the activity declined. The K-promoted catalyst showed a similar decline in activity as the unpromoted catalyst, but here the catalyst retained the carbide structure as the activity declined. Thus, retention of the carbide phase is not adequate for the retention of activity. The silicate-promoted catalyst did not have as much of the carbide phase as the two samples mentioned above, but did not lose the carbide phase as rapidly as the unpromoted material and also its activity at about the same rate. The phases for the

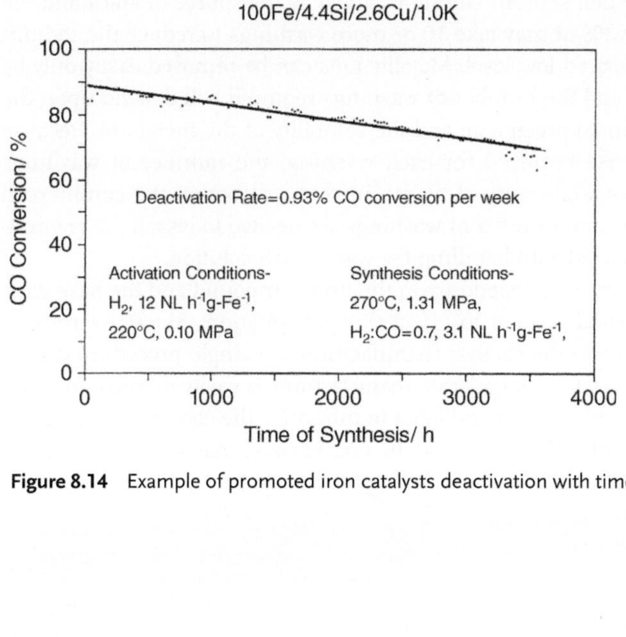

Figure 8.14 Example of promoted iron catalysts deactivation with time on stream.

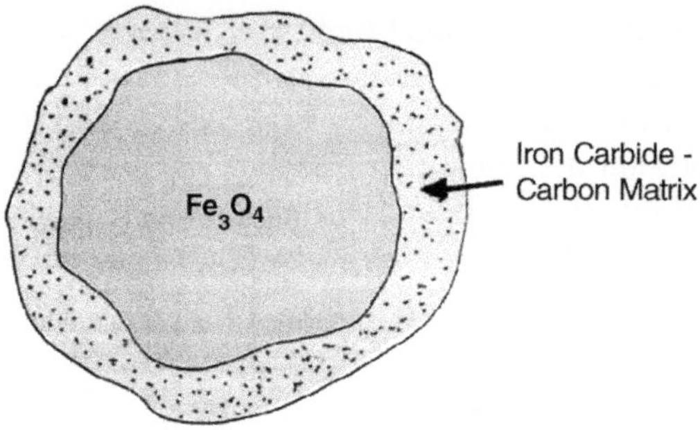

Figure 8.15 CAER model of working iron catalyst during FT synthesis.

stable doubly promoted (K + silicate) catalyst changed differently. During the initial period of the reaction, about half of the carbide was transformed into the Fe_3O_4 phase after which the ratio of the two phases, Fe_3O_4 and iron carbides, remained essentially constant. This led to the hypothesis that the structure of the doubly promoted catalyst was composed of a core of Fe_3O_4 and an outer rim of iron carbides (Figure 8.15). It therefore appears that the stability is determined by the ability of the catalyst to maintain a certain thickness of an outer shell of iron carbides and that the double promotion is needed to accomplish this.

8.4.3
Precipitate Washing

The washing step is usually the most time-consuming and the most labor-intensive part of the catalyst preparation. Unless base is added to the wash water, the pH of the wash water will decrease as washing proceeds. Unfortunately, as the pH is lowered, the iron precipitate becomes more flocculent and this makes the filtration slower. For example, when sodium silicate is added as the source of silica and the SiO_2 level is about 25 wt%, it may take 10 or more washings to reduce the sodium concentration to the desired low level. Metallic ions can be removed easily only by washing and the time and the number of washings required will depend upon the concentration in the initial precipitate and the solubility of the metal ion. Because of the time and expense required for each washing, the number of washings required in catalyst preparation must be minimized, at least on the commercial scale. Likewise, a minimum number of washings are needed to lessen the environmental problems associated with handling the waste wash solution.

The washing procedure will depend upon the iron compound and the base used as well as on the method chosen to effect the precipitation. Since a variety of approaches are available to the catalyst manufacturer, no single procedure can be given. The goal, however, for a large-scale manufacturer is easily defined: to minimize the amount of washing water and labor to minimize the cost and the environmental impact of this step. It should also be emphasized that what can be done easily at the lab scale cannot always be scaled up to the commercial level.

8.4.4
An Environmentally Greener Process

As indicated above, one of the difficulties in using the precipitation approaches is that severe environmental problems are encountered. In addition, there is a high cost due to the labor involved in the precipitation and the many washing cycles required. An innovative approach has been developed by Süd-Chemie, Inc. that overcomes most of these problems [21–24]. Metallic iron can be treated with an aqueous organic acid solution (e.g., acetic (ethanoic) acid), which is then subjected to a flow of oxygen. This produces an X-ray amorphous mixture of hydroxide, oxyhydroxide, and/or oxides. It has been reported that the surface area of the calcined catalyst can be in excess of $200\,m^2/g$ and is free of nitrates, sulfates, and so on that may be incorporated during the precipitation step. This catalyst preparation method received the US EPA Green Chemistry Award in 2003.

8.4.5
Chemical Promoters

The alkali promoter is usually added by impregnation using a nitrate salt solution. It is desirable to add the alkali promoter by an incipient wetness approach where no more solution is added than is required to fill the pores of the solid. If excess solution is present, the alkali ions exhibit a Henry's Law-type adsorption. Thus, only a portion of the alkali is added to the solid and the rest remains in the solution. The incipient wetness approach eliminates the adsorption problem. Analogously, if excess solvent is present after the alkali promoter is added and spray drying is utilized, essentially all the added alkali will be present in the dried solid.

8.4.6
Copper Promoters

For many years, copper has been added to iron FT catalysts, despite most reports indicating that Cu, added at the promoter level, has no impact upon the products produced or upon the activity of the finished catalyst. In fact, the presence of copper lowers the reduction temperature and decreases sintering during the activation of the catalyst with hydrogen. Surprisingly, the presence of copper also enhances the activation rate even when CO is used to activate the catalyst. Kölbel indicated that only about 0.05 wt% Cu is needed for the activation, although most workers utilize a loading of 1 or 2 wt% Cu.

Copper is usually added to the dried or calcined iron oxide by impregnation. Copper nitrate is usually chosen to prepare the impregnation solution. Although the impregnation may be carried out adding only copper, it is frequently carried out with both the copper and the potassium being added in a single impregnation.

8.4.7
Phase Changes

In going from the iron oxide to the carbide during the activation step, the volume changes by about 80% and the same change in volume occurs when the catalyst is oxidized during synthesis. This volume change places severe strains upon the catalyst particles. One approach to overcome the volume change and the attrition associated with the strains that this creates on the catalyst particles is to surround the iron by a silica shield. This is the major reason for Ruhrchemie incorporating 25% silica in the catalysts they developed during the early German work; this approach is still utilized by Sasol for their low-temperature processes. However, it appears that the carbiding destroys the bonding between iron and the oxide support. Thus, the interfacial binding that causes the iron oxide to adhere to the oxide support is lacking and the iron carbide particles are prone to attrition. The separation of wax from the catalyst slurry is one of the major problems still to be overcome in the use of the iron catalyst.

8.4.8
Other Iron Catalysts

There are many other routes to producing iron catalysts. However, these processes need much effort before they can compete with the ones described above. For example, Exxon Mobil has described a laser pyrolysis approach to form iron carbides by the reaction of iron carbonyl with ethylene that is induced by the energy adsorbed by the reactants. However, a catalyst prepared by this method does not contain the promoters and loses the high activity within 400 h, far too short for consideration for a commercial process. Thus, for a process to become viable, one must be able to add the promoters during or following the carbide preparation step and this is a demanding task. Iron carbonyl can be reacted with water in the gas phase and thereby produces very small (2–3 nm) iron oxide particles. Again, the oxide must be converted into the carbide and here particle growth occurs. In addition, the small particle sizes greatly increase the viscosity of the slurry and this limits the catalyst loading that can be achieved.

A number of attempts have been made to prepare supported iron catalysts. In one sense, the Sasol catalyst containing 25% silica is a giant step in that direction. However, because of its lower activity and the need to limit catalyst loading to avoid a too high viscosity, one is limited by how much silica support can be tolerated in the finished catalyst.

References

1 Schwertmann, U. and Cornell, R.M. (1991) *Iron Oxides in the Laboratory*, Wiley-VCH Verlag GmbH, Weinheim.

2 Cornell, R.M. and Schwertmann, U. (1996) *The Iron Oxides: Structure, Properties, Reactions, Occurrence and Uses*, Wiley-VCH Verlag GmbH, Weinheim.

3 Sasol (2011) Sasol Facts 2010. http://sasol.investoreports.com/sasol_sf_2010/

technology-and-production/our-main-south-african-production-processes (accessed July 2011).
4 Bromfield, T.C. and Vosloo, A.C. (2003) *Macromol. Symp.*, **193**, 29–34.
5 Veazey, M.V. (2010) *Downstream Today, February 15.*
6 Satterfield, C.N., Hanlon, R.T., Matsumoto, D.K., Donnelly, T.J., and Yates, I.C. (1989) Report DOE/PC/80015-T5, 51.
7 Bromfield, T.C. and Vosloo, A.C. (2003) *Macromol. Symp.*, **193**, 29–34.
8 Crous, R. and Bromfield, T.C. (2010) EP 2193842 A1.
9 Crous, R. and Bromfield, T.C. (2010) EP 2193841 A1.
10 Dry, M.E. (1981) *Catalysis* (eds J.R. Anderson and M. Boudart), Springer, Berlin, pp. 159–255.
11 Bukur, D.B., Carreto-Vazquez, V.H., and Ma, W. (2010) *Appl. Catal. A – Gen.*, **388**, 240–247.
12 Hu, X.D., O'Brien, R.J., Tuell, R., Conca, E., Rubini, C., and Petrini, G. (2007) US Patent No. 7,199,077 B2, April 3.
13 Srinivasan, R., Lin, R., Spicer, R.L., and Davis, B.H. (1996) *Colloids Surf. A*, **113**, 97–105.
14 Sarkar, A., Dozier, A.K., Graham, U.M., Thomas, T., O'Brien, R.J., and Davis, B.H. (2007) *Appl. Catal. A – Gen.*, **326**, 55–64.
15 Davis, B.H. (1992) DOE Final Report DOE/PC/90049-T7.
16 Kölbel, H. and Ralek, M. (1980) *Catal. Rev. Sci. Eng.*, **21**, 225–274.
17 Milburn, D.R., Chary, K.V.R., O'Brien, R.J., and Davis, B.H. (1996) *Appl. Catal. A – Gen.*, **144**, 133–146.
18 Milburn, D.R., O'Brien, R.J., Chary, K., and Davis, B.H. (1994) *Stud. Surf. Sci. Catal.*, **87**, 753–761.
19 Milburn, D.R., Chary, K.V.R., and Davis, B.H. (1996) *Appl. Catal. A – Gen.*, **144**, 121–132.
20 Sing, K.S.W., Everett, D.H., Haul, R.A.W., Moscou, L., Pierotti, R.A., Rouquerol, J., and Sieminelewska, T. (1985) *Pure Appl. Chem.*, **57**, 603–619.
21 Petrini, G., Conca, E., O'Brien, R.J., Hu, X.D., and Sargent, S. (2009) US Patent No. 7,566,680 B2, July 28.
22 Hu, X.D., O'Brien, R.J., Tuell, R., Conca, E., Rubini, C., and Petrini, G. (2007) US. Patent No. 7,199,077 B2, April 3.
23 Hu, X.D., Loi, P.J., and O'Brien, R.J. (2008) US Patent No. 7,452,844 B2, November 18.
24 O'Brien, R.J., Sargent, S.E., Petrini, G., and Conco, E. (2011) US Patent No. 7,939,463 B1, May 10.

9
Cobalt FT Catalysts

Burtron H. Davis

Synopsis

Industrial cobalt FT catalysts generally require a support. These are mostly based on alumina or silica, but titania has also recently been used. Details of the supports used and the methods of preparing the catalysts are summarized.

9.1
Introduction

The preparation and characterization of cobalt catalysts are complicated by the presence of a support. While unsupported cobalt catalysts have been utilized in a few instances, the activity is low and the catalyst lifetime short. Thus, only supported catalysts have found use in commercial operations. The first step in making a good catalyst then requires the careful preparation and characterization of the support material. Addition of the cobalt and other promoters to the support, calcination, and reduction produces the finished catalyst.

9.2
Early German Work

The initial German commercial plants utilized thoria-promoted cobalt–kieselguhr catalysts, first at 1 atm and then in a middle pressure range (about 20 atm). Compared to today's catalysts, these early catalysts were relatively inactive. One of the reasons for this is the support used: kieselguhr, while primarily silica, has a low surface area, usually less than $20\,m^2/g$. Despite this, the early workers defined a number of factors that are important for the preparation of today's catalysts. They demonstrated the need to use a support for the cobalt catalyst; they also identified the fact that a nonreducible oxide could be added, which decreased the reaction between the cobalt and the support, and showed that the addition of an easily

Greener Fischer-Tropsch Processes for Fuels and Feedstocks, First Edition. Edited by Peter M. Maitlis and Arno de Klerk.

reducible metal oxide could decrease the temperature needed to reduce the cobalt to the metallic state.

9.3
Support Preparation

To date, only three supports have been employed to prepare catalysts that have been utilized in commercial or large pilot plant reactors: alumina, silica, and titania. Many other supports have been used to make catalysts for scientific studies, but these have so far not advanced to the pilot plant or commercial stage.

As a major role of the support is to provide a high surface area for the metal, it is usually necessary for the support also to have a high surface area. The supports are usually metal oxides of two general classes: those where the surface area decreases continuously as the temperature is increased and those which retain the surface area to a high temperature and then rapidly lose surface area (Figure 9.1). Both silica and alumina (and also carbon) supports belong to this latter class. Other oxides tend to lose surface area as the temperature is increased. For this reason, silica and alumina are the oxides most often chosen as supports since they retain their high surface area upon calcination to reasonably high temperatures, and are therefore used for most FT catalysts. However, it appears that ExxonMobil has chosen titania as the support for their commercial catalyst since they have many patents on the preparation and use of this support.

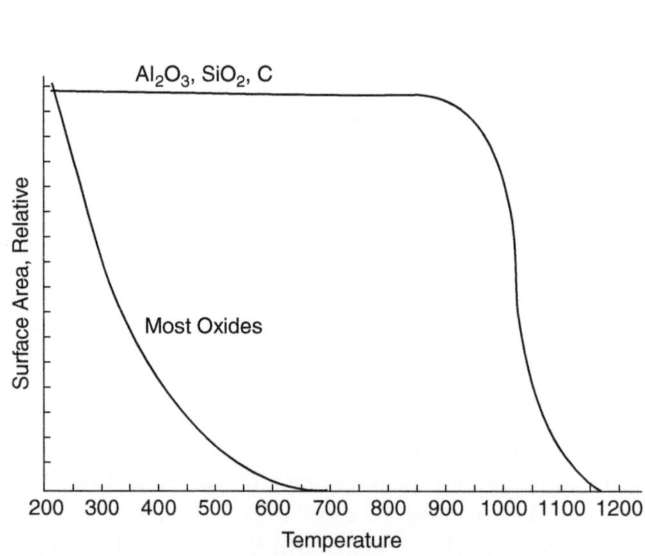

Figure 9.1 Schematic representation of surface area loss for metal oxides as they are heated to higher temperatures (unpublished CAER work).

9.3.1
Alumina Supports

Aluminum forms a range of hydroxides. Well-crystallized trihydroxide forms include gibbsite, bayerite, and nordstrandite. The more common oxyhydroxide forms are boehmite and diaspore. The most important form is gibbsite, but bayerite and boehmite are also available commercially. The phases available commercially are usually offered in a range of physical properties, such as surface area, pore size, and so on, for each phase, allowing the preparation of a variety of supported Co–alumina catalysts using commercially available alumina. Therefore, in preparing alumina-supported catalysts, it is wise to search for commercial aluminas that have properties that are desired and use these to prepare the catalyst.

Alumina may exist in several crystal structures (including phases denoted as α, χ, η, δ, γ κ, θ, and ϱ) that arise during the heat treatment of the hydroxide or oxyhydroxide. The crystal phase of the as-prepared material may exist in three forms that depend upon the method used for the preparation: gibbsite, boehmite, and bayerite. Calcination changes these phases as illustrated in Figure 9.2 [1]. The gamma

Decomposition Sequence of Aluminum Hydroxides

Conditions	Conditions Favoring Transformation① Paths a	Path b
Pressure	>1 atm	1 atm
Atmosphere	moist air	dry air
Heating Rate	>1°C/min	< 1°C/min
Particle Size	> 100 micrometer	<10 micrometer

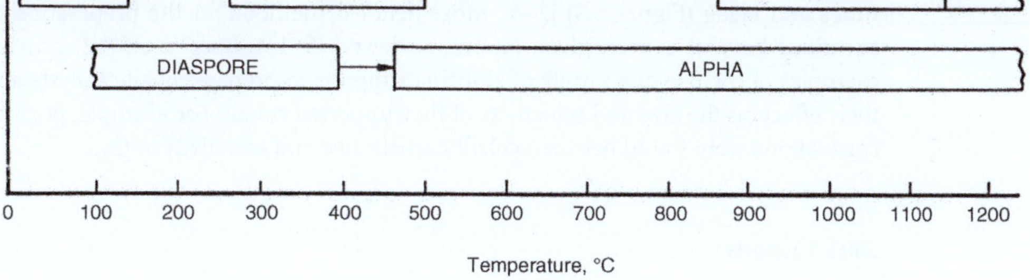

Figure 9.2 Decomposition sequence of aluminum hydroxides (enclosed area indicates range of occurrence; open area indicates range of transition).

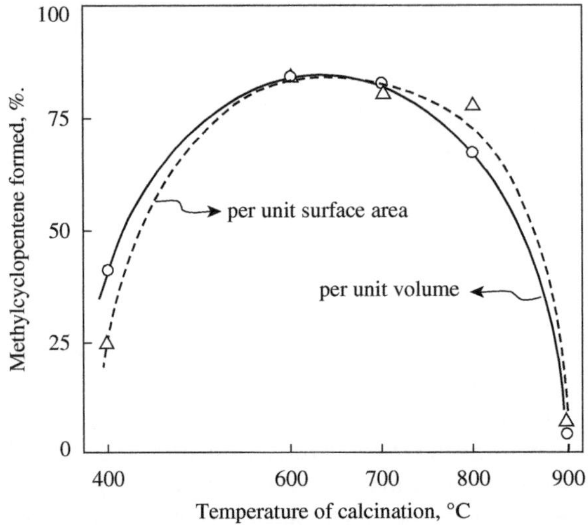

Figure 9.3 Isomerization of cyclohexane (410 °C, HLSV 2.0) as a function of calcination temperature of alumina. *Solid line:* Isomerization activity per unit volume. *Dashed line:* Isomerization activity per unit surface area [2].

phase is the most widely used as a catalyst support, but the other phases are also employed. For the cobalt FT catalyst, one usually calcines the support in the 400–600 °C range so that higher temperature phase changes do not occur during use. These aluminas will have a high (100–200 m^2/g and greater) surface area.

In spite of the commercial availability of alumina, many laboratory studies utilize supports made in-house. Aluminas of especially high purity can be prepared by the hydrolysis of an aluminum alkoxide and calcination of the product. The alkoxide can be prepared by reacting high-purity aluminum with an appropriate alcohol, such as isopropanol. Spherical aluminas with a high surface area can be prepared by the hydrolysis of aluminum chloride in a flame process and this form is available commercially, usually with a surface area of about 200 m^2/g. Laboratory preparations are available to make both acidic and nonacidic aluminas, and a series of investigations on the preparation and properties of these aluminas was initiated by Pines and Haag (Figure 9.3) [2–4]. More detailed methods for the preparation of aluminas have also been given [1] and reviewed [5–11]. Borg *et al.* [12] provide examples of the use of a variety of alumina supports to prepare cobalt catalysts and their effect on the size and selectivity of the supported cobalt. For example, positive correlations were found between cobalt particle size and selectivity to C_{5+}.

9.3.2
Silica Supports

Silicas were originally classified in one of two types: gels, which have very high surface areas with very small average pore sizes, and fused silica, which generally

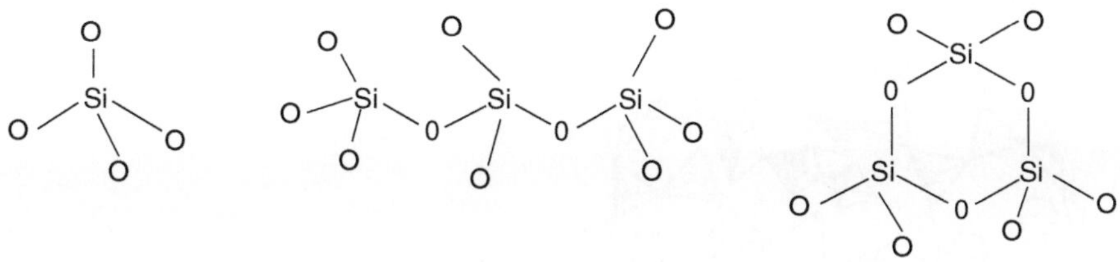

Figure 9.4 Linear and planar cyclic silicate anions (http://www.silicates.com/leading-speciation.asp?bhjs=1&bhsw=1280&bhsh=1024&bhswi=1003&bhshi=1092&bhflver=5&bhdir=0&bhje=1&bhcold=32&bhrl=-1&bhqt=-1&bhmp=-1&bhab=-1&bhmpex=&bhflex=&bhdirex=&bhcont=lan).

has a low surface area with large average pore sizes. The early FT catalysts utilized kieselguhrs that were composed of amorphous silica plus small amounts of alumina, iron oxides, and traces of other oxides. Anderson *et al.* [13] have summarized data for a number of kieselguhrs and reported surface areas less than 40 m^2/g.

The widespread use of silicas has led to the development of materials with a nearly continuous range of surface areas and pore sizes. For example, the Grace Division of W.R. Grace provides a wide range of silicas with surface areas in the range of 25–750 m^2/g and pore volumes in the range of 0.1–2.8 cm^2/g. In addition, these silicas can be made in a variety of shapes as well as in forms that have been surface treated to modify the properties.

Silicates have a range of complex structures in solution that depend upon the concentration and the ratio of base to silica. The structures of the monomer and of two trimers are shown in Figure 9.4; these can combine to form a variety of polymers, some of which are illustrated in Figure 9.5.

The anion distribution is mainly controlled by the concentration of the solids and the silica/alkali ratio (Figure 9.5).

Changes in either factor (Figures 9.5, 9.6 and 9.7) will upset the equilibrium and the system will adjust to the new position within minutes or days, the rates depending on the rearrangements involved.

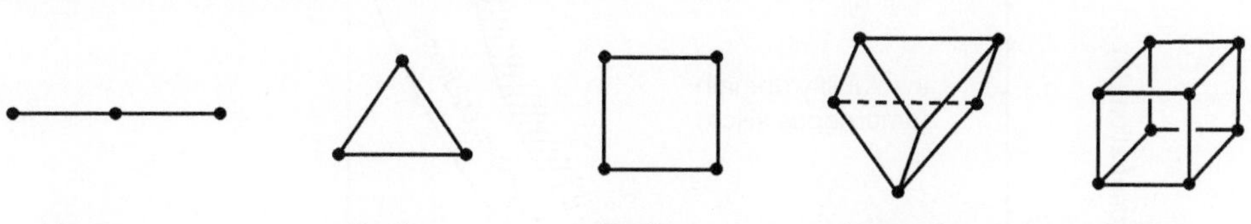

Figure 9.5 Linear, planar, cyclic, and three-dimensional silicate anion structures, with dots denoting silicon atoms. The oxygens linking the silicon atoms lie between the dots and are not shown (http://www.silicates.com/leading-speciation.asp?bhjs=1&bhsw=1280&bhsh=1024&bhswi=1003&bhshi=1092&bhflver=5&bhdir=0&bhje=1&bhcold=32&bhrl=-1&bhqt=-1&bhmp=-1&bhab=-1&bhmpex=&bhflex=&bhdirex=&bhcont=lan).

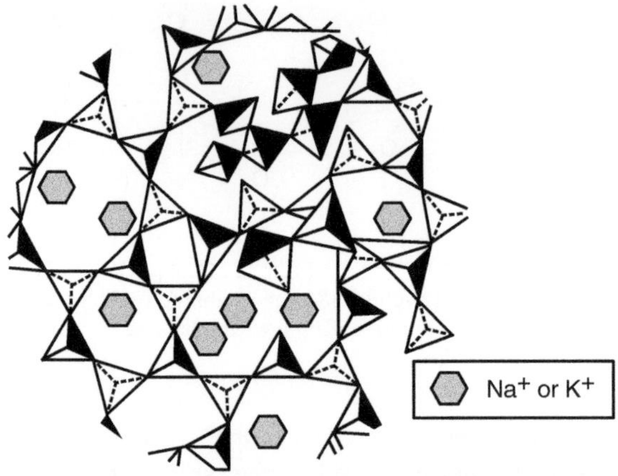

Figure 9.6 Silicate anion structure in amorphous silicate glass.

The discovery of mesoporous molecular sieves offered another form of silica in which the pore structure was uniform in size and whose size could be controlled reasonably closely [14, 15]. Two possible pathways to make the mesoporous MCM-41 are illustrated in Figure 9.8, one initiated by a liquid crystal phase, the other by a silicate anion. The sizes of the pores depend upon the organic ion used to template the synthesis as depicted by the Mobil workers in Figure 9.8. The unit cell parameter has been shown to depend upon the number of carbon atoms in the surfactant chain used to prepare the material (Figure 9.9) [16].

These mesoporous supports attracted attention for FT synthesis and a number of studies for their preparation have been reported. Khodakov *et al.* [17, 18] found that the conversion increased in line with the cobalt surface density ([Co] per m^2/g) for the mesoporous support but not for the silica-supported material (a wetted and dried Cab-o-Sil-M5) (Figure 9.10). The 35 h evaluations of the catalytic activity,

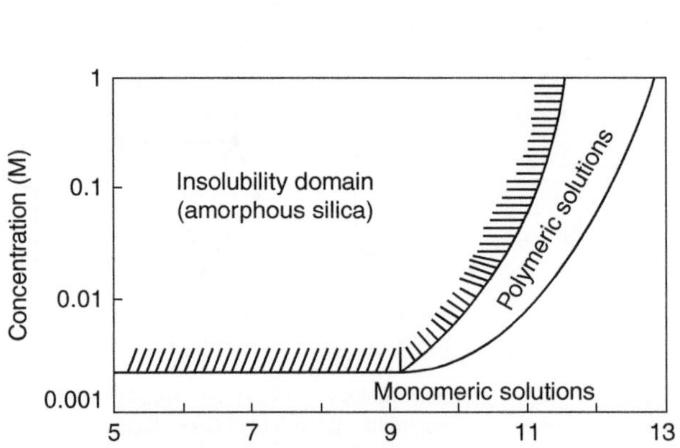

Figure 9.7 Soluble silicate speciation (http://www.inchem.org/documents/sids/sids/SolubleSilicates.pdf).

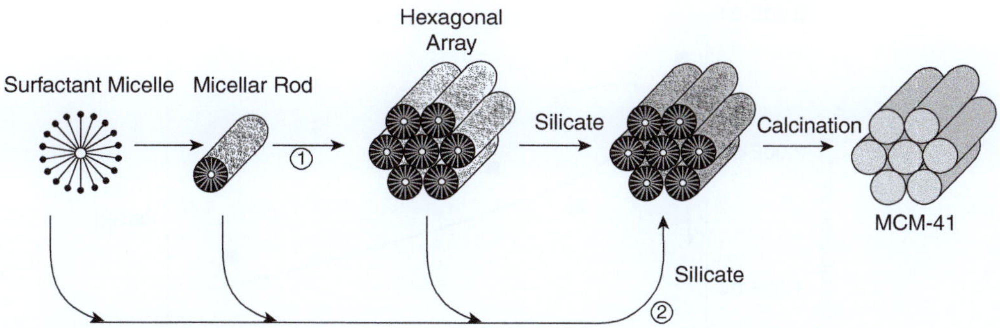

Figure 9.8 Possible mechanistic pathways for the formation of MCM-41. 1: liquid crystal phase initiated, and 2: silicate anion initiated.

however, were conducted at atmospheric pressure so that the methane production was high and the alpha-values for the catalysts were low.

A support with a unique bimodal structure has been claimed [19–21] Figure 9.11. A commercial silica with larger (about 50 nm) pores was impregnated with a silica sol and then the solvent was evaporated. The added silica was considered to develop a small pore structure within the larger pores. After calcining, the support is

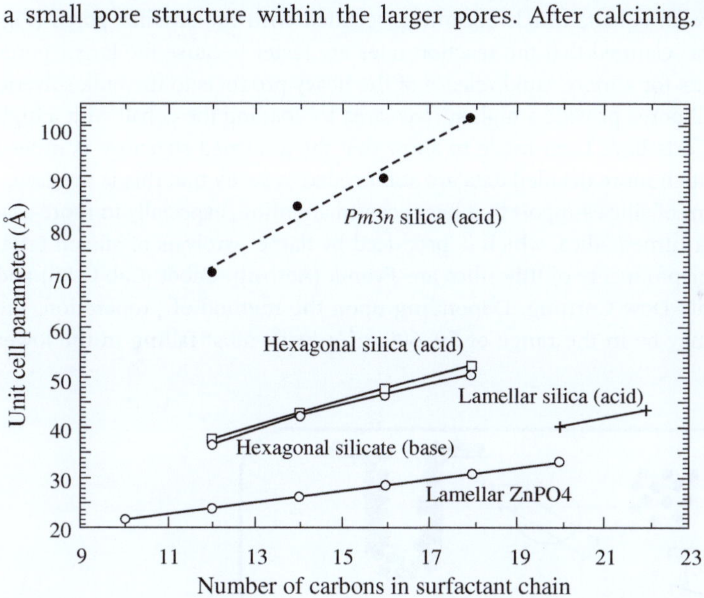

Figure 9.9 The relationship between unit cell parameter and the number of carbon atoms in the surfactant chain. A silica (Pm3n) phase was synthesized by using alkyltriethyl ammonium as template at room temperature for 1 h. Hexagonal silica and silicate phases were obtained in the presence of alkyltrimethyl ammonium at room temperature in acidic and basic media, respectively. Only $[C_{20-22}\ TMA]^+$ led to the formation of lamellar silica in acidic medium at room temperature. Lamellar $ZnPO_4$ was prepared by using alkyltrimethyl ammonium as template from acidic medium (pH 2.5). Estimated error is 1 Å in d-spacing of highest peak (d_{210} for Pm3n phase and d_{100} for hexagonal and lamellar phases).

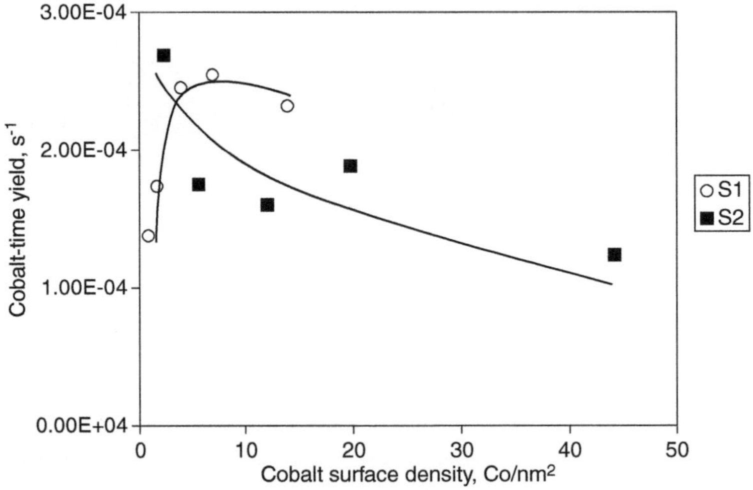

Figure 9.10 Cobalt–time yields as functions of cobalt surface densities in CoS1 and CoS2 catalysts.

impregnated with cobalt, which became associated with the smaller pores (Figure 9.11). It was claimed that the reaction rates are faster because the larger pores provide a means for a more rapid release of the heavy products to the bulk solvent, while the small pores provide a high surface area for loading the cobalt with a high dispersion. Efforts have been made to show that the assumed structure is indeed formed, but much more detailed data are still needed to verify that this is the case.

Another form of silica support that has received attention, especially in more academic work, is fumed silica, which is produced by flame pyrolysis of silicon tetrachloride. Major producers of this silica are Evonik (Aerosil), Cabot (Cab-O-Sil) and Wacker Chemie-Dow Corning. Depending upon the method of preparation, the surface area may be in the range of 50–600 m^2/g, with most falling in the lower

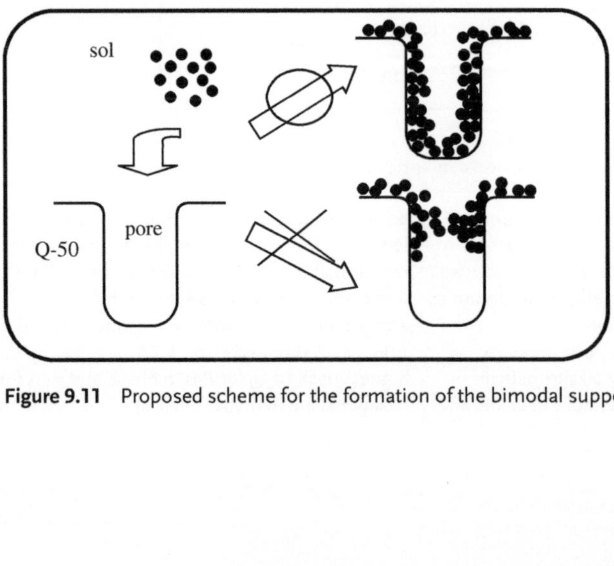

Figure 9.11 Proposed scheme for the formation of the bimodal support.

half of the range. These materials are generally composed of nonporous spheres and the overall porosity results from the packing of the spheres [22].

9.3.3
Titanium Dioxide Support

Titania received much attention as a support after it was demonstrated that it could be involved in strong metal support interaction (SMSI) [23]. As it was associated with unusual catalytic activity, the SMSI phenomenon led to much interest and many papers appeared on the subject in a short time (Section 11.3 and Box 11.4). These included a number of patents and publications by Exxon workers using titania as a catalyst support, many of which were for the FT-S.

Compared to silica and alumina, titanium dioxide has a low surface area. Early commercial samples had areas in the range of 20–50 m^2/g, but recently commercial samples have been reported that have surface areas in the 200 m^2/g range. Titania exists in three crystal forms: rutile, anatase, and brookite, the first two being important for catalysis. Conventional catalysts are based upon the rutile form of titania because of its better attrition resistance compared to the anatase form. A commercial form that has been utilized in many studies is the Degussa P25 titania. This has a rather low surface area; thus, a P25 sample was recently reported to have a surface area of 50.8 m^2/g and was about 83% anatase, with the remainder being rutile. Higher surface area titania supports are now available that may have areas greater than 300 m^2/g [24]. However, titania is one of the oxides that lose surface area rapidly as the calcination temperature is increased, as shown in Figure 9.12 [25].

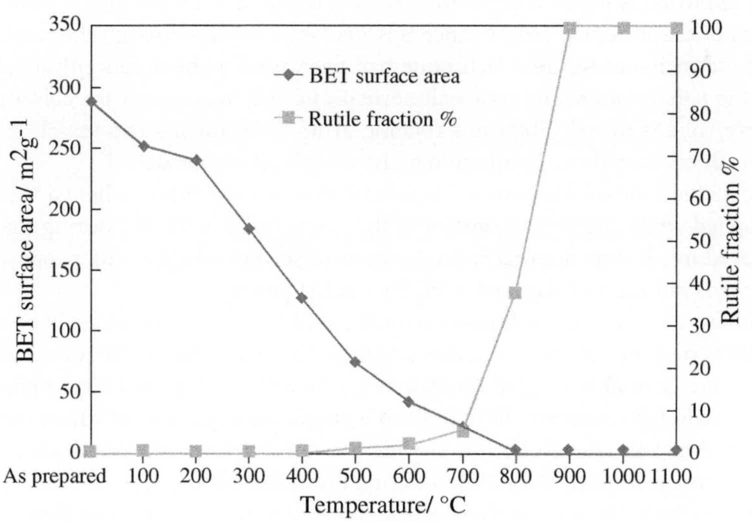

Figure 9.12 The effect of heat-treatment temperature (in air) on "as-prepared" nano-TiO_2 powder (i.e., 100% anatase) on the BET surface area and the transformed rutile weight fraction%.

In order to improve the attrition resistance and to provide better resistance to sintering and loss of surface area, a number of mixed oxides have been investigated. For example, both ExxonMobil and Shell have patents for a range of silica–titania compositions and their use as supports for cobalt catalysts. With an alumina–titania ratio of 1: 0.5, it was reported that the material had a surface area of 41 m^2/g even after calcining at 1000 °C [26]. While there are a large number of options for the preparation of these mixed oxides, so far they do not appear to have been utilized in a commercial operation.

9.4
Addition of Cobalt and Promoters

In the laboratory there are a number of sophisticated ways to add cobalt to the support; however, at the commercial scale, impregnation with an aqueous solution appears to be preferred. The chief impregnation techniques used are summarized.

A common approach uses the incipient wetness method. Here, a salt of the material to be added is dissolved in just sufficient water to fill the pores of the support without wetting the solid; thus, this approach avoids excess water so that the added salt is present in a uniform distribution in the catalyst pore system. Prior to the impregnation by this method, a preliminary run is made with pure water to determine the amount of water that the solid pore volume can hold before the "dry" solid begins to form a wet paste. This defines the volume of solution that can be added so that the pores retain all the added solution. However, as drying may concentrate the salt at those locations where the solvent last evaporates, one must also be concerned with the drying method utilized.

Another approach is to use more solvent than is required to fill the pores. Here, the removal of the solvent is critical since it is necessary to pass through the same stage as if the incipient wetness technique had been used without concentrating the salt at the pore mouths. The approach normally used is to evaporate the solvent as the slurry/solid is mixed, either in a rotating drum or by stirring in a vessel. As the slurry is dried, a uniform distribution of the added salt is approached.

Another method for solvent removal is described in a Sasol patent where a vacuum is applied while the vessel containing the slurry/solid is rotated during the drying procedure. In this approach, the amount of solvent added to the support usually exceeds the amount required to fill the catalyst pores.

The solubility of cobalt salts is limited, so only 10–12 wt% Co can be added in one impregnation step; hence, two or more impregnation steps will be required to obtain the more normal loading of 20–25 wt% Co. In spite of this solubility limitation, some catalyst preparations have utilized a single impregnation in which the solid is placed in enough solution to provide the desired Co loading and which then forms a supersaturated salt solution during the drying step.

Another approach that has received considerable attention is to deviate from a uniform distribution of the metal within the support. In this impregnation approach, the active catalytic material can be deposited either on the outer rim of

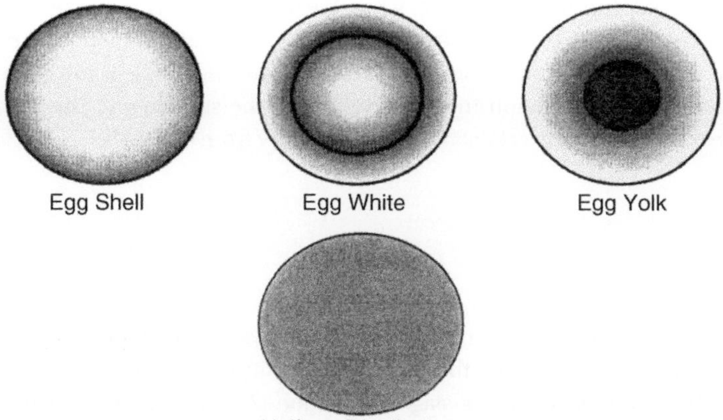

Figure 9.13 Different distributions of catalytically active components within support bodies.

the support (*egg shell*), in the middle (*egg white*), or at the center (*egg yolk*), as illustrated in Figure 9.13 [27].

The chemical promoters are normally added during the final impregnation with the cobalt or after the cobalt addition and calcination. The promoter is generally added as a nitrate salt dissolved in water. Although the halide or sulfate salts are more common and cheaper, both halide and sulfur are poisons for the cobalt catalyst. As an example involving another metal, an aqueous solution of $Pt(NH_3)_4(NO_3)_2$ is frequently used to add platinum to the catalyst [28]. If the promoter is to be added to the calcined catalyst, the material should equilibrate with water vapor and the incipient wetness approach is utilized. EXAFS data show that many, if not all, of the metals of Groups 9–11 (formerly known as Group VIII) act as promoters that associate with or form alloys with cobalt during reduction.

9.5 Calcination

Gases are released during the heating, the soaking, and at the calcination stages: notably water and decomposition products of the anion of the cobalt salt used to prepare the catalyst. During the calcination, the release of these gases needs to be controlled and the approaches used in the laboratory and in the commercial setting may differ.

In the laboratory, the sample is commonly heated in an oven in which a gas flows over the catalyst mass that is spread in a thin layer on a ceramic tray. The calcination is usually conducted in air or even pure oxygen at atmospheric pressure. The sample is usually heated from room temperature to the final calcination temperature (350–400 °C) during an hour or more for a small laboratory size sample. The calcination may also be conducted by placing the catalyst mass in a tube; here a high gas flow is passed through the catalyst, while the temperature is increased at a rate of about 0.1–1.0 °C/min. Again, the heating rate and gas flow need to be adjusted so that the concentration of the gases issuing from the sample is low (usually below 0.1 atm). The calcination time will vary with the sample and the salts used to prepare the catalyst. The calcination time may vary from 4 to 24 h.

9.6 Reduction

The most widely used approach to activating a cobalt catalyst is reduction in hydrogen. Depending upon the support and the catalyst composition, there are a variety of techniques for effecting this, such as a staged reduction [29]. Among the variables that one attempts to control is the partial pressure of water in the reducing gas. In a fixed bed reactor, a preferred approach is to add the hydrogen from the bottom of the reactor and to conduct the reduction at a slow rate that limits the partial pressure of water in the exit gas. In the Bureau of Mines work, the water partial pressure was limited to less than 0.2 atm. Thus, the heating rate needs to be adjusted so that it may take many hours to attain the final temperature and for the impregnate to attain a reasonably high degree of reduction. When cobalt nitrate is used for the impregnation, ammonia is formed during the reduction. The level of ammonia in the exit gas is recommended to be kept below certain levels, for example, 250 volume ppm [30].

One of the major reasons for adding a platinum metal to the catalyst is to cause a more rapid reduction and to attain a much higher degree of reduction. This effect of the promoter is illustrated in the laboratory-scale reduction of an unpromoted catalyst and one containing different levels of promoter, as illustrated in Figure 9.14. In general, the promoter increases the fraction of reduced cobalt by about a factor of 2, with the actual amount depending upon the cobalt loading (Figure 9.14).

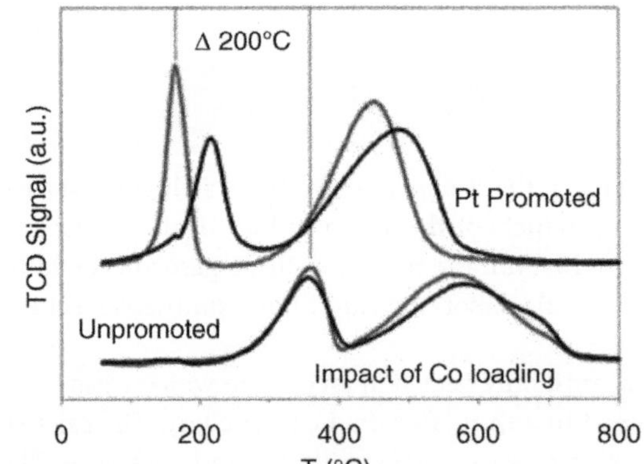

Figure 9.14 The impact of a Pt promoter (0.5 wt%) on the reduction of 15 and 25 wt% cobalt–alumina catalyst (unpublished CAER work).

Gulf workers provided an improved process for the reduction of supported cobalt catalysts that involved three steps: reduction, reoxidation, and reduction. The second reduction step added 2–5% CO conversion compared to that after the first reduction. This type of activation procedure appears to have been adopted elsewhere too [31].

9.7 Catalyst Transfer

In South Africa, Sasol makes their iron catalyst on-site; however, for the cobalt catalyst used in Qatar, they formed a venture with Engelhard (now BASF). To transport their finished catalyst from their preparation plant in the Netherlands to Qatar, the freshly reduced catalyst was reported to be dumped into molten wax, which then cooled and solidified, thereby protecting the reduced catalyst from exposure to air. At the plant site, the wax is melted and the catalyst–wax slurry transferred to the reactor.

9.8 Catalyst Attrition

Fines resulting from the cobalt/alumina catalyst used in a slurry bed reactor in the Oryx plant in Qatar were dealt with through modifications of the FT reactor operating parameters, and some small modifications made during the first shutdown [32].

9.9 Addendum Recent Literature Summary

A number of papers have appeared that describe some recent aspects of cobalt catalysts, their preparation [33–38], and catalyst preparation in general [39, 40]. Oukaci *et al.* [41] have reviewed the patent literature of FT cobalt catalysts; Zhang *et al.* [34] have reviewed the design of Co catalysts, while Zhang *et al.* [42] reviewed novel catalysts for the Fischer–Tropsch catalysts.

References

1 Gitzen, W.L. (ed.) (1970) *Alumina as a Ceramic Material*, The American Ceramic Society, Columbus, OH.
2 Pines, H. and Haag, W.O. (1960) *J. Am. Chem. Soc.*, **82**, 2471–2483.
3 Haag, W.O. and Pines, H. (1960) *J. Am. Chem. Soc.*, **82**, 2488–2494.
4 Pines, H. and Brown, S.M. (1971) *J. Catal.*, **20**, 74–87.
5 Truebam, M. and Trasatti, S.P. (2005) *Eur. J. Inorg. Chem.*, 3393–3403.
6 Cuenya, B.R. (2010) *Thin Solid Films*, **518** (12), 3127–3150.

7 Chen, M.S. and Goodman, D.W. (2007) *Chem. Phys. Solid Surf.*, **12**, 201–269.
8 Chen, M.S. and Goodman, D.W. (2008) *J. Phys.: Condens. Mater.*, **20** (26), 264013/1–2064013/11.
9 Balcar, H. and Čejka, J. (2007) *Phys. Chem.*, **243**, 151–166.
10 Čejka, J. (2003) *Appl. Catal. A – Gen.*, **254**, 327–338.
11 Seki, T. and Onaka, M. (eds) (2010) Mesoporous alumina: synthesis, characterization, and catalysts, in *Advanced Nanomaterials*, Wiley-VCH Verlag GmbH, Wienheim, pp. 481–521.
12 Borg, Ø., Eri, S., Blekkan, E.A., Storsæter, S., Wigum, H., Rytter, E., and Holmen, A. (2007) *J. Catal.*, **248**, 89–100.
13 Anderson, R.B., McCartney, J.T., Hall, W.K., and Hofer, L.J.E. (1947) *Ind. Eng. Chem.*, **39**, 1618–1628.
14 Kresge, C.T., Leonowicz, M.E., Roth, W.J., Vartuli, J.C., and Beck, J.S. (1992) *Nature*, **359**, 710–712.
15 Beck, J.S., Vartuli, J.C., Roth, W.J., Leonowicz, M.E., Kresge, C.T., Schmitt, K.D., Chu, C.T.W., Olsek, D.H., Sheppard, E.W., McCullen, S.B., Higgins, J.B., and Schlenker, J.L. (1992) *J. Am. Chem. Soc.*, **114**, 10834–10843.
16 Huo, Q., Margolese, D.I., Ciesia, U., Feng, P., Gier, T.E., Sieger, P., Leon, R., Petroff, P.M., Schüth, F., and Stucky, G.D. (1994) *Nature*, **368**, 317–321.
17 Khodakova, A.Y., Bechara, R., and Griboval-Constant, A. (2003) *Appl. Catal. A – Gen.*, **254**, 273.
18 Martinez, A. and Prieto, G. (2009) *Top. Catal.*, **52**, 75–90.
19 Xu, B., Fan, Y., Zhang, Y., and Tsubaki, N. (2005) *AIChE J.*, **51**, 2068–2076.
20 Zhang, Y., Yoneyama, Y., Fujimoto, K., and Tsubaki, N. (2003) *Top. Catal.*, **26**, 129–137.
21 Zhang, Y., Shinoda, M., and Tsubaki, N. (2004) *Catal. Today*, **93–95**, 55–63.
22 Brenner, A.M., Adkins, B.D., Spooner, S., and Davis, B.H. (1995) *J. Non-Cryst. Solids*, **185**, 73.
23 Tauster, S.J., Fung, S.C., and Garten, R.L. (1978) *J. Am. Chem. Soc.*, **100**, 170–175.
24 Lok, C.M. (2004) Process for preparing cobalt catalysts on titania support, WO 204/028687 A1, April 8.
25 Zhang, Z., Brown, S., Goodall, J.B.M., Weng, X., Thompson, K., Gong, K., Kellici, S., Clark, R.J.H., Evans, J.R.G., and Darr, J.A. (2000) *J. Alloys Compd.*, **476**, 451–456.
26 Padmaja, P., Warrier, K.G.K., Padmanabhan, M., and Wunderlich, W. (2009) *J. Sol.-Gel. Sci. Technol.*, **52**, 88–96.
27 Geus, J.W. (2007) Production of supported catalysts by impregnation and (viscous) drying, in *Catalyst Preparation: Science and Engineering*, CRC Press, Boca Raton.
28 Jacobs, G., Das, T.K., Zhang, Y., Li, J., Racoillet, G., and Davis, B.H. (2002) *Appl. Catal. A – Gen.*, **233**, 263–281.
29 Vissage, J.L., Botha, J.M., Koortzen, J.G., Datt, M.S., Bohmer, A., van de Loosdrecht, J., and Saib, A.M. (2008) Catalysts, WO 2008/135939 A2, November 13.
30 Vissage, J.L. and Veltman, H.M. (2006) Producing supported cobalt catalysts for the Fischer–Tropsch synthesis, WO 2006/075216 A1, July 20.
31 Mart, C.J. and Nesklora, D.R. (2001) Slurry hydrocarbon synthesis with fresh catalyst activity increase during hydrocarbon production, US Patent 6,323,248 B1, November 27.
32 Venables, S. (2008) Oryx Technology status feedback, www.sasol.com/sasol_internet/downloads/oryx_site_visit24November2008_technology.pdf (November 24).
33 Ducreux, O., Rebours, B., Lynch, J., Roy-Auberger, M., and Bazin, D. (2009) *Oil Gas Sci. Technol.*, **64**, 49–62.
34 Zhang, J., Chen, J., Li, Y., and Sun, Y. (2002) *J. Nat. Gas Chem.*, **11**, 99–108.
35 Karaca, H., Safoniva, O.V., Chambrey, S., Fongarland, P., Roussel, P., Griboval-Constant, A., Lacroix, M., and Khodakov, A.Y. (2011) *J. Catal.*, **277**, 14–26.
36 Viswanathan, B. and Gopalakrishnan, R. (1986) *J. Catal.*, **99**, 342–348.
37 Ataloglou, T., Bourikas, K., Vakros, J., Kordulis, C., and Lycourghiotis, A. (2005) *J. Phys. Chem. B*, **109**, 4599–4607.
38 Borg, Ø., Eri, S., Blekkan, E.A., Stoorsaeter, S., Wigum, H., Rytter, E., and Holmen, A. (2007) *J. Catal.*, **248**, 89–100.

39 Bourikas, K., Kordulis, C., Vakros, J., and Lycourghiotis, A. (2004) *Adv. Coll. Int. Sci.*, **110**, 97–120.

40 Bourikas, K., Kordulis, C., and Lycourghiotis, A. (2006) *Catal. Rev.*, **48**, 363–444.

41 Oukaci, R., Singleton, A.H., and Goodman, J.G., Jr. (1999) *Appl. Catal. A – Gen.*, **186**, 129–144.

42 Zhang, Q., Kang, J., and Wang, Y. (2010) *Chem. Catal. Chem.*, **2**, 1030–1058.

10
Other FT Catalysts

Burtron H. Davis and Peter M. Maitlis

Synopsis

Only catalysts based on Fe, Co, and Ru are sufficiently active and/or selective for the production of higher hydrocarbons from the hydrogenation of CO in the FT-S. This chapter deals with other less-active metals. Rh can also catalyze FT-S when suitably promoted; this reaction also gives oxygenates. Ni is generally a methanation catalyst, though the synthesis of higher hydrocarbons has been reported in certain situations. Detailed investigations indicated that Co and Ru appear only to be active when present in the metallic state. However, metallic Fe is not stable under FT reaction conditions, but converts into iron carbides; this adds significant complexity to the characterization of the active iron catalyst. Although ruthenium is very active for FT-S, because of its expense and limited supply, it has been used primarily in laboratory studies to investigate mechanisms. In addition to the single-metal catalysts, studies on some mixed metal systems have also been reported: manganese is often a component of the mixed-metal catalysts.

10.1
Introduction

The most commonly used industrial FT catalysts are those based on Fe and Co, which were discussed in Chapters 8 and 9, respectively. Other metals that have been found to be active for FT-S include ruthenium, rhodium, and nickel. Salts of platinum, manganese, and some other metals have also been reported to be efficacious as promoters or additives. Catalysts for some CO hydrogenation reactions other than FT-S are covered in Chapter 6.

The early history is described in Section 1.7.2, when Sabatier and Senderens reported the formation of methane by CO hydrogenation over a nickel catalyst [1]. In the 1940s, Pichler [2] attempted a more general survey of FT catalysts and the temperature and pressure ranges for their use. The correlation in Figure 10.1 shows the catalysts, as both the metals and the oxides, as well as the products that

Greener Fischer-Tropsch Processes for Fuels and Feedstocks, First Edition. Edited by Peter M. Maitlis and Arno de Klerk.

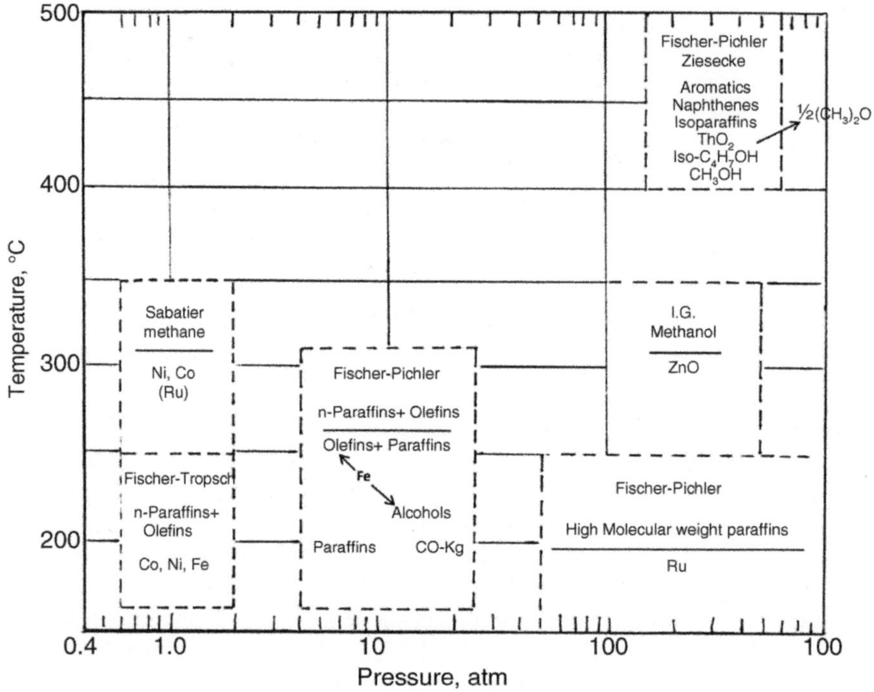

Figure 10.1 Relationship of catalyst type with respect to temperature and pressure. Redrawn from Ref. [1].

were reported as a function of operating conditions. This diagram has been reproduced in various forms in subsequent years largely without attribution.

10.2
Ni Catalysts

The nickel-catalyzed hydrogenation of CO to give methane or *methanation* was first reported by Sabatier and Senderens in 1902 [1]. In 1979, Vannice and Garten [3] working at Exxon found that the hydrogenation activity was modified if the nickel was supported on titania (Box 12.7). Methanation was not suppressed, but the activity was increased by comparison with nickel on alumina and the selectivity was changed in that more higher hydrocarbons (especially C_3) were formed. The difference in catalytic behavior between nickel on titania after reduction and Ni on alumina was complemented in the suppression of chemisorption of H_2 on titania. Other metals and some other oxides (mainly those that are more easily reducible) showed similar behavior, which came to be known as strong metal–support interactions (SMSI) (Chapter 12) [4]. Attempts made to improve the catalytic selectivity for higher carbon number products by alloying with a second metal such as Co or Fe have had little success and only low ($<C_5$) products were obtained.

10.3
Ruthenium Catalysts

10.3.1
Historical

Although not used industrially, ruthenium catalysts have received considerable attention in laboratory studies. The Ru FT catalysts were originally prepared by fusing 1 part commercial ruthenium powder with a mixture of 10 parts KOH and 1 part KNO_3. The fused mass was dissolved in water, methanol was added, and the solution was heated. The crude oxide precipitated and was dried, and then activated by reduction to the metal in flowing hydrogen.

Much of the early work was due to Pichler [5] who reported that Ru could be more active than either Fe or Co and gave large quantities of very high molecular weight products (≤100 000 Da) [5]. Very high pressures (1000 atm) but relatively mild temperatures (≤140 °C) were used [6]. At 300 °C and atmospheric pressure, the activity of the Ru catalyst was high, but only methane was formed [6]. At 180 °C, the conversion was zero at atmospheric pressure, but increased with increasing pressure so that at 1000 atm the CO conversion was 92%; furthermore, as the pressure was increased, the average molecular weight of the products increased (Table 10.1).

In the late 1970s, extensive studies were carried out by both industrial and academic groups on CO hydrogenations homogeneously catalyzed by Ru carbonyls in solution and heterogeneously by metallic Ru on a support [7–16]. This was driven by a search for direct routes from syngas to oxygenate products (such as ethylene glycol (1,2-dihydroxyethane)) that offered higher added value than the normal FT hydrocarbon synthesis products. In part these endeavors were successful, but the high pressure/high temperature conditions that were required inhibited development of a commercial process.

Following the 1970s energy crisis, there was a large increase in interest on Ru catalysts (as shown by the jump in *Chemical Abstracts* citations) (Figure 10.2), which have remained at a reasonably high level since then.

Table 10.1 Synthesis of hydrocarbons on ruthenium catalysts at 180 °C and different pressures [4].

Pressure (atm)	CO conversion (%)	% Converted CO to		
		Paraffin	Liquid hydrocarbons	Gaseous hydrocarbons
1	0	—	—	—
50	48	46	33	21
100	68	53	31	16
1000	92	59	26	15

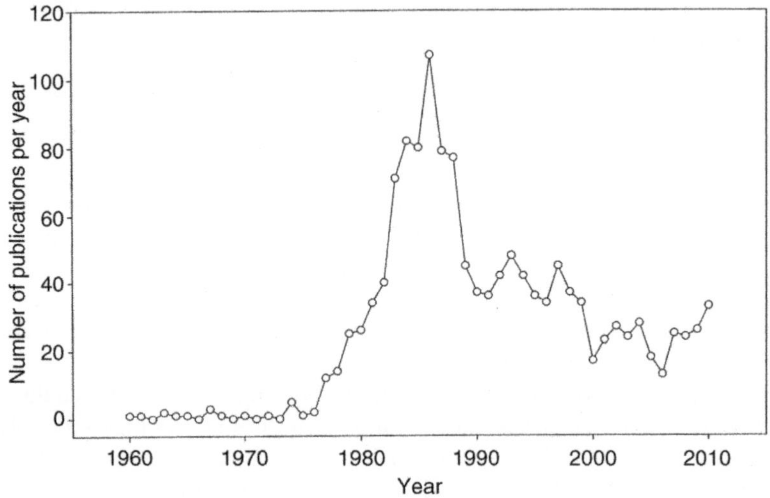

Figure 10.2 Number of publications per year for Ru catalysts reported for FT synthesis.

Acetaldehyde was found to be the principal oxygenated product formed over a silica-supported Ru catalyst, but methanol was the principal oxygenated species formed over an alumina-supported catalyst. IR spectroscopic studies showed that when the molecular ruthenium carbonyl, $Ru(CO)_5$, was present in THF solution, a *homogeneously* catalyzed CO hydrogenation reaction occurred under stringent conditions (CO: H_2, 2: 3, 268 °C, 1300 atm) giving methanol and methyl formate with >99% selectivities [11]. Carbon dioxide could also be converted at below 200 °C, but only methane was formed; however, CO appeared to inhibit that reaction [5].

10.3.2
Studies on Ru Catalysts

Although there have been many publications on Ru catalysts, there have unfortunately been few attempts to carry out wider studies. Publications in both the patent and the open literature have generally described reactions over individual metal catalysts and there have been rather few attempts to systematically explore conditions or optimize products. Some studies supplementing those described in Chapters 11 and 12 are summarized here.

Ru is an active FT catalyst and, as with the other catalysts, the FT activity seems largely to reside in the metal itself. CO hydrogenations based on various ruthenium starting materials have been investigated experimentally and theoretically [6–16]. Ligand complexes, other than the ruthenium carbonyls that decompose thermally to give the metal, show little or no FT activity. IR spectroscopic studies showed that when only $Ru(CO)_5$ was present in THF solution, a *homogeneously* catalyzed CO hydrogenation reaction occurred (CO: H_2, 2: 3, 268 °C,

1300 atm). However, only methanol and methyl formate were obtained (selectivities >99%) [6]. Decomposition of the homogeneous catalyst to metallic Ru was immediately accompanied by the formation of typical FT hydrocarbons and water among the products [9]. In summary, the data seem to imply that Ru complexes are easily reduced to metal under hydrogen and that metallic Ru is required for the FT reaction.

Some comparisons of FT-S reactivities at moderately elevated temperatures and 1 atm using metallic Fe, Co, Ru, and Rh catalysts on silica supports have been reported by the Sheffield group [17]. It was found that, to a first approximation, all four metals gave comparable distributions of organic products: largely methane, and *n*-1-alkenes plus some internal *n*-alkenes and some *n*-alkanes (Figure 12.2). This suggested that basically similar processes occurred on all the metals. The temperatures needed for roughly comparable product formation rates (pfr's) were Ru (150 °C), Co (180 °C), Rh (190 °C), and Fe (220 °C) indicating that Ru was indeed very active.

In important early mechanistic studies by the Shell group, Biloen and coworkers [18, 19] investigated the FT reaction using a labeled $^{12}CO/H_2$ feed over ^{13}C layers that had been deposited on Ni, Co, and Ru catalyst surfaces by the Boudouard disproportionation reaction ($2CO \rightleftharpoons CO_2 + C$). They found that the FT-S products included $^{13}CH_4$ as well as hydrocarbons containing several ^{13}C within one molecule. From this they concluded that oxygen-free species $\{CH_x\}$ ($x = 0$–3) were intermediates for methane production and were also incorporated into the growing hydrocarbon chains. Other studies in which the feed was rapidly switched between isotopomers, from $^{12}CO/H_2$ to $^{13}CO/H_2$, allowed them to calculate that the rate of C—C bond formation varied from $\geq 1\,s^{-1}$ for Ru to $\leq 0.1\,s^{-1}$ for Co. The data also indicated surface heterogeneity. These and similar results by others led to a decrease of enthusiasm for an oxygenate mechanism that had previously been advanced by Emmett and by Storch and the wider acceptance of a surface carbide mechanism.

Ekerdt and Bell found that under steady-state operation over Ru, the catalyst maintained a carbon reservoir with the magnitude dictated by temperature, the CO partial pressure, and the H_2/CO ratio [20, 21]. A major part of the carbon may be present on the support or attached to the Ru crystallites in the form of filaments. Ekerdt and Bell also found that alkenes cofed with the syngas were incorporated into the FT products: thus, ethene addition enhanced the formation of propene, while the addition of cyclohexene led to the formation of bicycloheptanes and alkylcyclohexanes (see also Section 12.3.3). The results implied that the added alkene reacted with species involved in chain propagation.

Henrici-Olivé and Olivé [22] summarized the results of CO hydrogenation studies involving ruthenium by various groups. Both hydrocarbons and oxygenates were found; however, the experimental conditions did not seem to be comparable and thus any conclusions had to be very tentative. These authors also presented some Exxon data [23] that showed that a Poisson curve fitted the data much better than the typical ASF distribution (Figure 10.3) characteristic of a C_1 polymerization mechanism.

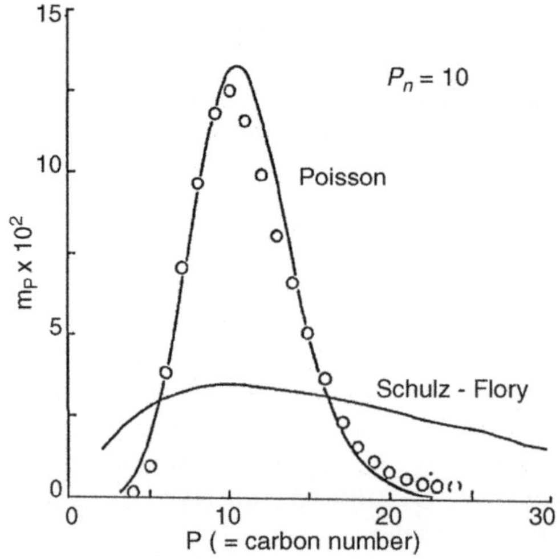

Figure 10.3 Theoretical Poisson and Schulz–Flory distributions for $P_n = 10$ (solid lines); (○) experimental data obtained at Exxon for a Ru catalyst, 30 atm, 241 °C. Reprinted with permission from Ref. [23].

IR Spectra Bell and his colleagues also noted that *in situ* IR spectra of CO on a Ru-SiO_2 catalyst showed that the surface is essentially completely covered by adsorbed CO [24]. Physisorbed and chemisorbed CO exchanged very rapidly compared to the steady-state rate of CO hydrogenation and, under reaction conditions, nearly a monolayer of hydrogen was adsorbed on the Ru surface [25, 26]. Two forms of surface carbon were identified under reaction conditions: C_α and C_β. The coverage of C_β grows during an initial period to become much larger than C_α. Steady-state methane and C_{2+} hydrocarbons formation were directly related to the surface concentration of C_α.

Particle Size Effects Co, Rh, and Fe crystallites below a certain size in the nanometer range have been reported to display a lower metal surface area-specific activity and a higher methane selectivity, smaller particles having lower activity per surface atom than larger particles [25–27]. González Carballo *et al.* [28] found that FT with Ru-based catalysts is a highly structure-sensitive reaction when Ru particles are less than 10 nm. The lower activity for particles below 10 nm may be related to stronger CO adsorption and partial blocking of surface sites.

Effects of Water on Ru-Catalyzed FT Synthesis Early work on Ru catalysts already showed that they were active in water and even in dilute acids [2]. It was also shown that the addition of water to the syngas feed can decrease the methane selectivity and increase chain growth and CO conversion (Figures 10.4–10.6) [29]. The effects in FT-S of Ru, Co, and Fe nanoparticles [30] that show higher activities and selectivities in water, ionic liquids, and high boiling organic solvents than can be obtained using conventional supported catalysts have been described. Xiao *et al.* [31]

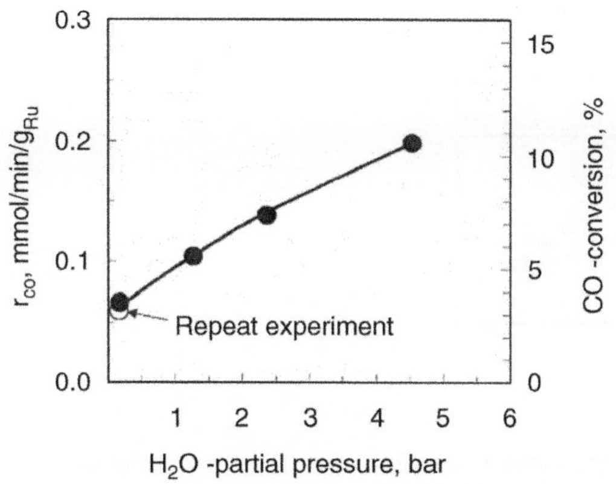

Figure 10.4 Rate of CO consumption (r_{CO}) and CO conversion as a function of water partial pressure. From Ref. [30].

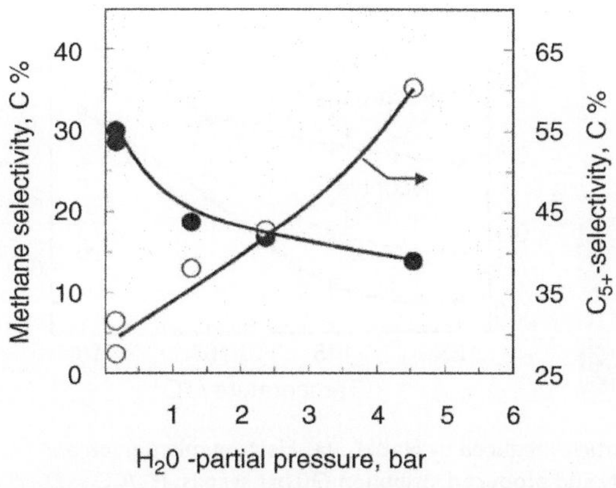

Figure 10.5 Methane selectivity and C_{5+} selectivity as a function of water partial pressure. From ref. [30].

reported FT-S using nanocluster Ru catalysts in an aqueous medium, where they obtained a rate that was 35 times higher than for a Ru-SiO_2 catalyst at 150 °C. Quek *et al.* [32] found a very high oxygenate selectivity (Figure 10.7) in a FT synthesis using poly-vinylpyrrolidone (PVP)-stabilized Ru nanoparticles made by aqueous borohydride reduction.

Liu *et al.* [33] found that in the presence of water, addition of Cl^-, OH^-, $H_2PO_4^-$, or HCO_3^- enhanced the activity for FT-S, while the addition of F^- decreased the activity. The addition of bromide or iodide caused a dramatic increase in the oxygenates.

ASF Distribution Some workers have reported deviations from the normal ASF product distribution [34]. For example, in the catalysts with Ru supported on a zeolite, there is a sharp decline in the products above C_7. The authors suggested

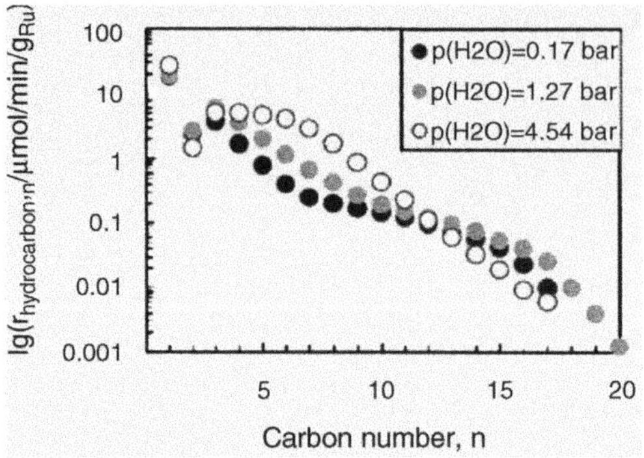

Figure 10.6 Anderson–Schulz–Flory plots: molar product formation rates (hydrocarbons) obtained at different water partial pressures. From Ref. [30].

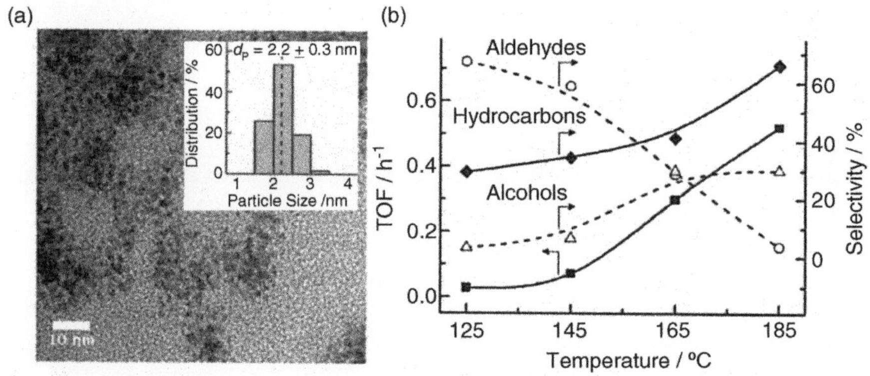

Figure 10.7 PVP-stabilized Ru nanoparticles reduced by $NaBH_4$. (a) Electron micrograph and particle size distribution. (b) FTS activity and product distribution (30 bar syngas; $H_2/CO = 2$). Reprinted with permission from Ref. [32].

that space constraints in the zeolite pores limited the chain growth to about seven carbon numbers. This has been disputed [35, 36] and it may be that the cutoff arises from some experimental shortcomings and not due to any particular catalyst or reactor design.

Bifunctional Catalysis Since oils heavier than transportation fuels are produced by many FT operations, researchers have sought to combine the FT and the hydrocracking reaction in the same reactor. An early experimental approach was conducted by workers at Mobil Oil. Caesar *et al.* [37] reported that the combination of an iron catalyst and a ZSM-5 zeolite could produce gasoline to over 60% of the total hydrocarbon product and essentially 100% of the liquid products. The Mobil workers interpreted their results as a polystep reaction following a bifunctional mechanism described by Weisz [38]. Thus, one function involved the generation of the ASF FT product distribution with the migration of these products through the gas/liquid phase to the acidic function where hydrocracking converted the FT primary product distribution into transportation range products. This approach that allowed two process units to be combined into a single stage unit was also used by Huang and Haag [39]. Higher temperatures favor aromatics production but also increase the fraction of the gaseous hydrocarbons.

The Mobil workers later switched to operating each catalyst in its own reactor to allow the conditions to be optimized for each reactor and for each catalyst. In addition, water and light products could be removed prior to adding the FT products to the acid-catalyzed reactor. Mobil evaluated their process at the pilot plant stage [40, 41], when they found that, in addition to the problems of water for the single reactor, alkali migrated from the FT catalyst to the acidic zeolite, which decreased the catalytic activity of the zeolite component with time on stream. This problem was also noted by Sasol workers [42]. Martínez *et al.* [43] recently reported that the CO also inhibited the acid-catalyzed reaction in the single reactor operation. In most recent reports of the bifunctional catalyst operation, the run length was only a few hours, which does not permit one to define whether the long-term operation is viable. Thus, while there continues to be interest in combining the synthesis and upgrading reactions using bifunctional catalysts in a single reactor, there remain considerable problems. Subiranas [44] considered the results and concluded that although there was potential for combining the two reactions in one reactor, there were many difficulties. Unexpectedly the most deleterious effect was that of CO on the cracking reactions, requiring recycle of the unconverted gas. Thus, it seems likely that the final product will be a mixture of gasoline and diesel until a diesel-selective hydroprocessing catalyst is found.

10.4 Rhodium Catalysts

In contrast to the situation with ruthenium where numerous papers have been published on different aspects but which seem unable to illuminate solutions that could inspire a new industrial process, work on rhodium-catalyzed CO hydrogenation reactions has been much more focused, despite the huge expense of the metal. This is largely because early studies showed that ethanol (methanol and other oxygenates) were important products in addition to the FT hydrocarbons [45, 46]. The search for high-selectivity reactions has also spurred some serious mechanistic studies, most recently by Goodwin and his colleagues [47] using “multiproduct steady-state isotope transient kinetic analysis.” These led to the conclusion that quite different surface sites were active for the production of methane, methanol, and ethanol. Rhodium by itself seems to have rather low activity for CO hydrogenation and, therefore, many promoters have been used, including cobalt, iron, vanadium, and rare earth oxides such as ceria.

10.5
Other Catalysts and Promoters

Manganese Kölbel reported that FeMn catalysts had a high selectivity for light olefins [48] when greater than 50% Mn was present in the catalyst. Subsequent studies with FeMn indicated that the H_2-activated catalyst is a mixture of spinels $Mn_xFe_{(3-x)}O_4$ and wistites (Fe, Mn)O, as well as metallic iron. Lohithharn and Goodwin [49] showed that the addition of Mn and/or K increased the concentration of active surface intermediates leading to product, which appeared to be a primary cause for the high catalytic activity observed in the K-promoted Fe and FeMn catalysts [50].

Kugler *et al.* [51] reported that supporting Ru on manganese oxide or other manganese-containing oxides gave an improved catalyst for the selective synthesis of low molecular weight, particularly C_2 and C_3, olefins. Nurunnabi *et al.* [52] found that Ru-Mn-γ-Al_2O_3 catalysts exhibited a high resistance to deactivation. These authors reported that the Mn removed Cl^- from the $RuCl_3$ used to prepare the catalyst and this led to easier formation of Ru with moderate particle size. Murata *et al.* [53] indicated that the addition of Mn and Na to a Ru–alumina catalyst was effective in raising the initial activity and C_{5+} selectivity, but after 20 h, the performance of the promoted catalyst was similar to that of an unpromoted catalyst. They demonstrated that agglomeration of metallic Ru occurred during the reaction.

While it has been demonstrated that Mn improves the activity of the Ru–alumina catalyst, the reports do not provide sufficient data to permit a decision on whether there is a promotion effect or a partial or complete elimination of chloride poisoning of the catalyst.

Den Breejen *et al.* [54] reported a highly active and selective Mn-promoted Co–silica catalyst, while Kinse *et al.* [55] found that for a Co–silica catalyst, the addition of Mn led to an increase in the primary formation of olefins and a decrease in the formation of paraffins. C_2–C_4 olefins underwent reincorporation into hydrocarbon chain growth at 10 atm operating conditions.

In summary, Mn has a significant impact on the three main FT metal catalysts, but the way in which Mn imparts this influence still needs to be defined.

References

1 Sabatier, P.S. and Senderens, J.D. (1902) *C. R. Chim*, **134**, 514–689.
2 Pichler, H. (1947) *Synthesis of hydrocarbons from carbon monoxide and hydrogen* (translated by R. Brinkley), US Department of Interior, Bureau of Mines Special Report, Pittsburgh, PA.
3 Vannice, M.A. and Garten, R.L. (1979) *J. Catal.*, **56**, 236.
4 Tauster, S.J. (1987) *Acc. Chem. Res.*, **20**, 389.
5 Pichler, H. (1938) *Brennst. Chem.*, **19**, 226.
6 Pichler, H., Firnhaber, B., Kioussis, D., and Dawallu, A. (1964) *Macromol. Chem.*, **70**, 12.
7 Pichler, H. (1952) *Adv. Catal.*, **4**, 271.
8 Fischer, F. and Pichler, H. (1933) *Brennst. Chem.*, **14**, 306.

9 Bradley, J.S. (1979) *J. Am. Chem. Soc.*, **101**, 7419.
10 Dombek, B. (1980) *J. Am. Chem. Soc.*, **102**, 6855.
11 Keim, W., Berger, M., and Schlupp, J. (1980) *J. Catal.*, **61**, 359.
12 Kellner, C.S. and Bell, A.T. (1981) *J. Catal.*, **67**, 175.
13 Kellner, C.S. and Bell, A.T. (1981) *J. Catal.*, **71**, 288.
14 Kellner, C.S. (1981) *The mechanism and kinetics of Fischer–Tropsch synthesis over supported ruthenium catalysts*. PhD thesis. University of California, Berkeley.
15 Bell, A.T. (1981) *Catal. Rev. Sci. Eng.*, **23**, 203.
16 Knifton, J.F. (1981) *J. Am. Chem. Soc.*, **103**, 3960.
17 Turner, M.L., Long, H.C., Shenton, A., Byers, P.K., and Maitlis, P.M. (1995) *Chem. Eur. J.*, **1**, 549.
18 Biloen, P., Helle, J.N., and Sachtler, W.M. H. (1979) *J. Catal.*, **58**, 95.
19 Zhang, X. and Biloen, P. (1986) *J. Catal.*, **98**, 468.
20 Ekerdt, J.G. and Bell, A.T. (1979) *J. Catal.*, **58**, 170.
21 Ekerdt, J.G. and Bell, A.T. (1980) *J. Catal.*, **62**, 19.
22 Henrici-Olivé, G. and Olivé, S. (1984) *J. Mol. Catal.*, **24**, 7.
23 Madon, R.J. (1979) *J. Catal.*, **57**, 183.
24 Kellner, C.S. and Bell, A.T. (1981) *J. Catal.*, **71**, 296.
25 Winslow, P. and Bell, A.T. (1984) *J. Catal.*, **86**, 158.
26 Winslow, P. and Bell, A.T. (1984) *J. Catal.*, **86**, 142.
27 Cant, N.W. and Bell, A.T. (1982) *J. Catal.*, **75**, 257.
28 González Carballo, J.M., Yang, J., Holmen, A., García-Rodríguez, S., Rojas, S., Ojeda, M., and Fierro, J.L.G. (2011) *J. Catal.*, **284**, 102–108, 29/37.
29 Fischer, F., Pichler, H., and Lohmar, W. (1939) *Brennst. Chem.*, **20**, 247–250.
30 Claeys, M. and van Steen, E. (2002) *Catal. Today*, **71**, 419.
31 Xiao, C.-X., Cai, Z.-P., Wang, T., Kou, Y., and Yan, N. (2008) *Angew. Chem., Int. Ed.*, **47**, 746.
32 Quek, X.-Y., Guan, Y., van Santen, R.A., and Hensen, E.J.M. (2011) *ChemCatChem*, **3**, 1735.
33 Liu, L., Sun, G., Wang, C., Yang, J., Xiao, C., Wang, H., Ma, D., and Kou, Y. (2012) *Catal. Today*, **183**, 136.
34 Wang, C., Zhao, H., Wang, H., Liu, L., Xiao, C., and Ma, D. (2012) *Catal. Today*, **183**, 143.
35 Gual, A., Godard, C., Castillón, S., Curella-Ferré, C., and Claver, C. (2012) *Catal. Today*, **183**, 154.
36 Nicolaiu, J., D'Hont, M., and Jungers, J.C. (1946) *Bull. Soc. Chim. Belg.*, **55**, 160.
37 Caesar, P.D., Brennan, J.A., Garwood, W.E., and Ciric, J. (1979) *J. Catal.*, **56**, 274–278.
38 Weisz, P.B. (1962) *Adv. Catal.*, **13**, 137–190.
39 Huang, T.J. and Haag, W.O. (1981) *Catalytic Activation of Carbon Monoxide*, *ACS Symposium Series*, vol. **152** (ed. P.C. Ford), American Chemical Society, pp. 306–323.
40 Kuo, J.C. (1985) *Two-stage process for conversion of synthesis gas to high quality transportation fuels*. Final Report, DOE/PC/60019-9, October 1985.
41 Kuo, J.C. (1985) *Two-stage process for conversion of synthesis gas to high quality transportation fuels*. Final Report, Appendix, DOE/PC/60019-9, October 1985.
42 Botes, F.G. and Böhringer, W. (2004) *Appl. Catal. A – Gen.*, **267**, 217–225.
43 Martínez, A., Rollán, J., Arribas, M.A., Cerqueira, H.S., Costa, A.F., and Falabella, E. Aguiar, S. (2007) *J. Catal.*, **249**, 162.
44 Subiranas, A.M. (2009) *Combining Fischer–Tropsch synthesis (FTS) and hydrocarbon reactions in one reactor*. PhD thesis. University of Karlsruhe.
45 Ichikawa, M.J. (1978) *Chem. Soc. Chem. Commun.*, 566.
46 Bowker, M. (1992) *Catal. Today*, **15**, 77.
47 Gao, J., Mo, X., and Goodwin, J.G. (2010) *J. Catal.*, **275**, 211.
48 Kölbel, H. and Tillmetz, K.D. (1979) Hydrocarbons and oxygen-containing compounds and catalysts thereof, US Patent 4,177,203.
49 Lohitharn, N. and Goodwin, J.G., Jr., (2008) *J. Catal.*, **260**, 7–16.

50 Ribeiro, M.C., Jacobs, G., Pendyala, R., Davis, B.H., Cronauer, D.C., Kropf, A.J., and Marshall, C.L. (2011) *J. Phys. Chem. C*, **115**, 4783–4792.

51 Kugler, E.L., Tauster, S.J., and Fung, S.C. (1980) US Patent 4,206,134.

52 Nurunnabi, M., Murata, K., Okabe, K., Inaba, M., and Takahara, I. (2008) *Appl. Catal. A – Gen.*, **340**, 203–211.

53 Murata, K., Okabe, K., Takahara, I., Inaba, M., and Saito, M. (2007) *React. Kinet. Catal. Lett.*, **90**, 275–283.

54 den Breejen, J.P., Frey, A.M., Yang, J., Holmen, A., van Schooneveld, M.M., de Groot, F.M.F., Stephan, O., Bitter, J.H., and de Jong, K.P. (2011) *Top. Catal.*, **545**, 768.

55 Kinse, A., Aigner, M., Ulbrich, M., Johnson, G.R., and Bell, A.T. (2012) *J. Catal.*, **288**, 104s.

11
Surface Science Studies Related to Fischer–Tropsch Reactions

Peter M. Maitlis

Synopsis

An overview is given of various techniques, many of them spectroscopic, that have been used to investigate the properties of surface species, especially those arising from the chemisorption of carbon monoxide and hydrogen. The formation of the catalytically active (chemisorbed) surface species, their spectroscopic properties, structures, and reactivities are considered. Since the surface species have only been partially defined, they are written in curly brackets, for example, {CH}, {H}, and so on.

11.1
Introduction: Surfaces in Catalysts and Catalytic Cycles

The Fischer–Tropsch reaction, as we use the term here, converts carbon monoxide and hydrogen into linear 1-alkenes and alkanes. It is commonly practiced by passing the reactants (CO and H_2) over a solid metallic catalyst, in either a two-phase gas–solid or in a three-phase gas–liquid–solid system. The catalyst is normally in the solid phase and thus investigations of the overall process must at some stage require an understanding of what goes on at the interface of the solid and the gas or liquid phase containing the reactants. Achieving a deeper understanding of the Fischer–Tropsch and other heterogeneously metal-catalyzed reactions has been slow because the tools to study such phenomena in depth have only become available in recent years.

The reason why metal-catalyzed processes are more difficult to study than many other chemical reactions is twofold: they involve at least two and often three or more phases, and the catalyst itself is often quite non-uniform, composed of surfaces of varying compositions and varying structures that may be present only in small amount. Thus, the first task facing an investigation is to ensure that the conditions are carefully defined so that reproducible catalytic reactions can be obtained. This is not nearly as straightforward as might be supposed since minute traces of impurities – sometimes at the barely detectable ppb level – can have

Greener Fischer-Tropsch Processes for Fuels and Feedstocks, First Edition. Edited by Peter M. Maitlis and Arno de Klerk.

dramatic effects on rates and selectivities. As a consequence, the early literature especially sometimes cites reactions that can be difficult to reproduce.

Heterogeneously catalyzed reactions have been classified into those where the rate is structure sensitive and those where it is not. Different sized nanoparticles will have many different surfaces available for adsorption and reaction. Since surface studies have shown that different surfaces (effectively crystal faces) have different activities, on the macroscale one may expect variations in activity with particle size. Structure-sensitive reactions are significantly affected by changes in the physical state of the catalyst, such as the particle size. For example, the hydrogenolysis of ethane to methane ($C_2H_6 + H_2 \rightarrow 2CH_4$) is structure sensitive, but the hydrogenation of ethene to ethane ($C_2H_4 + H_2 \rightarrow C_2H_6$) is not. The situation for CO hydrogenation is not clear, but it is usually regarded as structure insensitive.

There is good evidence, usually spectroscopic, identifying some of the organic species involved in catalytic processes that have been detected on metal surfaces and this is discussed in sections 11.2–11.3. However, definitive characterizations of the species involved in a catalytic cycle are still rare and this is an area where research is urgently required (see Sections 12.12 and 16.5).

11.2 Heterogeneous Catalyst Characterization

A whole raft of specialist techniques has been developed to try and cope with the problems involved in defining a heterogeneous catalyst. Most use spectroscopic or diffraction techniques, but quite a lot of attention has been focused on theoretical modeling, using either the *ab initio* or the DFT (*density functional theory*) approaches. A further frequently used tool involves models based on the structures, properties, properties and reactivities of molecular metal complexes, which are very much easier to define and study. These approaches will be briefly reviewed and their uses described in specific cases in Sections 11.3.1–11.3.5.

11.2.1 *Diffraction Methods*

Diffraction methods represent the ultimate techniques for defining a surface and species adsorbed thereon. If the material to be investigated is crystalline and uniform, XRD (X-ray diffraction), neutron diffraction (ND), and LEED (low-energy electron diffraction) can be used to find out where the atoms are in a crystalline bulk solid.

11.2.2 Spectroscopic Methods

A number of spectroscopic techniques have been developed to study surfaces:

Vibrational spectroscopy Infrared (IR) transmission spectra can easily be obtained for liquid (solution)- or gas-phase systems, today they are usually measured by Fourier transform infrared spectroscopy (FTIR). This relatively straightforward technique can also be used for some solid-state samples. However, the examination of surfaces is best carried out using reflected radiation; thus, diffuse reflection infrared spectroscopy (DRIFTS) is used for the *in situ* IR analysis of non-transparent materials and also for measurements at higher temperatures and pressures.

Raman spectra measure the radiation scattered at angles to the incident radiation and can provide information complementary to IR spectra. Raman spectroscopy of solid materials has been used to define bulk solids, but it suffers from the disadvantages of relatively low sensitivity since it depends on scattered radiation. Modifications to overcome this are being investigated.

New techniques are also constantly being developed in order to probe more fully what goes on at a catalyst surface [4, 5]. One of these combines atomic force microscopy (AFM) and vibrational spectroscopy as in AFM tip-enhanced Raman spectroscopy. Another, known as sum frequency generation (SFG), is especially useful under the conditions of much higher temperatures and pressures used for an operating catalyst (as in "sum frequency generation and polarization–modulation infrared reflection absorption spectroscopy"). This is in contrast to the ultrahigh vacuum (UHV) (10^{-9} Torr, $\sim 10^{-12}$ atm) that is normally needed for many techniques.

HREELS (high-resolution electron energy loss spectroscopy) The inelastic scattering of electrons from surfaces is utilized to study electronic excitations or vibrational modes of the surface or of molecules adsorbed at a surface. HREELS deals with small energy losses in the range of 10^{-3} to 1 eV, in contrast to other electron energy loss spectroscopies (EELS).

Photoelectron spectroscopy (PES) or X-ray photoelectron spectroscopy (XPS), also known as ESCA (electron spectroscopy for chemical analysis), is a quantitative surface chemical analysis technique that can determine the elemental composition, empirical formula, and chemical and electronic states of the elements present in a material.

The more sophisticated EXAFS (extended X-ray absorption fine structure) spectroscopy analyzes the X-ray absorption spectrum using very high-energy synchroton radiation. The chemical environment of an element can be measured in terms of the number and type of its neighbors, and the interatomic distances and structural disorders, within a 4–8 Å radius from the element, can be calculated. The nature of the surface is then extrapolated from the knowledge of the structure of the bulk material. EXAFS can be used for amorphous powders. Auger electron spectroscopy (AES) is based on the analysis of energetic electrons emitted from an excited atom after a series of internal relaxation events AES is used for analyzing different atoms in a surface.

11.2.3
Microscopy Techniques

The above spectroscopic methods give total or average information about a surface. To investigate specific areas, especially of nonuniform surfaces, techniques involving *microscopy* are commonly used. They allow examination of particular aspects of a surface, frequently pinpointing where reaction is believed to occur.

They include high-resolution electron microscopy (HREM) that yields images obtained with a transmission electron microscope (TEM). Scanning tunneling microscopy (STM) is used for direct real-space imaging of atoms, molecules, and adsorbate structures on surfaces [6]. Atomic force microscopy (AFM) or scanning force microscopy (SFM) is a very high-resolution scanning probe microscopy, with resolution of fractions of a nanometer, more than 1000 times better than the optical diffraction limit. AFM is one of the foremost tools for imaging, measuring, and manipulating matter at the nanoscale. The information is gathered by "feeling" the surface with a mechanical probe. Piezoelectric elements that facilitate tiny but accurate movements on (electronic) command enable the very precise scanning.

11.2.4 Molecular Metal Complexes as Models

Mechanisms of surface reactions such as those involved in heterogeneous catalyses are still very much the subjects for experimentation and discussion with only a few clear signposts. By contrast, many metal-based catalytic reactions in *homogeneous* (liquid and solution) phase are now reasonably well understood, founded on the knowledge of the structures of the molecular or ionic reactants involved and the kinetics of the processes. Though even these have not been easy to study as the highly reactive intermediates are normally present only in small amounts and are short-lived, their investigation has provided a good basis for mechanistic discussion in a chemical sense. Even though the results cannot be directly applied to complex multistep reactions such as the Fischer–Tropsch, they offer valuable guidance concerning the reaction pathways.

It should also be noted that the actual active catalytic entity is normally *not* the material added to the reaction mix initially (sometimes called the *precatalyst*), but arises from it by some preliminary transformation, often involving the loss of one or more (protecting) ligands. However, the precatalyst can be defined, and knowledge of its basic chemistry will indicate how the actual catalytic intermediates may be derived from it.

Studies of what goes on between the first step, when the reagents are introduced to the reaction, and the final step, when the products are liberated, are aided by the presence of good "handles" on the reactants and the intermediates, such as strong signature bands for spectroscopy. Some of the best-understood catalyzed reactions are the carbonylations in which carbon monoxide (CO) is added to an organic substrate mediated by a soluble transition metal complex.

As an example, we will consider the carbonylation of methanol to acetic acid. This is known as the Monsanto-BP process (Box 11.1), which is carried out on an industrial scale using a rhodium/iodide catalyst at about 20–50 atm CO and at 160–180 °C. The related BP-Cativa process uses an iridium–ruthenium/iodide catalyst under similar conditions and has the advantage that it can be run at very low water

Box 11.1 Simplified representation of the cycles involved in the rhodium and iodide homogeneously catalyzed carbonylation of methanol to acetic acid (Monsanto-BP process) [7, 8]

The cycle starts with the reaction of methanol with HI to give MeI (iodomethane, methyl iodide) (Equation 11.1); this then adds in an *oxidative addition* reaction to a low-valent rhodium(I) anionic complex $[Rh(CO)_2I_2]^-$ to give the anionic methyl-rhodium(III) complex $[MeRh(CO)_2I_3]^-$ (Equation 11.2). In the next step, the dicarbonyl complex $[MeRh(CO)_2I_3]^-$ reacts with CO present in the reaction solution to give the acetyl complex $[MeCORh(CO)_2I_3]^-$ (Equation 11.3), which then loses acetyl iodide (*reductive elimination*) (Equation 11.5) to regenerate the initial complex $[Rh(CO)_2I_2]^-$ that can start the cycle again. The acetyl iodide is hydrolyzed to acetic acid and HI (Equation 11.4). It may be noted that steps represented by Equations 11.1 and 11.5 are organic, while steps Equations 11.2–11.4 involve the active anionic metal intermediate species.

$$HI + MeOH \rightarrow MeI + H_2O \tag{11.1}$$

$$[Rh(CO)_2I_2]^- + MeI \rightarrow [Rh(Me)(CO)_2I_3]^- \tag{11.2}$$

$$[Rh(Me)(CO)_2I_3]^- \rightarrow [Rh(COMe)(CO)I_3]^- \tag{11.3}$$

$$[Rh(CO)Me(CO)I_3]^- + CO \rightarrow [Rh(CO)Me_2I_3]^- \rightarrow [Rh(CO)_2I_2]^- + MeCOI \tag{11.4}$$

$$MeCOI + H_2O \rightarrow MeCOOH + HI \tag{11.5}$$

concentrations, thus avoiding energy-demanding and expensive distillations to isolate the desired anhydrous product.

The mechanistic studies were greatly helped by Fourier transform infrared spectroscopy (FTIR) that allowed the monitoring of the reaction *in situ*. This was possible since CO itself shows a very strong band at 2143 cm^{-1} in the IR spectrum and the various metal carbonyl complexes also show strong absorption bands in a region where not many other resonances occur, 1700–2150 cm^{-1}. The positions, multiplicities, and intensities of these ν(CO) bands are due to different species, which can be assigned to individual molecules using isotopic labeling. This information on the environment in which the CO finds itself can then lead to the construction of a reaction scheme featuring plausible initial, intermediate, and final species, which can then be used with kinetic analyses to make testable predictions of the reaction under consideration.

Almost by definition, the active species in a catalytic cycle are present in low concentrations, thus the high intensity of the ν(CO) vibrations is particularly useful since it allows the detection of species present only in a very low amount. A simplified version of the Monsanto-BP cycle is presented in Box 11.1.

Other unsaturated species with characteristic infrared signatures similar to that of CO, including NO, acetylenes, some olefins, CS, and so on, have also been used to investigate catalytic reactions. Another very powerful technique for defining organic species in solution is nuclear magnetic resonance (NMR). This also has its counterpart in *solid state NMR* that allows the analysis of bulk solids. In both, the surroundings of specific atomic nuclei (such as ^{1}H, ^{13}C, ^{19}F, ^{31}P, and even ^{103}Rh) can be probed. However, for surfaces, NMR is a rather less sensitive technique and other methods are often preferred (see below).

It is now recognized that heterogeneously catalyzed reactions can also be analyzed in terms of cycles of elementary steps, in this case involving surface species. However, the actual structures and binding of the surface intermediates at the atomic level are still debated.

11.3
Species Detected on Surfaces

Metal surfaces are important in catalysis, electronics, environmental protection, and energy conversion processes. Nanoparticles of the metals are used in order to expose a large surface area to the reacting molecules from the gas or liquid phase. Nanoparticulate surfaces are often microcrystalline, containing well-defined planes together with steps, kinks, edges, and corners. Adsorbing molecules will bind to the various surface sites in different ways and with different energies.

A common way to find out what happens in a heterogeneous reaction is to use surface science measurements to define the surface structures and the species present thereon. The investigations often involve using techniques such as measuring the quantitative adsorption of nitrogen or of an inert gas to estimate total surface areas.

An important aspect of a surface reaction is the strength of binding of the adsorbate, and a number of methods have been developed to determine this. One is temperature programmed desorption (TPD), which relies on the fact that different species are adsorbed on surfaces with different binding strengths. Thus, they will desorb from the surface at different temperatures, which are characteristic for the species and surfaces concerned [9–11].

Many of the techniques used to probe solid-state structures have counterparts in solution- or gas-phase investigations, for example, FTIR and DRIFTS (see above). However, the surface techniques often require an ultrahigh vacuum as they rely on the detection of short-range electrons or ions emanating from the surface.

Many catalysts are microcrystalline, multicomponent, and nonuniform with surfaces made up of steps, kinks, and other dislocations. Figure 11.1 gives a representation of two ruthenium surfaces, one smooth and the other slightly corrugated with step, and kink dislocations. In addition, surfaces can also contain pores and channels within which species adsorb and reactions can occur. There has been

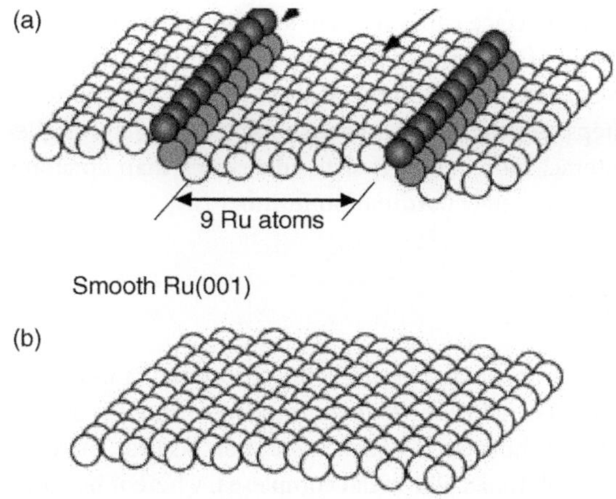

Figure 11.1 Representations of two metal surfaces. (a) Stepped (Ru109). (b) Smooth (planar, Ru(001)). After Ref. [12].

much interest in defining exactly where adsorbates bind and react. Fischer–Tropsch catalysts typically consist of a metal (Fe, Co, Ru, etc.) on an oxide support (such as alumina or silica) and there is considerable evidence that key reactions take place at metal–oxide interfaces, which must therefore be investigated.

Supported metallic nanoparticles in a high surface area catalyst can be viewed as consisting of a distribution of surfaces with different local geometries – different facets or edges, corners, steps, and kinks. It is important to know the number of *catalytically active* sites within a surface. The question of how many *active* sites exist within a given surface area can be addressed using model systems and also by measuring the uptake of some non-inert gases (such as CO) that adsorb in defined active sites. There has been a substantial increase in our understanding of the natures of active sites and of the structure dependence of surface reactions. Most significant, studies of single crystals have shown that the chemical activity at steps can be several orders of magnitude higher than that on close-packed (effectively *flat*) surfaces [13].

Two types of interaction between the adsorbate and the surface are commonly defined: *physisorption* and *chemisorption*. Physisorbed species are rather weakly bound (energies $\leq$30 kJ/mol) and are held on the surface largely by van der Waals or electrostatic forces, while in chemisorption true chemical bonds ($\geq$100 kJ/mol) are formed between the adsorbed molecules and atoms in the surface. The chemisorbed species are also activated toward reaction. In practice however, a continuum of bonding types is identified by spectroscopic and diffraction techniques.

It is also worth pointing out that the interaction between a given adsorbate and a surface is not a constant, but rather depends on the nature (topography) of the site at which adsorption occurs. Adsorption often changes the nature of the adsorbate, as shown, for example, by the decrease in bond order of carbon monoxide on coordination (see below). The opposite phenomenon is also found, since an adsorbate can modify the structure of a metal, giving a dynamical surface.

There are two ways in which the surface can affect the stability of reaction intermediates and the activation energy of a chemical reaction. One effect is electronic and the other effect is geometrical. Late transition metal atoms with a low coordination number (open surfaces, steps, edges, kinks, and corners) tend to have higher lying d-states and, therefore, interact more strongly with adsorbates than do atoms on close-packed surfaces with a high metal coordination number [14].

11.3.1
Carbon Monoxide on Surfaces {CO}

A frequently used probe for the study of metal surfaces is the adsorption of CO. Spectroscopic methods allow many useful conclusions to be drawn concerning the state of the CO and, in turn, of the nature of the sites at which adsorption occurs. The basis for this comes from the study of small molecular metal complexes, where it has been established by detailed single-crystal X-ray structural determinations that CO can be coordinated ("attached") to transition metals, M, in various ways that give rise to characteristic fingerprint bands in the carbonyl region of the infrared spectrum. Since the positions, intensities, and multiplicities can be correlated with specific structural features, the spectra have been used to probe the binding of CO on surfaces.

The simplest is a linear binding via the carbon of CO to a single-metal atom M–CO as in octahedral $Cr(CO)_6$ (Figure 11.2a). Here the infrared absorption band is typically in the region of 1850–2125 cm^{-1}, indicating a lower CO bond order in

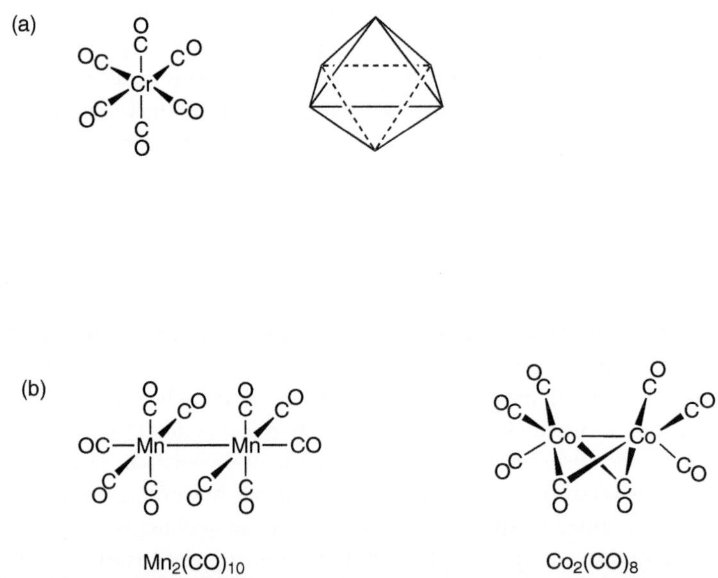

Figure 11.2 Structures of some molecular metal carbonyl complexes (diffraction determinations). (a) Mononuclear $Cr(CO)_6$, octahedral geometry, terminal carbonyls. (b) Dinuclear $Mn_2(CO)_{10}$ (octahedral metal atoms held by a metal–metal bond, terminal carbonyls); and dinuclear $Co_2(CO)_8$ with six terminal carbonyls and two carbonyls bridging the two cobalt atoms.

the metallic complex than that in CO itself where the triply bonded C≡O has ν(CO) at 2143 cm^{-1}. In complexes such as $Co_2(CO)_8$ (Figure 11.2b, right) where two CO's are found to bridge two metal atoms through the carbon, the IR band is shifted down still further (1750–1900 cm^{-1}). Organic carbonyl compounds, conventionally represented by a double bond (such as acetone, $Me_2C{=}O$), show ν(CO) lower still 1705–1725 cm^{-1}. Care must be taken in discussing these numbers as the precise positions of the bands depend on the environment of the CO and on the charge, as well as on the medium, as there can be differences between measurements carried out in the gas phase, solution, and solid. Pictorial representations of the bonding between the metal and CO are shown in Figure 11.3.

Very similar bands can be observed in the vibrational spectra when a metal surface is exposed to gaseous CO and bands in the region 2000–2130 cm^{-1} are ascribed to CO linearly bonded to surface atoms, while those in the regions 1960–2000 cm^{-1} and 1800–1920 cm^{-1} indicate CO bridging two and three surface metal atoms, respectively. Again, these are only ranges and the actual numbers will depend on the metal and also on the extent of surface coverage, which will depend on neighboring atoms.

11.3.2
Activation of CO

Chemisorbed {CO} readily cleaves on many metal surfaces (Box 11.2). By contrast, although there have been many thousands of publications on carbon monoxide as a

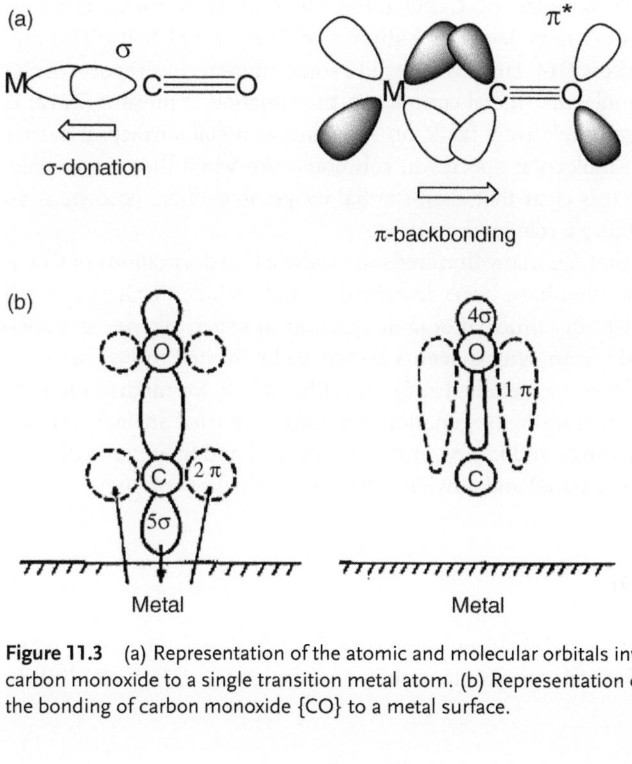

Figure 11.3 (a) Representation of the atomic and molecular orbitals involved in the bonding of carbon monoxide to a single transition metal atom. (b) Representation of the Blyholder model of the bonding of carbon monoxide {CO} to a metal surface.

Box 11.2 CO cleavage on metal surfaces

CO binds quite strongly to many d-block transition metals. On the atomic scale, the main binding is through the carbon, but binding to a second metal through the oxygen can also take place.

The adsorption of CO at surfaces generally leads to a medium–weak interaction in which the CO bond order is reduced and some metal–C bonding is established. In some cases, this can eventually lead to scission of the C—O bond and the establishment of metal–O and metal–C bonds. These stages are readily followed by changes in the vibrational spectra (IR, HREELS, etc.).

Other diatomic molecules, CO, N_2, NO, and O_2, can behave similarly on metal surfaces; the more to the left the metal lies in a given d-block transition series, the greater the tendency for such molecules to dissociate. Furthermore, the approximate borderline between dissociative and molecular adsorption moves to the left from the 3d to the 4d to the 5d series. Thus, CO dissociates readily on the 3d metals, Ti, V, Cr, and Mn; on the 4d metals, Zr, Nb, and Mo; and on the 5d metals, Hf and Ta. Surfaces of the 3d metals Fe and Co, the 4d metals Tc and Ru, and the 5d metals W and Re show intermediate character. CO is largely undissociated on the metals further to the right in those transition series. Similar situations have been evaluated for the adsorption of the diatomics, CO, N_2, NO, and O_2.

ligand, relatively few examples of C—O bond cleavage in a metal complex have been reported. Most notable are the studies of Shriver and Sailor [15] and Wolczanski and coworkers [16]. However, despite some more recent reports of CO cleavage reactions in molecular metal complexes, the situation at present (2012) is that while CO cleavage is relatively facile on a transition metal surface, it occurs only with difficulty on molecular species in solution, even when they are in polyatomic metal clusters. It is clear that a substantial cooperative effect (known as an ensemble effect) must play a role on the surface.

It is also significant that the many hundreds of catalytic transformations of CO in homogeneous solutions that have been described do not include Fischer–Tropsch reactions. The otherwise very unusual organic reaction in which alkenes and alkanes are formed directly from syngas seems largely to be limited to *heterogeneous* surface processes catalyzed by metallic Fe, Co, Ru, Rh, and Ni, for methanation. By contrast, if carbon monoxide hydrogenation reactions are tried in *homogeneous* solution, very high pressures and temperatures are needed and then generally give oxygenates, especially methanol and ethylene glycol (1,2-dihydroxyethane).

11.3.3 Transformations of {CO}

There have been several investigations exploring the CO cleavage process, many of which were performed on single-metal crystals under UHV conditions.

Furthermore, some of the metals that have been investigated (e.g., platinum) are not normally associated with FT activity, and it remains to be seen just how closely the results can be applied to catalytic reactions under ambient conditions or above.

It was noted early on that the susceptibility of the CO molecule to dissociation upon chemisorption varies systematically depending on the position of the metal in the Periodic Table. Similar results were obtained for other diatomic molecules (Box 11.2) [17, 18]. Thus, O_2 dissociates on all transition metals at room temperature and the close-packed surfaces are least active for dissociation, while steps and kinks may have extra activity.

Using SFG (sum frequency generation surface vibrational spectroscopy), CO was found to dissociate on Pt(111), Pt(557), and Pt(100) at 673, 548, and 500 K, respectively, under 40 atm. The CO on top site frequency shifted as a function of temperature and also with time, indicating the surface was being modified, probably due to the formation of volatile platinum carbonyls that moved platinum atoms around the surface lattice. For both the (111) and (100) surfaces of platinum, the crystal needed to be heated to produce the step and kink sites required for dissociation, with the Pt(111) surface exhibiting a much higher CO dissociation temperature compared to Pt(100) because it was more stable. Since the Pt(557) is essentially a (111) surface with steps already present in the structure, no extra heating was needed to produce the step and kink sites for CO dissociation [19].

By contrast, CO adsorbed on palladium (111) surfaces showed no CO dissociation in the range of 300–400 K [20, 21]. It was also found that the FT active metals, ruthenium and cobalt, needed pronounced crystallographic corrugation before becoming active for dissociation [22, 23].

While there is not necessarily any correlation between CO dissociation on a clean single-metal crystal and FT activity of the supported metal, the question as to whether the chemisorbed {CO} splits to give {C}, a surface carbide, and {O}, a surface oxide, or participates in a more complex process, with the {CO} being attacked by adsorbed {H}, has been hotly debated for a long time. In the dissociative path, both the {C} and the {O} then react with {H} to form methylidyne {CH} and {OH} and hydroxo (or water), respectively. By contrast, in the associative path, species such as {COH} and {CHOH} are postulated as intermediates that can then react further (see Sections 12.5.2 and 12.6).

There are organometallic model systems that can be related to each pathway (see Chapter 12). But while there is considerable evidence for the dissociative path on surfaces, reliable evidence for surface species such as {COH} and {CHOH} arising from an associative path does not seem to have been obtained as yet, and indeed some attempts to find them have failed [24, 25].

11.3.4
Hydrogen on Surfaces {H_2} and {H}

The modification of a molecule on adsorption varies from a small perturbation to complete breakup. A good example of the cleavage of a molecule on adsorption is provided by molecular hydrogen H_2, which is also readily adsorbed by metals.

Box 11.3 Activation of H_2

In the physisorption of H_2, the molecular orbital of the H_2 interacts with the surface electronic states to generate bonding and antibonding orbitals. The H_2 1σ and $2\sigma^*$ orbitals shift and broaden as they approach the surface; electron transfer from the metal to the H_2 occurs because the $2\sigma^*$ orbital drops in energy and broadens as the H_2 approaches the surface. As it drops, the Fermi energy electrons begin to populate the orbital and the H_2 bond weakens and eventually breaks, while the metal–H bonds become stronger. For optimum orbital overlap, the axis of the H_2 should be parallel to the surface [27]. The picture that emerges is very similar to that found for hydrogen in molecular metal complexes where two bonding patterns have been established: those involving the bonding of molecular dihydrogen to a metal atom, written as $M(\kappa^2{-}H_2)$, analogous to physisorption, and the σ-bonded dihydride, $M(\kappa^1{-}H)_2$, where the H—H bond has been replaced by two M—H σ-bonds; that is, close to the picture of chemisorption.

$$M(\kappa^2 - H_2) \rightleftarrows M(k^1 - H)_2$$

The change from $M(\kappa^2{-}H_2)$ to $M(\kappa^1{-}H)_2$ can also formally be regarded as equivalent to the oxidative addition of H_2 to the metal, and is in many cases quite readily reversible [28].

Initially it is physisorbed, which may imply a state similar to that binding a *dihydrogen* ligand in a metal complex where H_2 is sideways bonded to a transition metal in what is termed a σ (sigma)-complex. The next step can be cleavage of the coordinated $\{H_2\}$ to give two hydrides $\{H\}$ (Box 11.3). A very detailed physical analysis has been given [26].

These surface metal hydrides are generally regarded as responsible for the activity of metals in hydrogen exchange and hydrogenation reactions. A form of hydrogen known as spillover hydrogen also plays an important role [29]. Its character depends on the substrate to which it is attached and it can behave as $H^\bullet$ (atomic hydrogen), H^- (hydride ion), or $H^\oplus$ proton [28, 30].

11.3.5 Transformations of {H}

As the metal surfaces are dynamic on the nanoscale, adsorbed species will migrate easily. It has also been found experimentally that on surfaces, the hydrogen in H_2 exchanges very readily with D_2 (giving mixed HD) and also with C—H(D) and other molecules containing X—H(D) bonds. These reactions proceed via dissociation, $\{H_2\} \rightleftharpoons 2\{H\}$. While it is generally accepted that such reactions take place between two surface species, some workers have postulated direct reactions between a gaseous molecule and an adsorbed species.

Box 11.4 Strong metal–support interactions (SMSI)

Many catalytic systems consist of nanosized metal particles supported on oxides. It has been found that the interactions between metals and oxide supports, so-called *metal–support interactions*, are of great importance in heterogeneous catalysis. Strong metal–support interactions were first suggested by Tauster *et al.* to explain the suppression of both H_2 and CO chemisorption capacity of metal clusters supported on TiO_2 after reduction at high temperatures. Later, SMSI was widely observed in many metal/oxide catalytic systems. Two major factors contribute to the SMSI states, one is electronic and the other is geometric. The electronic factor is determined by a perturbation of the electronic structure of the metal catalyst, which originates from charge transfer between the metal and the oxide, while the geometric factor results from a thin layer of reduced oxide support physically covering the metal particles (called the *encapsulation* or *decoration* model), which blocks active catalytic sites at the metal surface (see Box 12.7) [31].

11.3.6 Reactions of {CO} and {H}

The reactions undergone by {CO} and {H} in surfaces are discussed in Chapter 12. One point that must be mentioned is the role that the support plays in reactions that occur at the metal surface. Since these are quite difficult situations to probe experimentally, there have been rather few studies. However, the effect of the support on the catalytic activity of the metal has been recognized for a long time; one of the first attempts to quantify the question led to the introduction of the concept of strong metal–support interactions (SMSI; Box 11.4), first suggested by Tauster to explain the suppression of both the H_2 and the CO chemisorption capacity of metal clusters supported on TiO_2 surfaces, which had been reduced at high temperatures. It was also found that SMSI effects could be used to promote FT reactions [31].

11.4 Theoretical Calculations

The great difficulties facing experimentalists studying the Fischer–Tropsch and other heterogeneous reactions have led many researchers to apply various theoretical approaches to the problem, in particular *DFT* (density functional theory) [32]. Aspects of the FT hydrocarbon synthesis that have been emphasized include (i) the natures of the reactive sites for hydrogenation and for C—C coupling steps; (ii) mechanisms involving methylidynes (surface carbynes), methylenes (surface carbenes), CO insertion, and hydroxyl-carbenes; (iii) methane formation, α-olefin selectivity, and chain growth probability; and (iv) catalytic activity.

Van Santen [33] has proposed a molecular theory of the structure sensitivity of catalytic reactions based on the computed activation energies of corresponding elementary reaction steps on transition metal surfaces. Thus, cleavage of molecules such as CO or N_2 requires a reaction center with a unique configuration of several metal atoms and step-edge sites, which physically cannot be present on transition metal particles less than 2 nm. This is called class I surface sensitivity, and the rate of reaction will sharply decrease when the particle size decreases below the critical size [33].

An alternative view based on DFT calculations has been offered by Jenkins and King [34]. They have computed the existence of a remarkably strong alkali-induced polarization of the C—O bond and have analyzed the consequences for understanding of the FT hydrocarbons synthesis (see also Section 12.6.2) [35].

References

1 Zaera, F. (2009) *Acc. Chem. Res.*, **42**, 1152–1160.
2 Ma, Z. and Zaera, F. (2006) *Surf. Sci. Rep.*, **61**, 229–281.
3 Zaera, F. (2003) *Catal. Lett.*, **91**, 1–10.
4 Vidal, F. and Tadjeddine, A. (2005) *Rep. Prog. Phys.*, **68**, 1095–1127.
5 Rupprechter, G. (2007) *Adv. Catal.*, **51**, 133–263.
6 Bowker, M. (2007) *Chem. Soc. Rev.*, **36**, 1656–1673.
7 Maitlis, P.M., Haynes, A., Sunley, G.J., and Howard, M.J. (1996) *J. Chem. Soc., Dalton Trans.*, 2187–2196.
8 Haynes, A., Maitlis, P.M., Morris, G.E., Sunley, G.J. *et al.* (2004) *J. Am. Chem. Soc.*, **125**, 2847–2661.
9 Thomas, J.M. and Thomas, W.J. (1997) *Principles and Practice of Heterogeneous Catalysis*, Wiley-VCH Verlag GmbH, Weinheim.
10 Somorjai, G.A. (1994) *Introduction to Surface Chemistry and Catalysis*, Wiley-Interscience, New York.
11 Kolasinski, K.W. (2002) *Surface Science*, John Wiley & Sons, Inc., New York.
12 Zubkov, T., Morgan, G.A., and Yates, J.T. (2002) *Chem. Phys. Lett.*, **362**, 181.
13 Nørskov, J.K., Bligaard, T., Hvolbæk, B., Abild-Pedersen, F., Chorkendorff, I., and Christensen, C.H. (2008) *Chem. Soc. Rev.*, **37**, 2163–2171.
14 Nørskov, J.K., Bligaard, T., Logadottir, A., Bahn, S., Hansen, L.B., Bollinger, M., Bengaard, H., Hammer, B., Sljivancanin, Z., Mavrikakis, M., Xu, Y., Dahl, S., and Jacobsen, C.J.H. (2002) *J. Catal.*, **209**, 275.
15 Shriver, D.F. and Sailor, M.F. (1988) *Acc. Chem. Res.*, **21**, 374–379.
16 Miller, R.I., Wolczanski, P.T., and Rheingold, A.L. (1993) *J. Am. Chem. Soc.*, **115**, 10422–10423.
17 Broden, G., Rhodin, T.N., Brucker, C., Benbow, R., and Hurych, Z. (1976) *Surf. Sci.*, **59**, 593–611.
18 Andreoni, W. and Varma, C.M. (1981) *Phys. Rev.*, **B23**, 437.
19 McCrea, K., Parker, J.S., Chen, P.L., and Somorjai, G. (2001) *Surf. Sci.*, **494**, 238–250.
20 Kaichev, V., Morkel, M., Unterhalt, H., Prosvirin, I.P., Bukhtiyarov, V.I., Rupprechter, G., and Freund, H.J. (2004) *Surf. Sci.*, **566**, 1024–1029.
21 Kaichev, V.V., Bukhtiyarov, V.I., Rupprechter, G., and Freund, H.J. (2005) *Kinet. Catal.*, **46**, 269–281.
22 Fan, C.Y., Bonzel, H.P., and Jacobi, K. (2003) *J. Chem. Phys.*, **118**, 9773–9782.
23 Kim, Y.K., Morgan, G.A., and Yates, J.T. (2007) *J. Phys. Chem. C*, **111**, 3366.
24 Zubkov, T., Morgan, G.A., Yates, J.T., Kuehlert, O., Lisowski, M., Schillinger, R., Fick, D., and Jaensch, H.J. (2003) *Surf. Sci.*, **526**, 57–71.
25 Morgan, G.A., Sorescu, D.C., Zubkov, T., and Yates, J.T. (2004) *J. Phys. Chem. B*, **108**, 3614–3624.

26 Christmann, K. (1988) *Surf. Sci. Rep.*, **9**, 1–163.

27 Kolasinski, K.E. (2004) Chemisorption, physisorption and dynamics, in *Surface Science*, John Wiley & Sons, Inc., New York.

28 Morris, R.H. (2008) *Coord. Chem. Rev.*, **252**, 2381–2394.

29 Conner, W.C. and Falconer, J.L. (1995) *Chem. Rev.*, **95**, 759–788.

30 Roland, U., Braunschweig, T., and Roessner, F. (1997) *J. Mol. Catal. A*, **127**, 61–84.

31 Fu, Q. and Wagner, T. (2007) *Surf. Sci. Rep.*, **62**, 431–498.

32 Cheng, J., Hu, P., Ellis, P., French, S., Kelly, G., and Lok, C.M. (2010) *Top. Catal.*, **53**, 326–337.

33 van Santen, R.A. (2009) *Acc. Chem. Res.*, **42**, 57–66.

34 Jenkins, S.J. and King, D.A. (2000) *J. Am. Chem. Soc.*, **122**, 10610–10614.

35 Inderwildi, O.R., Jenkins, S.J., and King, D.A. (2008) *J. Phys. Chem. C.*, **112**, 1305–1307.

12
Mechanistic Studies Related to the Fischer–Tropsch Hydrocarbon Synthesis and Some Cognate Processes

Peter M. Maitlis

Synopsis

Mechanisms are discussed to account for the hydrocarbon products and the way they are formed in the FT reaction. Recent results have led to the conclusion that dual mechanisms occur and that at least two pathways coexist. One, termed *dissociative,* is where the initial chemisorption step leads to dissociation of CO on the surface. In the other, termed *associative,* a direct reaction takes place between chemisorbed CO and hydrogen. It is proposed that the precise steps eventually leading to free hydrocarbon formation depend on the interactions between the surface hydrocarbyl species $\{C_xH_y\}$, the metal, and any support (or additive) (see Chapter 11). Polar surfaces may be expected to facilitate associative reaction paths involving electrophiles and/or nucleophiles, while steps involving more neutral species will be favored on nonpolar interfaces. The associative path can also account for the coproduction of some higher oxygenates in the Fischer–Tropsch synthesis (FT-S). However, the "low-pressure" methanol synthesis reaction seems to proceed by a quite different path (see Chapter 6).

12.1
Introduction

The Fischer–Tropsch synthesis (FT-S) is very unusual in the context of normal solution or gas-phase organic chemistry and must involve rather unique steps. As we will use the term here, the FT-S refers to the organic compounds that are formed when *syngas*, a mixture of carbon monoxide and hydrogen (see Chapter 2), is passed over certain late d-block transition metal catalysts. The chief products are *n*-alkenes, *n*-alkanes, methane, and *water.* The formation of water can be considered the thermodynamic "driving force" of the reaction. Depending on the catalyst, the support and promoters, temperatures and pressures, and even the type of reactor used, the FT-S also gives rise to small amounts of other organic compounds, including branched chain hydrocarbons (mainly monomethyl alkenes and monomethyl alkanes), oxygenates including higher alcohols, and even carboxylic acids (see Table 4.1). Several other metal-catalyzed reactions involving hydrogen and carbon monoxide,

Greener Fischer-Tropsch Processes for Fuels and Feedstocks, First Edition. Edited by Peter M. Maitlis and Arno de Klerk.

but giving quite different products, can also occur. These include methanol synthesis (Equation 12.3, and Chapter 6) and the water-gas shift reaction (WGSR) (Equation 12.4) (also see Section 2.3).

The thermodynamics of the CO hydrogenation reactions (Equations 12.1–12.3) indicate that the most favorable path is that leading to methane and water (Equation 12.1). It is therefore quite remarkable that the course of the reaction can be efficiently diverted to give either long-chain hydrocarbons and water (Equation 12.2) or methanol (Equation 12.3)

$$3H_2 + CO = H_2O + CH_4, \quad \Delta G_{(227°C)} - 96\,kJ/mol \tag{12.1}$$

$$2H_2 + CO = H_2O + 1/3(C_3H_6), \quad \Delta G_{(227°C)} - 31\,kJ/mol \tag{12.2}$$

$$2H_2 + CO = CH_3OH, \quad \Delta G_{(227°C)} + 27\,kJ/mol \tag{12.3}$$

$$H_2O + CO \rightleftharpoons H_2 + CO_2, \quad \Delta G_{(25°C)} - 28\,kJ/mol \tag{12.4}$$

The conditions under which all these reactions are carried out are broadly similar: they involve passing syngas over the heated metallic catalyst at a temperature usually between 200 °C and 350 °C and under moderate pressure. A catalyst giving largely methane (Equation 12.1) is nickel on alumina, while the classical FT long-chain hydrocarbon synthesis (Equation 12.2) is catalyzed by iron, cobalt, ruthenium, or rhodium, often on a silica or alumina support. The production of small amounts of higher oxygenates may also be part of that reaction matrix.

Methanol synthesis, also formed from syngas in a heterogeneously catalyzed process (over Cu + ZnO on alumina), could be written as part of this scheme (Equation 12.3). However, detailed investigations have indicated that it largely arises from carbon *di*oxide in a quite different sequence (see Chapter 6). The water-gas shift reaction (Equation 12.4) that relates hydrogen, water, carbon monoxide, and carbon dioxide is also very important since it is used industrially to make hydrogen.

Why should we study the mechanism? Since the FT-S is an integral part of the CTL, GTL, and BTL conversion technologies (Chapter 2), a better understanding of how the reaction works allows changes in parameters to be made that can improve the activity and selectivity.

Researchers in the field, right back to Fischer and Tropsch, have speculated on the mechanisms, and many experimental and theoretical investigations have been carried out with the aim of clarifying what happens on the catalyst surface. Classically, most information has been derived from kinetics measurements and full reaction product analyses, combined with evaluations of the changes in products arising from the addition of various probe molecules to the syngas stream.

However, these are rather indirect methods and some results, especially those reported in the early work, have been quite difficult to reproduce independently, in many cases due to small variations in the composition or morphology of the catalysts. Thus, the literature sometimes contains different (and even contradictory) results.

Current research in FT-S has the target of making the process more environmentally acceptable and sustainable. One aspect is to make the FT-S more selective, so that the major products are, for example, 1-alkenes of a very narrow range of sizes. Thus considerable effort has been devoted to optimizing the catalyst and reactor performance. To some extent, emphasis on engineering the FT-S has come at the cost of fundamental mechanistic studies.

Since there is so much information on the FT-S now available both in the open and in the patent literatures, we have had to be quite selective for this review of mechanisms; hence, we will be concerned mostly with offering a broad-brush picture of recent developments. Section 12.11 as well as Chapter 16 contains some thoughts for the future and a consideration of the directions that may most usefully be followed.

12.1.1 A Brief Background: Classical Views of the Mechanism

Since the original discovery of the FT-S by Fischer and Tropsch in 1925, there have been many publications on the reaction, and the ISI Web of Knowledge lists over 3000 scientific papers in the open literature alone between 1939 and 2011, many of which have included consideration of the mechanism. Because of the great industrial interest, there are also hundreds of patents on FT processes. Several monographs [1–4] have been written on the FT-S and although these deal mainly with the practical (engineering) side (see also Chapters 4 and 5), some useful mechanistic insights are also offered.

The complexity and heterogeneity of the FT-S system and some difficulties in consistently repeating reactions over accurately defined catalysts have limited unambiguous mechanistic conclusions. In addition to the overall kinetics and reaction product analyses, many studies have focused on what are presumed to be the individual steps of the process; these include surface science investigations, the use of organometallic reactions in solution as models, and most recently, *ab initio* and density functional theory (DFT) theoretical calculations.

Even some basic facts are not universally accepted. Thus, while most workers agree that 1-*n*-alkenes are the kinetically formed first products of the FT hydrocarbon synthesis under normal conditions, many have also reported that *n*-alkanes are *primary* products, notably over cobalt catalysts. However, cobalt is a good hydrogenation catalyst and it is difficult to ensure that the alkanes are indeed primary products of the FT-S and are not formed from the 1-*n*-alkenes in secondary reactions over the catalyst and the support. Oxygenates, especially *n*-1-alkanols, formed in small amounts, have also been described as primary products. These considerations also highlight the difficulty of defining a primary product (see below).

Although the main aim of research is to improve the activity and selectivity of the FT process, since catalysts are expensive and do deactivate over time, it is also important to minimize the loss or alteration of catalyst during the preparation, pretreatment, and actual reaction. Significant modifications of catalyst structure can and do occur during the reactor start-up and during FT synthesis since the performance of FT catalysts evolves with time on stream [5].

12.2
Basic FT Reaction: Dissociative and Associative Paths

The first ideas concerning the formation of long-chain hydrocarbons were advanced by Fischer and Tropsch in 1925 [6, 7], and later expanded notably by Biloen, Pettit, and others [8–11]. Although a number of serious problems with the Fischer–Tropsch–Pettit–Biloen (FTPB) theory have been highlighted [12], it is still widely quoted and elements have been incorporated into present-day thinking. The main steps involve CO cleavage to a surface carbide {C}, hydrogenation (eventually to methane), and the formation and coupling of surface hydrocarbyl $\{C_nH_x\}$ species, from which 1-alkenes are liberated by β-elimination.

To distinguish the FTPB from paths not involving initial CO cleavage, we refer to it as the *dissociative* (or *carbide*) path. Figures 12.1a–c are representations of the FTPB scheme for CO hydrogenation on metal surfaces; Figure 12.1a is a pictorial display of the activation of CO to give surface carbide and the formation of various

(a)

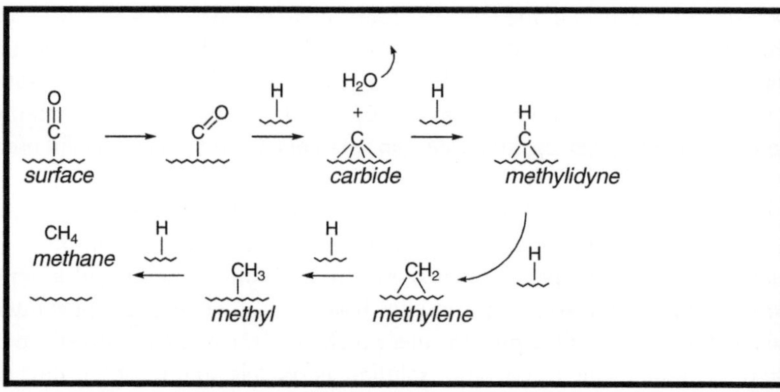

(b)

$\{CO\} + \{H\} \rightarrow \{C\}$ (carbide) + $H_2O \uparrow$ (water)
$\{C\} + \{H\} \rightarrow \{CH\}$ (methylidyne)
$\{CH\} + \{H\} \rightarrow \{CH_2\}$ (methylene)
$\{CH_2\} + \{H\} \rightarrow \{CH_3\}$ (methyl)
$\{CH_3\} + \{H\} \rightarrow CH_4 \uparrow$ (methane)

(c)

$\{CH_3\} + \{CH_2\} + \{CH_2\} \rightarrow \{CH_3CH_2CH_2\}$
$\{CH_3CH_2CH_2\} \rightarrow \{CH_3CHCH_2\} + \{H\}$
$\{CH_3CHCH_2\} \rightarrow CH_3CHCH_2 \uparrow$ (propene)

Figure 12.1 (a) A pictorial representation of the first steps in the classical Fischer–Tropsch–Pettit–Biloen (FTPB) scheme illustrating the dissociation of CO to carbide followed by hydrogenation. (b) An alternative less picturesque version of Figure 12.1a. (c) The coupling of three surface C_1 hydrocarbyls to give a C_3 hydrocarbyl, which then β-eliminates the C_3 organic, propene, and returns {H} to the catalyst, as envisaged in the FTPB scheme.

hydrogenated C_1 species. Since surface species are rarely well defined, they are written in curly brackets, as shown in Figure 12.1b. Figure 12.1c shows the formation of propene by the coupling of $3 \times \{C_1\}$ species and the regeneration of {H}.

Several reviews of the current thinking on the mechanism of the FT hydrocarbon synthesis to make alkenes and alkanes from syngas have been published. The most informative approach today is based on the likely individual steps that occur on the surface (see Chapter 11). The reactions start with the physisorption of both H_2 and CO, followed by their chemisorption (activation) at metal centers.

12.2.1
Dissociative Activation of CO

It is generally agreed that activation of H_2 takes place by dissociation to give surface atomic hydrogen, written as {H} (Equation 12.5), which is the reactive entity in a nonpolar environment.

Chemisorption of CO also occurs, and two different modes for the activation of CO at a surface have been proposed: the *dissociative* (or "carbide" route) and the *associative* (or "oxygenate" path). Dissociation (cleavage of the CO bond) occurs to essentially give atomic {C}, carbide, and {O}, oxide (Equation 12.6) and Box 12.1, which can then react further with {H} giving hydrocarbyls $\{CH_x\}$ (Equation 12.7) and eventually methane (Equation 12.8), together with hydroxo- {OH} surface species and eventually water (Equation 12.9). In the *methanation* reaction, the final product CH_4, methane, desorbs and is released. Since the main FT-S reaction products are long-chain hydrocarbons, the overall reaction must involve a polymerization of $\{CH_x\}$ and other hydrocarbyl–metal intermediates on the surfaces. The structures of these last species and how they arise and then react further are only partly agreed. The most usual explanation is that in the FT hydrocarbon synthesis, the partially hydrogenated monohydrocarbyl species $\{CH_x\}$ also couple to give further intermediates containing di- or trinuclear hydrocarbyls, $\{C_2H_y\}$, $\{C_3H_z\}$, and so on, which lead to higher oligomers and polymers that β-eliminate to give the alkenes. This is essentially the FTPB scheme.

Instead of chain termination by β-elimination or coupling, the hydrocarbyls can react with {H} to give alkanes (Equations 12.12a and 12.12b). It should be noted that in this mechanistic scheme, there is competition at all stages between the propagation and termination steps and that in order to account for the wide range of hydrocarbon product molecular sizes, their overall rates must be very comparable.

$$H_2 \rightleftarrows 2\{H\} \tag{12.5}$$

$$CO \rightleftarrows \{C\} + \{O\} \tag{12.6}$$

$$\{C\} + \{H\} \rightarrow \{CH\}; \quad \{CH\} + \{H\} \rightarrow \{CH_2\}; \quad \{CH_2\} + \{H\} \rightarrow \{CH_3\} \tag{12.7}$$

$$\{CH_3\} + \{H\} \rightarrow CH_4 \uparrow \tag{12.8}$$

$$\{O\} + \{H\} \rightleftarrows \{OH\}; \quad \{OH\} + \{H\} \rightleftarrows H_2O \uparrow \tag{12.9}$$

Box 12.1 CO cleavage

Surface studies (Chapter 11) have provided evidence for the cleavage process described by Equation 12.6 and others on a variety of metals. One important additional result from measurements on single crystals is that the CO cleavage is greatly facilitated at imperfections in the surfaces such as edges and kinks. This also explains why nanoparticles that have irregular surfaces and high surface area to volume ratios are particularly active. There is also evidence that hydride {H} can facilitate CO dissociation [13].

Many hypotheses discuss the polymerization in terms of chain lengthening by surface monomers such as $\{CH_2\}$, *methylene*, or {CH}, *methylidyne*. Important aspects of the FTPB theory as illustrated in Figure 12.1 have been criticized: for example, the couplings of simple $\{C_{sp3}\} + \{C_{sp3}\}$ surface alkyls have been calculated to be high-energy processes and hence are problematic [12]. The final step of β-elimination of a 1-alkene from an {alkyl} on a surface covered with {H} may also be difficult as this will be an easily reversed step.

12.2.2 Associative Activation

The other popular view has been that CO is activated in an associative process [5, 14]. Since the initial carbon–oxygen bond is preserved, this is also known as the *oxygenate* theory in contrast to the *carbide* (or dissociative) mechanism. Several versions of this have been proposed, including reaction with another {CO} (i.e., a dimerization), but the initial step most commonly considered is that of {CO} with {H} (Equation 12.10):

$$\{CO\} + 3\{H\} \rightarrow \{CH_2OH\}; \quad \{CH_2OH\} + 2\{H\} \rightarrow \{CH_3\}, \quad \text{initiation} \tag{12.10}$$

This is then followed by propagation steps involving the insertion of CO into a surface alkyl followed by hydrogenation to give the hydrocarbyls (Equations 12.11a and 12.11b):

$$\{R\} + \{CO\} \rightarrow \{RCO\}; \quad \{RCO\} + 2\{H\} \rightarrow \{RCH(OH)\}: \\ \{RCH(OH)\} + 2\{H\} - H_2O \rightarrow \{RCH_2\}, \quad \text{propagation} \tag{12.11a}$$

$$\{RCH_2\} + \{CO\} \rightarrow \{RCH_2CO\}; \quad \{RCH_2CO\} + 2\{H\} \rightarrow \{RCH_2CH(OH)\}: \\ \{RCH_2\,CH(OH)\} + 2\{H\} - H_2O \rightarrow \{RCH_2\,CH_2\}, \quad \text{propagation} \tag{12.11b}$$

As in the dissociative mechanism, chain growth can be terminated in several ways, for example, by reaction with {H} to form *n*-alkanes (Equation 12.12a), by loss of H to give *n*-1-alkenes (Equation 12.12b), or by a dimerization to give larger alkanes (Equation 12.12c):

$$\{RCH_2CH_2\} + \{H\} \rightarrow RCH_2CH_3 \uparrow \tag{12.12a}$$

$$\{RCH_2CH_2\} - \{H\} \rightarrow RCH{=}CH_2 \uparrow \tag{12.12b}$$

$$2\{RCH_2CH_2\} \rightarrow RCH_2CH_2CH_2CH_2R \uparrow \tag{12.12c}$$

Other termination steps, giving oxygenates, are also possible, for example, Equation 12.12d to form 1-*n*-alcohols or Equation 12.12e to form 1-aldehydes:

$$\{RCH_2C(OH)\} + 2\{H\} \rightarrow RCH_2CH_2OH \uparrow \tag{12.12d}$$

$$\{RCH_2C(OH)\} \rightarrow RCH_2CHO \uparrow \tag{12.12e}$$

Overall, the associative reaction scheme is essentially a carbonylation mechanism in which CO is "inserted" into metal–carbon bonds in the surface (Box 12.3). An interesting proposal [16] that the combination of 2{H} and {CO} leads to the formation of formaldehyde ($H_2 + CO \rightarrow HCHO$), which can then react further, unfortunately suffers from very adverse thermodynamics.

One major problem with the associative approach is that while there are many experimental observations of CO on metal surfaces and of its dissociation, there appear to be few well-documented examples of CO insertions on metal surfaces. Virtually all the literature on such carbonylations involves the transformations of defined molecular species *in solution* (see Box 12.3).

That the CO hydrogenation reactions in solution and on surfaces are quite different is also illustrated by the fact that when CO hydrogenations catalyzed by carbonyls of the FT active metals (such as Ru, Fe, or Rh) are carried out in solution, very stringent conditions are required ($\gg$100 atm; >350 °C) and the products are essentially oxygenates (methanol and other alcohols, glycol, etc.). Virtually no hydrocarbons are formed [17, 18]. By contrast, the heterogeneously catalyzed FT synthesis occurs under much milder conditions (<100 atm. and ~250 °C) on metal surfaces very largely gives hydrocarbons for a different view see Box 12.2.

Box 12.2 Other views of the FT propagation steps

Other views of the FT mechanism have also been put forward, including the suggestion that chain lengthening occurs by a process involving carbene plus olefin steps analogous to those that are responsible for olefin metathesis [27].

Box 12.3 Solution carbonylation mechanisms

Many complexes bearing an alkyl (or aryl) ligand attached to a metal atom undergo carbonylation in homogeneous solution [19]. In this reaction, which is easily observed, for example, by following the carbonyl stretching bands in the IR spectra, the metal alkyl is converted into a metal acyl (ν_{CO} about 1400–1600 cm^{-1}; the exact value depending on the metal, its charge, and the other ligands present). An intermediate that can sometimes be seen but is often very short-lived is the alkyl metal carbonyl in which the alkyl and the carbonyl (ν_{CO} 1800–2100 cm^{-1}) are on the same metal atom and preferably adjacent (i.e., *cis*) to each other [20]:

$$\text{M-R} + \text{CO} \rightleftarrows \text{MR(CO)} \rightleftarrows \text{M(COR)}$$

When starting from a non-carbonyl-containing complex, it is usually necessary for one or more ligands to be labile and easily replaceable so that the alkyl metal carbonyl can be formed. More detailed investigations have shown that the reaction forming the acyl generally occurs by a *migration of the alkyl onto the coordinated CO* rather than by an insertion of the CO into the metal–alkyl bond. However, for simplicity, many authors use the term *insertion* to describe all such processes [21]. The CO insertion reactions can be promoted by a Lewis acid, usually an aluminum or boron compound of the type ZX_3; ($Z = B$ or Al). It has been shown that the promotion is due to the formation of a Lewis acid–Lewis base adduct $M(C(R)O–ZX_3)$ involving the acyl-oxygen.

12.2.3
Dual Mechanism Approaches

In order to explain the observed products from the broader FT-S, it is necessary to "square the circle?" How may we explain two apparently closely related processes, one of which gives largely methane, while the other favors long chain hydrocarbons?

The simplest answer is that several distinct and rather differently based mechanisms occur in parallel. Several authors have come to similar conclusions independently. For example, Gaube and Klein suggested that the Anderson–Schulz–Flory (ASF) plots (see below) obtained from analysis of FT hydrocarbon synthesis reactions arose from the participation of two "incompatible" mechanisms. Both involve surface alkyls, but in one case the chain growth takes place by insertion of CO, while in the other the reactive entity is a surface carbene $\{CH_2\}$ as in the FTPB scheme [22, 23]. We will offer an answer in Section 12.3, but first we discuss some further experiments in Section 12.2 that bear on the question.

12.3
Some Mechanisms-Related Experimental Studies

12.3.1
The Original Work of Fischer and Tropsch

The procedures that Fischer and Tropsch carried out at the Kaiser Wilhelm Coal Research Institute in Műlheim involved the hydrogenation of carbon monoxide over a wide variety of iron-containing catalysts and different supports at *atmospheric pressure* [6, 7]. The volatile products identified were ethane, propane, and butane: ethene, propene and butene were also formed but in more minor amounts. The higher boiling fractions (named *benzin* and *petroleum*) were analyzed and shown to be mixtures of hydrocarbons with an average C : H ratio of about 1 : 2.15, but the individual components were not separated or identified; a solid hydrocarbon paraffin wax with a melting point of 61 °C was also isolated. The original system described by Fischer and Tropsch used a catalyst containing iron and zinc oxide at 250–300 °C. However, other finely divided metals were active in making higher hydrocarbons, including cobalt (plus chromium oxide), which had much better activity already at 270 °C, and even nickel (under specific conditions). The use of any of a large number of different oxide supports was also reported. Promotion of the catalyst by addition of alkali led to increased activity. The reactions were generally carried out in glass tubes using *water gas* (a form of syngas), but the authors noted that other industrial $CO + H_2$ mixtures in metallic tubes also gave good results.

Fischer and Tropsch found that at the end of the reaction, the catalyst was in the form of finely divided solids, identified as metal carbides. On treatment with hydrogen, these solids gave only methane, though they gave the same higher hydrocarbon mixtures that were obtained from the original catalyst when they were reacted with syngas. It was also found that in contrast to the *high-pressure synthesis* (that gave products containing alcohols and other oxygenates, which they termed *synthol*), the low-pressure process gave only hydrocarbons.

We use the term *Fischer–Tropsch hydrocarbon synthesis* here to apply to the CO hydrogenation reaction largely giving linear hydrocarbons. This distinguishes it from the earlier *methanation* reaction of Sabatier and Senderens that gave methane and also from the methanol synthesis reactions (Chapter 6). As normally practiced in industry, the FT-S produces paraffins, methane, some aromatics, and oxygenates, as well as *n*-alkenes and water. Labeling studies indicate that the methane, the *n*-alkenes, and the *n*-alkanes have a common ancestry; this is also sometimes the case for oxygenate formation. The other large-scale industrial use of catalytic CO hydrogenation is methanol synthesis, but that takes place over Cu + ZnO/alumina and is quite different (Chapter 6).

Two regimes of the FT-S are commonly distinguished in industry: the low temperature Fischer–Tropsch and the high-temperature Fischer–Tropsch, known as LTFT and HTFT, respectively. The LTFT process, often using a slurry reactor, is designed to make high molecular weight alkanes (high wax production), while the HTFT process is aimed mainly at producing lighter alkene-rich products. The HTFT process uses an iron-based catalyst with the syngas passing through a fluidized bed of finely divided catalyst and is usually run below 250 °C in order to minimize unwanted methane production and to maximize selectivity. The catalyst in the HTFT process is much more productive than in the LTFT process, and so the gas throughput and hydrocarbon production rates are much higher [24]. Further details are in Chapters 3, 8 and 9.

Many mechanistic discussions of the FT hydrocarbon synthesis are based on the results from laboratory-scale experiments typically involving around 1 g of the oxide support containing 1–20% w/w of the active metal together with promoters in a microreactor. As the main products of interest are hydrocarbons, analyses are by gas chromatography (GC) for quantification, coupled to mass spectrometry (MS) for identification of the individual compounds. Further and more detailed information is now frequently obtained from proton or carbon nuclear magnetic resonance (^{1}H, ^{13}C, NMR etc.) and infrared (IR) spectroscopy of the products. Because of the difficulties in quantifying them, gaseous products such as carbon dioxide and water-soluble products have sometimes been ignored. Some laboratory-scale experiments described by the Sheffield group [25, 26] are summarized in Box 12.4.

Some questions arise over the simplification obtained by carrying out small-scale reactions: different reactor designs are not readily allowed for, and other aspects such as the times needed for the catalyst to equilibrate in the industrial system are

Box 12.4 Laboratory-scale experimental procedures

The reaction is often run with nanoparticles (very fine granules) of the metal dispersed on an oxide support such as silica, alumina, titania, or even a zeolite. These catalysts are generally prepared by the *incipient wetness technique* (see also Section 9.4) [25, 26].

A very pure, porous, and finely divided commercial oxide such as Davisil 645 or Alpha Aesar silica gel is used: the oxide support needs to have a high surface area for maximum activity, but must be sufficiently granular to allow the volatile reactants and products to pass through readily. The support is impregnated to incipient wetness, with an aqueous (acidic) solution of an appropriate salt of the metal catalyst (often the nitrate), together with solutions of any other promoter salts. The impregnated oxide is then heated to drive off moisture, first slowly to 100 °C and then to 200 °C. A measured amount of the dry catalyst is then transferred to a microreactor (often a glass tube) where it is heated under a stream of hydrogen gas (at say 400 °C) to ensure complete reduction to the metallic state. The microreactor tube is allowed to cool and the hydrogen gas flow replaced by syngas of defined CO: H_2 composition; the FT reaction then commences at a temperature appropriate for the catalyst used. The composition of the catalyst is also carefully monitored, and the metals (and other elements) contained therein measured to at least ppm levels using elemental analysis and techniques such as ESCA or Auger spectroscopy.

Most of the smaller molecule reaction products from the FT-S are volatile at the reactor temperature and are found in the exit gases where they are analyzed by GC-MS; the higher molecular mass products with lower vapor pressures are trapped prior to analysis. Isotopic labeling (usually ^{13}C or ^{14}C, but occasionally ^{17}O or ^{18}O) of the carbon monoxide and of added probe molecules has been successfully employed to define mechanisms; in this case, a defined amount of the labeled probe is added to the syngas stream. Deuterium (D_2, $^{2}H_2$) rather than hydrogen has also been tried, but the ease of H/D exchange under the conditions of the FT-S makes this a rather less attractive technique.

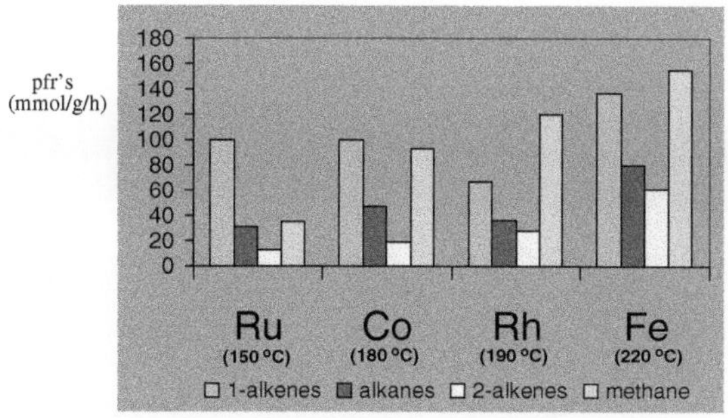

Figure 12.2 Bar graphs showing catalysts, products, and product formation rates (pfr's) found for comparative FT hydrocarbon synthesis reactions over Ru/150 °C, Co/180 °C, Rh/190 °C, and Fe/220 °C. Adapted from Ref. [25].

also quite difficult to mimick accurately. On the other hand, the use of microreactors does allow high-throughput screening, which is invaluable if new catalyst formulations are to be explored.

12.3.2
Laboratory-Scale Experimental Results

While the FT reactions are carried out on vast industrial scales, the mechanistic discussions of the FT-S given here are often based on the results from laboratory-scale microreactor experiments (Box 12.4), supplemented by catalyst studies (Chapter 11). However, the conditions are typically less stringent than in the industrial plants.

The bar graphs in Figure 12.2 show very much simplified comparisons based on the *product formation rates* (pfr's) of the four main classes of products that were measured over four catalysts (Box 12.5). In each case, methane and 1-*n*-alkenes predominated, with the methane increasing at higher temperatures. The similarity of product distributions (methane plus 1-*n*-alkenes and *n*-alkanes) and of their pfr's (μmol/($g_{catalyst}$ h)) should be noted. However, the reaction temperatures required to give comparable activity were different. The conditions used for the FT-S were chosen to minimize further reactions.

Box 12.5 Laboratory-scale catalyst case studies

Some of the chief results of laboratory-scale experiments by the Sheffield group are summarized in Figure 12.2 [25, 26]. Comparable reactions of syngas over four different metal catalysts, Fe, Co, Ru, and Rh plus ceria, all on silica, were examined. The temperatures required for efficient measurable pfr's were Ru, 150 °C; Co, 180 °C; Rh, 190 °C, and Fe, 220 °C at a total pressure of 1 atm and a CO: H_2 ratio of 1: 2; the other parameters were kept as similar as possible.

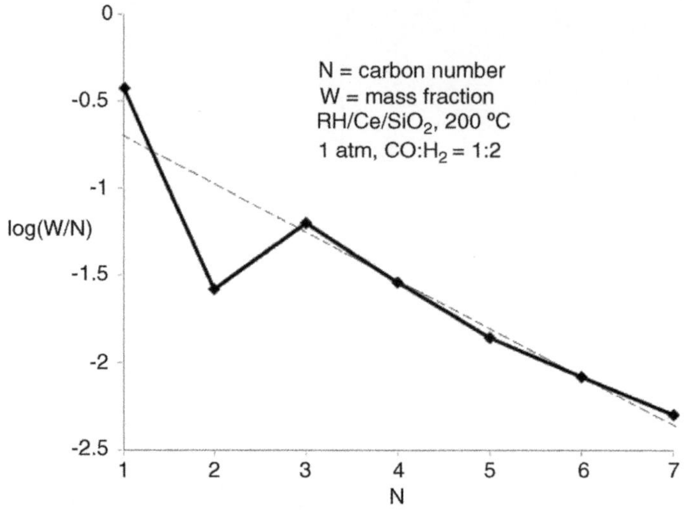

Figure 12.3 A typical Anderson–Schulz–Flory plot of FT hydrocarbon products showing the relation of the carbon number to the mass fraction (*conditions*: Rh + ceria on silica catalyst. CO : H_2, 1 : 2; 1 atm, 200 °C). From Ref. [25]

Measurements at different syngas flow rates over the Ru catalyst showed that at lower flow rates proportionately more alkanes and internal alkenes are generally found, while the reverse was true of the 1-alkenes and methane. As *lower* flow rates increase the residence times of the gases on the catalyst, this suggested that secondary reactions are involved in the formation of alkanes and 2-alkenes. The strong family resemblances between the results suggested that rather similar primary reactions occur over all the metals.

The amounts of the various hydrocarbon products are conveniently described in terms of the Anderson–Schulz–Flory (ASF) plot that defines the distribution of hydrocarbon products from a CO hydrogenation. A typical plot is shown in Figure 12.3, where W, the mass fraction of each carbon number (N), defined by $\log(W/N)$, is plotted against N. The dashed line is the theoretical ASF plot for step-growth polymerization involving C_1 monomer species, while the solid line represents an experimentally observed value here measured on a Rh plus ceria on silica catalyst after 1.5 h on stream using a CO: H_2 ratio of 1: 2 and total pressure of 1 atm at 200 °C [25]. The major deviations occur for C_1 (methane), of which much more is obtained than would be expected, and for the C_2 fraction, which is significantly smaller than expected. The dip at C_2 indicates a special feature about the C_2 species – either they are formed very slowly or, more likely, they are removed preferentially from the products. If the latter, it suggests a special activity is associated with the ethene. The ASF distribution has also been extensively modeled by Botes [61].

More detailed information concerning FT products is given in Chapter 4 and also in publications by Davis, Schulz, and their coworkers [28–30].

12.3.3
Probe Experiments and Isotopic Labeling

Valuable insights have been gained from the results obtained by adding small amounts of certain functionalized organics as probes to the syngas stream. Suitable probes are those that readily decompose at the reactor temperature to give organic fragments that are incorporated into the FT-S products; examples are ethene and other olefins, organic halides, alcohols, diazomethane, and nitromethane. GC-MS analysis can readily detect the incorporation of the resultant organic fragments into the FT-S products, and mechanistic conclusions can then be drawn from the changes observed. For example, the classical studies by Baker and Bell found that the hydrocarbon bicyclo[4.1.0]heptane (norcarane) was formed when cyclohexene was added to a syngas reaction (over Ru on silica at 225 °C); this was consistent with the presence of surface $\{CH_2\}$ species [31, 32].

The use of isotopically labeled probes has led to especially meaningful results (Box 12.4) [32, 33]. Some workers have used carbon monoxide enriched with either ^{14}CO or ^{13}CO, others used oxygen isotopes; D_2 in place of H_2 has also been employed. The products are then analyzed to determine the extent of isotope incorporation [33, 34]. A useful variant is provided by the experiments using steady-state isotopic transient kinetic analysis (SSITKA) in which the changes in products are determined when the gas flow is switched between ^{12}CO and ^{13}CO.

12.3.3.1 ^{13}C Labeling

Further information has come from studies of the ^{13}C NMR spectra of products after the addition of small amounts of $^{13}C_2H_4$ to a normal syngas stream ($^{12}CO + H_2$). The first experiments showed that under the quite stringent conditions initially used (Co/Al; 240 °C), the ^{13}C abundances were close to those seen when ^{13}CO was used as probe [35]. The presence of mostly mono-^{13}C labeled isotopomers indicated that there had been considerable cleavage of the C—C bond of the ethene probe added.

The ^{13}C NMR studies carried out a little later by the Sheffield group of the incorporation of $^{13}C_2$ moieties derived from the addition of small amounts of doubly labeled $^{13}C_2H_3X$ probes (X = H, Br, and SiR_3) to a syngas stream ($^{12}CO + H_2$) under milder conditions (Ru/150 °C, Co/180 °C; Fe/220 °C; or Rh/190 °C; all on silica; $CO + H_2 = 1:1$; 1 atm), together with GC-MS measurements were more informative as less cleavage of the C_2 probe had occurred. These showed that the *n*-alkene and *n*-alkane products contain terminal $^{13}CH_3$-$^{13}CH_2$- tails; for example, the label in the 1-butene was mainly $^{13}CH_3$-$^{13}CH_2$-$^{12}CH{=}^{12}CH_2$ with largely natural abundance labeling elsewhere (Figure 12.4) [25, 26, 36, 37].

Similar product labeling patterns were seen for the other *n*-1-alkene and *n*-alkane products ($^{13}CH_3{}^{13}CH_2(CH_2)_nCH_3$ ($n = 1$–4) over all the four catalysts. Thus, the main hydrocarbon products with $C_{\geq 3}$ had incorporated $^{13}C_2$ at the ends of the alkyl chains. This indicated that they were formed by successive additions of $\{C_1\}$ to a $\{C_2\}$ parent unit, a result consistent with the hydrocarbons being formed by a regiospecific polymerization of $\{C_1\}$ (derived

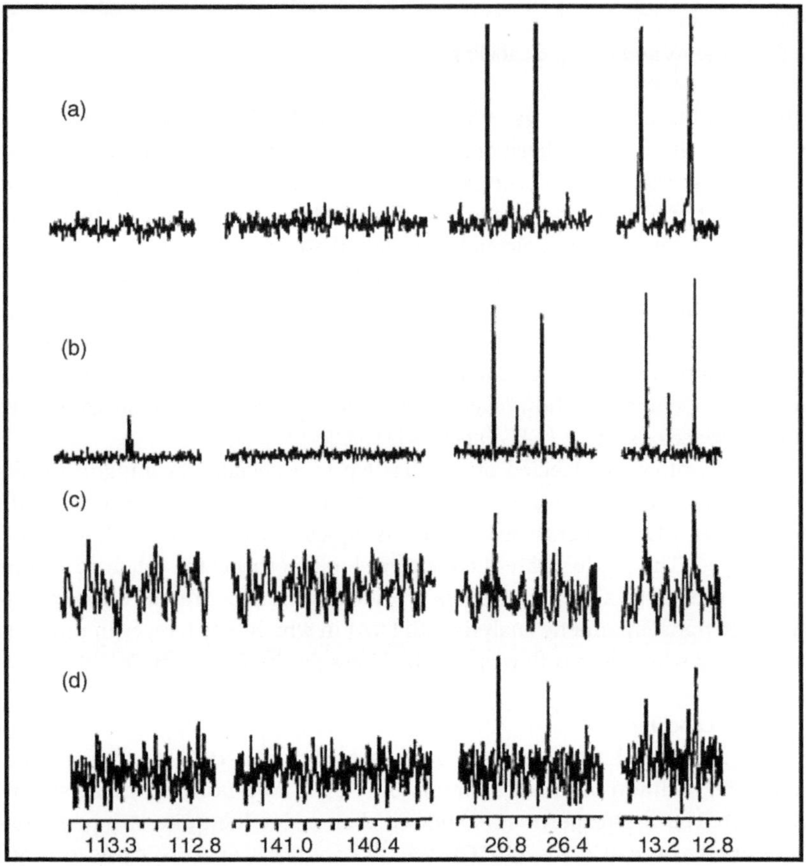

Figure 12.4 ^{13}C NMR spectra (δ; ^{1}H-decoupled) of 1-butene formed from $^{12}CO + H_2 + {}^{13}C_2H_4$ over (a) ruthenium at 150 °C, (b) cobalt at 180 °C, (c) rhodium at 190 °C, and (d) iron at 220 °C in the CH_2= (δ 113.1), =CH— (δ 140.6), —CH_2— (δ 26.7), and —CH_3 (δ 13.1) regions. All largely show the presence of single isotopomer $CH_2{=}CH^{13}CH_2{}^{13}CH_3$. Adapted from Ref. [36].

from CO) initiated by the $^{13}C_2$ probe. Blank experiments involving addition of small amounts of unlabeled ethene ($^{12}C_2H_4$) to the FT-S gave only minor perturbations in that the pfr's of hydrocarbons $C_{\geq 2}$ were increased and methane formation was repressed [36].

Under the temperature conditions stated, there was good $^{13}C_2$ incorporation in the hydrocarbon products and little sign of $^{13}C_1$ containing compounds. However, when the reactions were carried out at higher temperatures over Ru and Co catalysts $^{13}C_1$ containing products (as well as some $^{13}C_3$ and other isotopic species) were found [36]. This is ascribed to some cleavage of the added $^{13}C_2H_4$ probe molecule at higher temperatures, but may also in part arise from fragmentation of the $^{13}C_2$-labeled products in the reactor.

12.3.3.2 ^{14}C Labeling

Catalyst studies using ^{14}C labeling have been known since 1948 [38, 39], and have been used to evaluate the effects on activity and selectivity of small changes in catalyst and reaction conditions. Davis and coworkers at CAER have examined carefully the iron-promoted FT-S (see also Chapter 7). Under their conditions (260 °C; 7 atm; doubly promoted fused iron), they concluded that ^{14}C-ethene was incorporated into the FT-S and that alkanes were formed as primary products. The data were consistent with ethene acting to initiate chain growth in the FT-S, but for a different species to be involved in the chain propagation steps [40, 41]. They interpreted their results to support an oxygenate initiation step for the FT-S.

12.4 Current Views on the Mechanisms of the FT-S

We here offer a distillation of the mechanistic proposals that currently look the most reasonable and well based and that offer guidance for future studies.

12.4.1 The First Steps: H_2 and CO Activation

As described in Sections 11.3.2 and 11.3.3, surface studies indicate that the FT reaction starts when the CO and H_2 are activated. CO activation occurs preferentially at faults on the surfaces of late transition metals (Fe, Ru, Co, or Rh) and also at interfaces of the metal with "islands" of promoters, for example, Lewis acid oxides, such as alumina or titania. More details are given in Boxes 12.5 and 12.6.

It is found both experimentally and theoretically that late transition metal atoms at steps, edges, kinks, and corners (i.e., those with low coordination numbers) tend to have higher lying electronic d-states and, therefore, interact more strongly with adsorbates than the metal atoms on close-packed surfaces with higher coordination numbers [42].

The picture of H_2 activation at a surface that we now have (Box 11.3) is very similar to that found for hydrogen in molecular metal complexes where two bonding patterns have been established. In the first, molecular dihydrogen binds to a metal atom. This is written as $M(\kappa^2\text{-}H_2)$ and is analogous to physisorption at a surface. In the second, the H—H bond has been replaced by two M—H σ-bonds in a σ-bonded dihydride, $M(\kappa^1\text{-}H)_2$, which is close to the picture of chemisorbed hydrogen. The

Box 12.6 Activation of CO

The adsorption of CO at surfaces generally leads to a medium–weak interaction in which the CO bond order is reduced and some metal–C bonding is established (Figure 11.3, and [64]). This can eventually lead to scission of the C—O bond and the establishment of metal–O and metal–C bonds. The various stages can be followed by changes in the vibrational spectra (IR; HREELS, etc.).

path from $M(\kappa^2\text{-}H_2)$ to $M(\kappa^1\text{-}H)_2$ can also formally be regarded as equivalent to an oxidative addition of H_2 to the metal, and it is in many cases quite readily reversible (*reductive elimination*).

$$M(\kappa^2\text{-}H_2) \rightleftarrows M(\kappa^1\text{-}H)_2 \tag{12.13}$$

However, the picture for CO offered by Box 12.6 does not take into account the effect of the "promoters" often used in the catalysts. For example, addition of an alkali metal facilitates the ability of a metal surface to cleave CO; thus, although CO does not dissociate on Cu(111), HREELS measurements at room temperature showed that CO adsorption is dissociative on a sodium-precovered Cu(111) surface, and that the CO molecules occupy adsorption sites directly adjacent to those of alkali adatoms [43] see also Box 12.7.

An interesting theoretical analysis by Jenkins and King of the situation when potassium and CO are coadsorbed on a Co{10¯10} surface indicates that the C—O bond is weakened by the presence of the alkali metal. It is also suggested that the promoter effect of K in Fischer–Tropsch hydrogenation is in part due to the strong polarization of adsorbed CO molecules that thus become susceptible to nucleophilic and/or electrophilic attack [44].

While many surface studies had to be carried out in ultrahigh vacuum (UHV), CO dissociation has also been measured on several platinum single crystals, at high pressures and temperatures, using sum frequency generation (SFG) surface vibrational spectroscopy [45] as a model for CO activation in FT-S. CO was found to dissociate at higher temperatures over the more stable planar Pt(111) surface than over the Pt(557) and Pt(100) surfaces. The SFG spectra evolved with time, indicating the surface was being modified, probably due to the formation of platinum carbonyl species, which could move atoms from the surface lattice.

Box 12.7 SMSI (strong metal–surface interactions)

Vannice and Garten found that the activity of nickel in CO hydrogenation was modified from giving only methane to a broader spectrum of hydrocarbons (C_1—C_7) when the nickel was supported on titania [46, 47]. Other metals (e.g., rhodium) and other supports showed similar effects that became known as strong metal–surface interactions (SMSI) (see also Box 11.4). The studies concluded that the acceleration observed in CO hydrogenation arose from the interaction of nanoparticles of the metal with small amounts of oxides located in islands that "decorate" metal surfaces. Such metal–surface interactions are especially evident in the hydrogenation of CO and of organic carbonyl compounds. They may be understood in terms of a CO molecule bound (conventionally through C) on the surface to the transition metal, also being further activated by the interaction of the O to the Lewis acid of the oxide support, for example, titania (TiO_x), in an anion vacancy on a neighboring "island." Such interactions appear to be distinct from the activation caused by added alkali metals.

In addition to the experimental surface studies, there have been numerous computations concerning the mode by which CO is activated in the FT reaction on single-crystal surfaces. The results cover a range of possibilities, depending on the metal and the surface under consideration. It is however agreed that reactions occur preferentially at surface imperfections, such as steps or kinks, where the surface metal atoms have a low coordination and may be expected to be more reactive [48].

We may summarize the above by saying that the experiments and calculations indicate that under appropriate conditions, CO does dissociate on metal surfaces, and that dissociation is easier on "rough" surfaces containing atoms of lower coordination numbers, and is also facilitated by alkali metals and certain oxide supports. These results are important for our understanding of the initial steps of the FT-S.

12.4.2
Organometallic Models for CO Activation

In the 1980s and 1990s, there was considerable interest among chemists to model the CO activation step homogeneously in organotransition metal molecules. Both single metal and cluster complexes were explored and there are now a number of examples where CO is cleaved by a metal complex into carbide and oxide ligands. However, only the work of Shriver on the proton promoted cleavage of CO in an anionic Fe_4 cluster [49] seems directly relevant to the FT-S. The other examples of the cleavage of CO in a soluble metal complex are largely restricted to the earlier, more oxophilic transition metals (Zr, Nb, Ta, and W). However, once CO cleavage has occurred to form {C} and {O}, it is rarely possible to complete the remaining steps of a cycle and regenerate {CO} at a low-oxidation state metal site, and these metals are not active in Fischer–Tropsch synthesis [50]. Successful experiments to develop cycles could provide better models for catalysis in future.

Although there are many examples of the addition of "soft" nucleophiles (easily polarizable, with loosely held valence electrons, such as sulfides, iodide, or alkoxides) to metal carbonyls as models for the associative mechanism, only a few involve the addition of H [51].

12.5
Now: Toward a Consensus?

In order to see which mechanistic concepts are most appropriate now (in 2012), we first focus on the best aspects of the various proposals that have been made, based on the most widely accepted ideas, the dissociative (or *carbide*-based) and the associative (or *oxygenate*) theories, before offering a more comprehensive blend.

12.5.1
Routes Based on a Dissociative (Carbide) Mechanism

In the dissociative pathway, we propose refinements to the classical FTPB mechanism. These extend the view of the intermediates involved, namely, that they are essentially nonpolar and that the reaction proceeds on nonpolar surfaces as set out in Box 12.8. The first steps, cleavage of H_2 and CO to hydride (Equation 12.14a) and carbide (Equation 12.14b), respectively, are followed by hydrogenation (Equations 12.15 and 12.16), coupling to form di- and tri-mer units (Equations 12.17

Box 12.8 Alkenyl plus methylene chain propagation route [25]

The initiation steps follow the FTPB scheme. The main differences are in the chain propagation and termination. Propagation involves coupling between an alkenyl-like $\{C_{sp2}\}$ and a methylene rather than an alkyl plus methylene or a methylene plus methylene coupling. The alkenyl-like species can also be hydrogenated to give the alkene that is released (Equation 12.20); it can also undergo multiple hydrogenations with {H} to give the alkane.

Initiation:

$$H_2 \rightleftarrows 2\{H\} \tag{12.14a}$$

$$CO \rightleftarrows \{CO\} \rightleftarrows \{C\} + \{O\} \tag{12.14b}$$

$$\{O\} + \{H\} \rightleftarrows \{OH\} \rightleftarrows \{H_2O\} \rightleftarrows H_2O_{(g)} \tag{12.15}$$

$$\{C\} + \{H\} \rightarrow \{CH\} \rightarrow \{CH_n\} \rightarrow CH_4 \uparrow \tag{12.16}$$

Chain propagation:

$$\{CH\} + \{CH_2\} \rightarrow \{CHCH_2\} \tag{12.17}$$

$$\{CHCH_2\} + \{CH_2\} \rightarrow \{CH_2CHCH_2\} \tag{12.18}$$

$$\begin{aligned} &1,3\text{-H shift}\ \{CH_2CHCH_2\} \rightarrow \{CH_3CHCH\} \\ &\{CH_3CHCH\} + \{CH_2\} \rightarrow \{CH_3CHCHCH_2\};\ 1,3\text{-H shift} \rightleftarrows \{CH_3CH_2CHCH\} \\ &\text{etc} \end{aligned} \tag{12.19}$$

Chain termination:

$$\{CH_3CH_2CHCH\} + \{H\} \rightleftarrows CH_3CH_2CH{=}CH_2 \uparrow \tag{12.20}$$

$$\{CH_3CH_2CHCH\} + 3\{H\} \rightarrow CH_3CH_2CH_2CH_3 \uparrow \tag{12.21}$$

and 12.18). Polymerization of the surface $\{C_{1-3}H_x\}$ units then occurs (linking C_{sp2}'s on surface methylene-, alkenyl- [25], or methylidyne-like species [52]).

In the methylene + alkenyl C—C bond making step, a $\{C_{sp2}\}$ and a $\{C_{sp3}\}$ are formally linked; this is more energetically favorable than the $C_{sp3} + C_{sp3}$ coupling of the FTPB proposal. In order that C—C bonding steps involving $\{C_{sp2}\}$ can continue, 1,3-*H*-migrations also occur (Equation 12.19). Reasonable models for each of the individual steps have been reported in the literature [25, 36].

Another lower energy variation to the FTPB route is provided by the alkyl + methylidyne + hydride mechanism, illustrated in Box 12.9. After the initiation and $\{C_1\}$ hydrogenation steps (Equations 12.14a–12.16), chain propagation occurs in two steps: a surface alkyl reacts first with a methylidyne, {CH}, and then with{H} (Equations 12.22 and 12.23) [53]. This has the attraction that the {CH} species is well documented and can fit conveniently into the tetrahedral holes between the close-packed metal atoms on surfaces. Further steps proceed in the same manner between the higher surface alkyl $\{RCH_2\}$, methylidyne {CH}, and {H} in two stages (Equation 12.23). The reaction sequence can again terminate either with the loss of H to give an alkene or by addition of {H} to give the alkane (Equations 12.24a and 12.24b)

12.5.2
Routes Based on an Associative (or Oxygenate) Mechanism

The second group of mechanisms that is considered is based on an initial associative step where {CO} and {H} combine in some way, for example, as illustrated in Equation 12.25, and the {CO} is not cleaved. This can then be followed by further hydrogenation and loss of hydroxo or water (Equation 12.26) and carbonylation (Equation 12.27). Hydrogen transfer (Equation 12.28) and further hydrogenation and loss of OH or water (Equation 12.29) lead to alkyl surface species that can either lose an H, and release a 1-alkene, or gain an H and release the *n*-alkane.

Box 12.9 Alkyl + methylidyne + hydride chain propagation [53]

Initiation and hydrogenation: Equations 12.14a, 12.14b, 12.15, and 12.16

Chain propagation:

$$\{CH_3\} + \{CH\} + \{H\} \rightarrow \{CH_3CH_2\} \quad (12.22)$$

$$\{RCH_2\} + \{CH\} + \{H\} \rightarrow \{RCH_2CH_2\}, \text{etc.} \quad (12.23)$$

Chain termination:

$$\{RCH_2CHCH\} + \{H\} \rightarrow RCH_2CH{=}CH_2 \quad (12.24a)$$

$$\{RCH_2CHCH\} + 3\{H\} \rightarrow RCH_2CH_2CH_3 \quad (12.24b)$$

Many other variations have been suggested, such as couplings between two {CHOH} species, to initiate C—C bond formation:

$$\{CO\} + \{H\} \rightleftarrows \{CHO\}; \quad \{CHO\} + \{H\} \rightleftarrows \{CH_2O\} \tag{12.25}$$

$$\{CH_2O\} + 2\{H\} - \{OH\} \rightarrow \{CH_3\} \tag{12.26}$$

$$\{CH_3\} + \{CO\} \rightarrow \{CH_3CO\} \tag{12.27}$$

$$\{CH_3CO\} + 2\{H\} \rightarrow\rightarrow \{CH_3CH_2O\} \tag{12.28}$$

$$\{RCH_2CH_2O\} + 2\{H\} - \{OH\} \rightarrow \{RCH_2CH_2\}, \text{ etc.} \tag{12.29}$$

$$\{RCH_2CH_2\} - \{H\} \rightleftarrows RCH{=}CH_2 \uparrow \tag{12.30}$$

$$\{RCH_2CH_2\} + \{H\} \rightarrow RCH_2CH_3 \uparrow \tag{12.31}$$

Associative mechanisms have been supported by various studies. Thus, the investigations by Davis (at CAER) and others of FT-S on Fe catalysts (often the massive iron metal itself, doubly promoted with potassium oxide and carbonate) concluded that their results are best explained by a mechanism in which the C—O bond is *not* cleaved initially [54, 55].

They also found that ^{14}C-ethene (or ^{14}C-ethanol) could initiate chain growth and concluded that it took place via an oxygenate mechanism. The associative mechanisms have the advantage that they can easily be modified to account for the formation of some of the smaller by-products such as alcohols. However, surface studies have not really detected any intermediates corresponding to steps such as Equation 12.25.

12.6 Dual FT Mechanisms

We propose that the differing views can be reconciled by a "dual path mechanism." This has two components: a path that is followed on a largely nonpolar surface and involves uncharged intermediates, and another one that involves polar species (electrophiles and/or nucleophiles). The first path follows the dissociative route, illustrated in Figure 12.5. The other path involves an associative step driven by surface polarization effects, as illustrated in Figure 12.6.

12.6.1.1 Dual FT Mechanisms: The Nonpolar Path

The first part of the dual mechanism as outlined in Figure 12.5 takes into account the recent experimental and theoretical work on the intimate details of the catalytic steps, including the nature of the metal and of the support and the interactions between them. Since at least some of the reactions are site specific, we need to consider surface defects and the sizes and shapes (i.e., the morphology) of the nanoparticles that make up the real catalyst. In addition, we also draw attention to steps not *directly* involving the atomic transformations at the catalyst but that also play major roles in determining the nature of the products obtained.

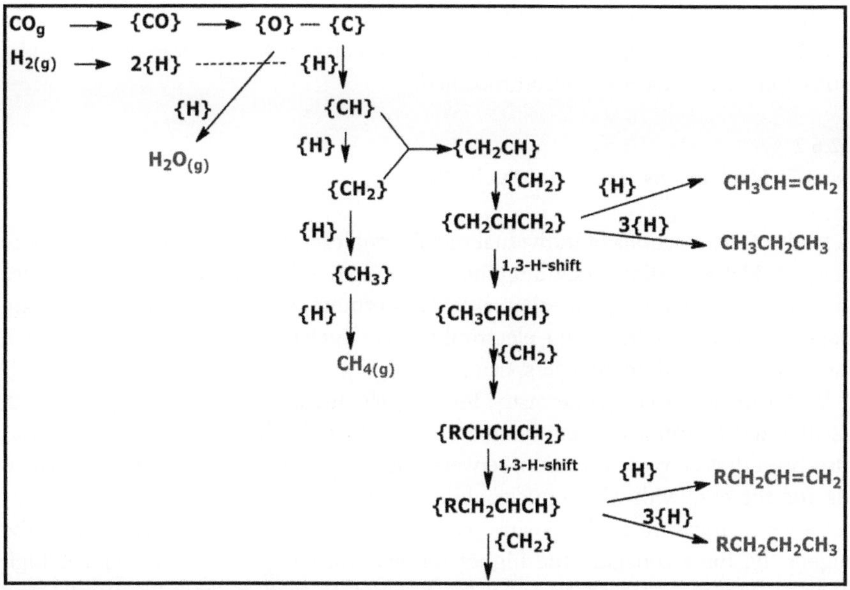

Figure 12.5 Representation of the dissociative FT path involving relatively nonpolar intermediates.

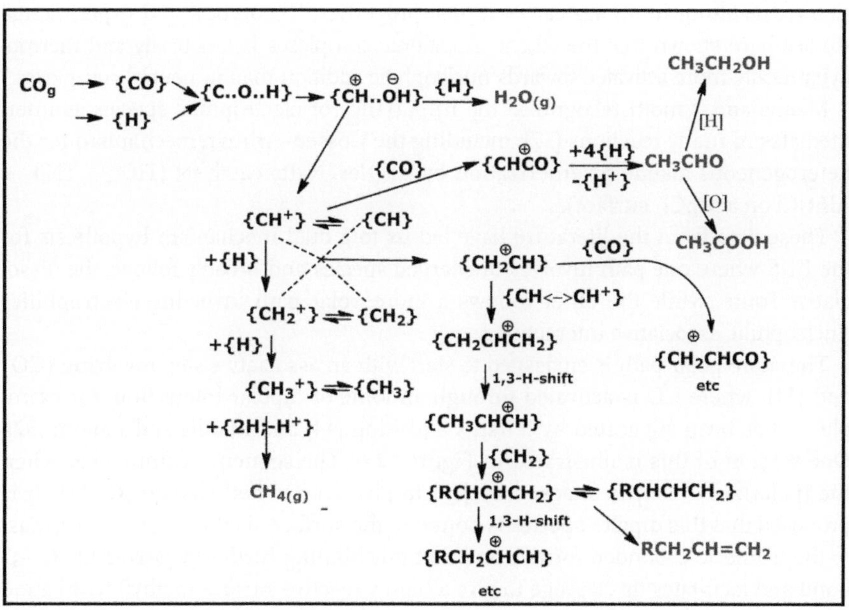

Figure 12.6 Representation of an associative FT path involving ionic and dipolar intermediates. *Note*: As usual in a consideration of organic reaction mechanisms, the charges shown are for illustration only; they will certainly be delocalized over several atoms of the substrate and adjacent atoms of the supports.

The nonpolar route (Figure 12.5) follows the dissociative path mapped out in Box 12.8 (or Box 12.9) and occurs on very nonpolar surfaces where electrophiles/nucleophiles are not easily accommodated.

12.6.2
Dual FT Mechanisms: The Ionic/Dipolar Path

A study of the reactions of individual metal complexes brings out the importance of charges. Many reactions including those forming C—C bonds involve polar species and proceed by ionic-type mechanisms; furthermore, where valid comparisons are possible, reactions involving electrophiles or nucleophiles are generally much faster than those of uncharged species.

Within metal carbonyl chemistry, for example, the addition of nucleophiles such as alkoxides or amines to coordinated CO is well established [21], and attention has also been drawn to a correlation between the CO force constants in metal carbonyls and the ease with which nucleophilic attack at CO occurs. The force constants represent a measure of the positive charge on the carbon of the CO group – the higher the force constant, the higher the positive charge on the carbon. A high positive charge is found to favor C—X bond formation, both kinetically and thermodynamically.

Similar situations occur in homogeneously catalyzed and other organometallic reactions. Thus, Hahn [56] noted that "the electrophilicity of coordinated alkenes in transition metal complexes can be strongly enhanced by increasing the positive net charge, resulting in strong carbocationic properties. Theoretical and experimental studies have shown that the alkene in cationic complexes is kinetically and thermodynamically more activated towards nucleophilic addition than in neutral complexes."

Maitlis and Zanotti recognized the importance of electrophilic species as intermediates in many reactions [57], including the Cossee–Arlman mechanism for the heterogeneous alkene polymerization by Ziegler–Natta catalysts ($TiCl_4 + TiCl_3 + AlEt_2Cl$ on a $MgCl_2$ surface).

These data from the literature have led us to a dual mechanism hypothesis for the FT-S where one path involves uncharged species and broadly follows the dissociative route, while the other follows a more polar path involving electrophilic/nucleophilic associative interactions.

The more polar path is envisaged to start with an associative step involving {CO} and {H}, where CO is activated through an ionic or dipolar interaction, for example, as has been suggested by Jenkins and King [44] and Maitlis and Zanotti [57]. One version of this is illustrated in Figure 12.6. The sequence commences when the (polarized) {CO} is attacked by {H} to give an oxomethylidyne {OCH}. It is proposed that this dipolar species is bound to the surface via the oxygen, in contrast to the isomeric C-bonded formyl, and that this binding further polarizes the C—O bond and facilitates its cleavage to give a highly reactive surface methylidyne cation $\{CH^+\}$ as shown.

We propose that the reactive chain carrier in the sequence is the highly electrophilic methylidyne cation $\{CH^+\}$. This can then react in various ways:

(i) with {H} to give $\{CH_2\}$, then $\{CH_3\}$, and eventually CH_4, methane; or (ii) with another $\{CH_x\}$ species forming a new reactive surface species containing a C—C bond, for example, $\{CHCH_2{}^{+}\}$. Alternatively, (iii) it can react with CO, in the manner established for other strongly cationic species [15] to build up oxygenate precursors. While the first path (i) gives methane and the second (ii) leads to long-chain alkenes and alkanes, the last (iii) can give higher oxygenates such as aldehydes or alcohols.

12.7
Cognate Processes: The Formation of Oxygenates in FT-S

In addition to the formation of *n*-1-alkenes and *n*-alkanes, the FT-S also produces smaller quantities (typically <10%) internal *n*-alkenes, branched chain alkenes and alkanes, aromatics, and oxygenates. Many of these products are formed from the initially produced *n*-1-alkenes in secondary processes.

The importance of secondary processes has been shown, for example, by the observation that in the alkene products, the ratio of internal to terminal olefin increases when the gas flow rates through the reactors are decreased. Since the supported FT catalysts are also able to hydrogenate and isomerize, it is to be expected that longer residence times and higher reaction temperatures will increase the amounts of secondary products.

Oxygenates such as alcohols, ketones, and carboxylic acids are formed in the FT-S as well as the hydrocarbons. In addition to "direct" paths, some can arise from the primary *n*-1-alkenes by secondary processes such as a metal-catalyzed hydroformylation in which 1-alkenes are converted into *n*- and *i*-aldehydes (Equation 12.32). Hydroformylation has been shown to take place very easily (1–10 atm syngas; 50–150 °C) in homogeneous media, particularly with cobalt or rhodium catalysts (see Section 6.7 and Box 6.3) [19]:

$$RCH{=}CH_2 + CO + H_2 \rightarrow RCH_2CH_2CHO + RCH(CHO)CH_3 \qquad (12.32)$$

By far, the other most important catalyzed processes undergone by syngas are the water-gas shift reaction (Equation 12.4) and methanol synthesis (Equation 12.3). Against all expectations, however, methanol synthesis in the major extent results from the hydrogenation of carbon *dioxide*, not carbon *monoxide*, as is explained in detail in Chapter 6. The key lies in the WGSR that links CO, H_2O, CO_2, and H_2 (see also Section 2.3). Current views on the mechanism of the WGSR (over copper catalysts) indicate that the steps outlined in Equations (12.33a–12.33g) are followed. A formate-mediated path (Equation 12.34) may also be significant.

$$CO_g \rightleftarrows \{CO\} \qquad (12.33a)$$

$$H_{2(g)} \rightleftarrows 2\{H\} \qquad (12.33b)$$

$$H_2O_{(g)} \rightleftarrows \{H\} + \{OH\} \qquad (12.33c)$$

$$\{OH\} \rightleftarrows \{H\} + \{O\} \qquad (12.33d)$$

$$\{CO\} + \{O\} \rightleftarrows \{CO_2\} \tag{12.33e}$$

$$\{CO_2\} \rightleftarrows CO_{2(g)} \tag{12.33f}$$

$$\{OH\} + \{OH\} \rightleftarrows \{H\} + \{H_2O\} \tag{12.33g}$$

$$\{CO_2\} + \{H\} \rightleftarrows \{HCOO\} \tag{12.34}$$

A close link between the WGSR and the FT hydrocarbon synthesis mechanisms has also been suggested [60]. While a direct mechanistic link between the FT hydrocarbon synthesis and the methanol synthesis can be written on paper, such links have not as yet been established and must therefore remain speculative, in particular as the catalysts involved are very different.

12.8
Dual Mechanisms Summary

As we have indicated in Section 12.6, a useful mechanistic rationalization of the products from the FT-S is provided by an approach involving Dual Mechanisms. One mechanism entails quite non-polar intermediates on surfaces of low polarity. The second mechanism involves the interaction of polar electrophiles and/or nucleophiles with polar surfaces. Both mechanisms can lead to similar hydrocarbon products though the reactions involving polar species may be expected to be faster.

12.9
Improvements by Catalyst Modifications

FT products are suitable as liquid fuels. However, in order to obtain products of higher added value such as those required by the chemicals industry, a greater specificity, preferably to one major product, is needed. Since the FT-S gives a wide range of products much effort has gone into evaluating the effects on activity and selectivity of small changes in reaction conditions or the addition of "promoters."

One recent publication has reported the rather selective conversion of syngas into C_2–C_4 olefins with up to 60 wt% selectivity over nanoparticulate iron catalysts (promoted by Na and S) on α-alumina or carbon nanofiber supports [65].

The role of alkali metals, especially potassium, in activating catalysts is well known and has already been commented on. In addition, it was found that magnesium and barium oxides also act as promoters as they lower sintering rates by decreasing metal atom mobility.

Water has a complex relation with FT-S. It is a major FT product and because water promotes the WGSR, it has profound effects on the reaction since it can radically change the hydrogen and carbon monoxide concentrations. Water can also deactivate the active metal catalyst itself, since at the temperatures of the FT-S, it acts as an oxidizing agent and converts catalytically active metals or carbides into inert oxides. For example, the hydrothermal degradation of cobalt on alumina and cobalt on silica catalysts to form aluminates and silicates is a serious problem in FT synthesis at the high-pressure, high-conversion conditions typical of commercial operation.

12.10 Catalyst Activation and Deactivation Processes

Vital aspects of every catalytic process are the catalyst start-up and its decay (see also Chapter 13). In many cases, the FT-S needs to be run for some time before it is fully activated and optimum activity is reached [57]. All catalysts eventually die; however, the longer they survive and function in the reactor, the more economical and practicable the process. Thus, mechanisms involved in the activation and the deactivation of FT-S catalysts are of great importance. Various forms of activation are used. A common practice is to add a promoter to the main catalyst, for example, alkali (K_2O) to iron catalysts.

Bartholomew's classical review lists six causes of catalyst deactivation of three classes – chemical, mechanical, and thermal deactivation: (i) poisoning, (ii) fouling, (iii) thermal degradation, (iv) vapor compound formation accompanied by transport, (v) vapor–solid and/or solid–solid reactions, and (vi) attrition/crushing [63]. A brief summary of the key features is given below.

The poisons most likely to affect typical FT metals (cobalt, iron, and ruthenium) are sulfur (and sulfur compounds such as H_2S), ammonia, water, and possibly metal carbonyls. It has been found that the sulfur tolerances of Ni, Co, Fe, and Ru are very low: each can suffer three to four orders of magnitude loss in activity at 15–100 ppb of H_2S. Sulfur and sulfur compounds usually act by binding strongly to the FT metal and thus preventing the reagent gases from interacting with the metal.

Another type of deactivation is shown by certain gases that can destroy the active phases; for example, at high temperatures, carbon dioxide can oxidize the active iron carbides to inactive iron oxide. A variation on this is the transformation of catalytically active carbides into inactive oxides on Fe/K/Cu catalysts. Another form of deactivation arises from loss of active metal atoms from the surface by reaction with carbon monoxide to form volatile carbonyls.

A major product of the FT-S reaction is water that can also act as an oxidant for the catalyst to convert it into (catalytically inert) oxide. The effect of water on the FT-S is still the subject of considerable study (see Section 12.7). In some cases, oxidation of model catalysts with water was found to be difficult and particle size dependent [59].

Fouling (the physical deposition of material onto the catalyst surface) blocks sites and/or pores and leads to loss of activity. It can be caused by the formation of high molecular weight waxes and cokes, and in the advanced stages it may result in disintegration of catalyst particles. Deposits of filamentous carbon due to CO disproportionation during operation at relatively high temperatures and/or at low H_2/CO or steam/C ratios can occur in FT-S. Another important cause of deactivation is the sintering of the catalyst so that fewer active sites are available.

12.11
Desorption and Displacement Effects

The above discussion has focused chiefly on the effects on the product distribution of operating conditions such as temperature, pressure, the nature of the catalyst, the composition of the feedstock, and the reactor type. There is also a further set of variables that must be considered, namely, the effects of secondary reactions on the product distributions [61, 62]. Thus, once a primary product has been formed, it is desorbed or displaced by other materials in the reactor mix, which can go on to influence the final products found. These effects can include growth and cleavage of the polymer products as well as hydrogenation and dehydrogenation and carbonylation reactions. The wide range of possible secondary products can sometimes make it hard to distinguish them from the primarily formed substances.

12.12
Directions for Future Researches

Although the broad outlines of the mechanisms involved in the FT-S are clear, many details still need to be resolved. We describe some of the important outstanding questions relating to the basic underlying science and how they may most usefully be addressed.

12.12.1
Surface Spectroscopic Studies

While the FT-S is a complex process with many interrelated parts, a fundamental understanding of the interactions between the surface metal atoms and the various adsorbates (CO, hydrogen, and also the often transient reaction intermediates) should be able to clarify the initial stages. As is explained in Chapter 11, more knowledge of the nature, structures, and reactivities of the surface intermediates will go a long way to giving a sound basis for mechanistic suggestions. In recent years, many sophisticated spectroscopic techniques have been developed for such studies such as EXAFS and SFG (sum frequency generation) and polarization–modulation infrared reflection absorption spectroscopy, and these can now be more widely applied.

One important reason for further studies is that techniques have been developed that can be applied under realistic conditions. Previously, many of the spectroscopic techniques were only viable under ultrahigh vacuum conditions, but more are now being developed that give useful results under the conditions of temperature and pressure ($\sim$200 °C, $\geq$10 atm) where heterogeneously catalyzed reactions normally occur.

12.12.2
Surface Microscopic Studies

In a similar vein as the spectroscopies mentioned in Section 12.11.1, various microscopic techniques are now available that allow the study of surfaces and adsorbed species under realistic catalytic conditions. These may even lead to our being able to "see" what happens on a surface at the atomic level. Since it is now clear that many of the relevant transformations during the catalysis take place at the interfaces between the metal and the support, frequently an oxide, this should be a key area for detailed investigation as it would be of considerable help if the precise surface structure responsible for a given catalysis could be defined. Such measurements can then lead to analysis of the effects of changing the catalyst composition on the products. For example, what features of the surface determine whether in the FT hydrocarbon synthesis reaction C + C coupling (leading to longer chains) or hydrogen-transfer (chain termination) is preferred. It will also be valuable to probe how and why some metal oxide supports exhibit SMSI (strong metal–surface interactions) (Section 11.3) that can significantly modify the hydrogenation properties of metals.

12.12.3 Labeling and Kinetic Studies

Labeling and kinetic studies have been at the forefront of catalytic investigations for many years, and they still can provide the answers to many questions. So far the most instructive results have come from the use of ^{13}C- and ^{14}C-labeled CO; extension of such studies to ^{18}O-enriched CO will be useful in shedding light on the routes by which some of the FT oxygenate products arise.

It is not yet clear if the use of deuterium in place of H_2 can also be useful as a label and this should be carefully investigated. If reproducible conditions can be found where D is not randomly scrambled during a catalytic reaction, it should provide immense and useful knowledge.

One conundrum in this area is how apparently small variations in the nature of the catalyst surface can change the whole reaction, for example, going from methanation to hydrocarbon synthesis to methanol synthesis. Such variations must ultimately be under electronic control, but we know very little about the whys and wherefores and thus how to tune the reactions effectively.

Further kinetic studies should also be carried out involving SSITKA (steady-state isotopic transient kinetic analysis) in which the gas flow over a catalyst is switched between ^{12}CO and ^{13}CO and the changes in products determined.

Considerable attention from many viewpoints is being focused on the preparation, characterization, and properties of nanoparticles since such materials have special characters and special properties. Many active catalysts are in nanoparticulate form and studies on the relationship between nanoparticles and heterogeneous catalysts will have considerable impact on the development of catalysts with novel and useful properties.

12.12.4 Theoretical Calculations

The application of *DFT* and *ab initio* calculations have already led to many new insights into what happens in reactions such as those occurring on surfaces.

Newer and more sophisticated programs are being implemented and will certainly lead to better understanding of the interactions between the surfaces and the adsorbates.

12.13 Caveat

As in all such discussions of mechanisms, it is important to remind ourselves that although many plausible schemes can be drawn, the factors that determine which path is followed depend on the relative rates of the several possibilities. These are sometimes hard to establish.

Nevertheless, we are confident that given sufficient encouragement from industry and governments for further basic studies, scientists and engineers will use the information now available to design and construct new and more effective catalytic processes.

References

1 Anderson, R.B. (1984) *The Fischer–Tropsch Synthesis*, Academic Press, Orlando, FL.
2 Steynberg, A. and Dry, M. (2004) *Fischer–Tropsch Technology*, Elsevier, Amsterdam.
3 Rhodes, C., Hutchings, G.J., and Ward, A.M. (1995) *Catal. Today*, **23**, 45–58.
4 Khodakov, A.Y., Chu, W., and Fongarland, P. (2007) *Chem. Rev.*, **107** (5), 1692–1744.
5 Khodakov, A.Y. (2009) *Catal. Today*, **144**, 251.
6 Fischer, F. and Tropsch, H. (1926) *Brennst. Chem.*, **7** (7), 97–104.
7 Fischer, F. and Tropsch, H. (1926) *Chem. Ber.*, **59**, 830–831.
8 Biloen, P., Helle, J.N., and Sachtler, W.M. H. (1979) *J. Catal.*, **58** (1), 95–107.
9 Biloen, P., Helle, J.N., and Sachtler, W.M. H. (1981) *Adv. Catal.*, **30**, 165–216.
10 Brady, R.C. and Pettit, R. (1980) *J. Am. Chem. Soc.*, **102**, 6181.
11 Brady, R.C. and Pettit, R. (1981) *J. Am. Chem. Soc.*, **103**, 1287.
12 Maitlis, P.M., Quyoum, R., Long, H.C., and Turner, M.L. (1999) *Appl. Catal. A – Gen.*, **186**, 363–374.
13 Shetty, S., Jansen, A.P.J., and van Santen, R.A. (2009) *J. Am. Chem. Soc.*, **131** (36), 12874.
14 Pichler, H. (1952) *Adv. Catal.*, **4**, 271–341.
15 Bagno, A., Bukala, J., and Olah, G.A. (1990) *J. Org. Chem.*, **55**, 4284–4289.
16 Fahey, D.H. (1981) *J. Am. Chem. Soc.*, **103**, 136–141.
17 Dombek, B. (1983) *Adv. Catal.*, **32**, 325–416.
18 Knifton, J.F., Lin, J.J., Storm, D.A., and Wong, S.F. (1993) *Catal. Today*, **18** (4), 355–384.
19 Chiusoli, G.P. and Maitlis, P.M. (2006) *Metal-Catalysis in Industrial Organic Processes*, RSC Publishing, Cambridge, p. 290.
20 Maitlis, P.M., Haynes, A., Sunley, G.J., and Howard, M.J. (1996) *J. Chem. Soc., Dalton Trans.*, 2187–2196.
21 Collman, J.P., Hegedus, L.S., Norton, J.R., and Finke, R.G. (1987) *Principles and Applications of Organotransition Metal Chemistry*, University Science Books, Mill Valley, CA.
22 Gaube, J. and Klein, H.-F. (2008) *J. Mol. Catal. A*, **283**, 60–68.
23 Gaube, J. and Klein, H.-F. (2010) *Appl. Catal. A – Gen.*, **374**, 120–125.
24 Dry, M. (1999) *Appl. Catal. A – Gen.*, **189**, 185–190.
25 Turner, M.L., Long, H.C., Shenton, A., Byers, P.K., and Maitlis, P.M. (1995) *Chem. Eur. J.*, **1**, 549–556.

26 Quyoum, R., Berdini, V., Turner, M.L., Long, H.C., and Maitlis, P.M. (1998) *J. Catal.*, **173**, 355–365.

27 Hugues, F., Besson, B., Bussiere, P., Dalmon, J.A., Basset, J.M., and Olivier, D. (1981) *Nouv. J. Chim.*, **5**, 207.

28 Davis, B. (2009) *Catal. Today*, **141**, 25–33.

29 Schulz, H. (1999) *Appl. Catal. A – Gen.*, **186**, 3–12.

30 Schulz, H., Erich, E., Gorre, H., and Steen, E.V. (1990) *Catal. Lett.*, **7**, 157–167.

31 Baker, J.A. and Bell, A.T. (1982) *J. Catal.*, **78**, 165–181.

32 Bell, A.T. (1995) *J. Mol. Catal. A*, **100**, 1–11.

33 Raje, A. and Davis, B.H. (1996) *RSC Special. Rep. Catal.*, **12**, 52–131.

34 Lohitharn, N. and Goodwin., J.J. (2009) *Catal. Commun.*, **10**, 758.

35 Percy, L.T. and Walter, R.I. (1990) *J. Catal.*, **121**, 228–235.

36 Turner, M.L., Marsih, N., Mann, B.E., Quyoum, R., Long, H.C., and Maitlis, P.M. (2002) *J. Am. Chem. Soc.*, **124**, 10456–10472.

37 Quyoum, R., Berdini, V., Turner, M.L., Long, H.C., and Maitlis, P.M. (1996) *J. Am. Chem. Soc.*, **118**, 10888–10889.

38 Kummer, J.T., Dewitt, T.W., and Emmett, P.H. (1948) *J. Am. Chem. Soc.*, **70**, 3632–3643.

39 Eidus, Y.T. (1967) *Russ. Chem. Rev.*, **36**, 338–349.

40 Tau, L.-M., Dabbagh, H.A., and Davis, B.H. (1990) *Energy Fuels*, **4**, 94–99.

41 Shi, B. and Davis, B.H. (2003) *Top. Catal.*, **26**, 157.

42 Nùrskov, J., Bligaard, T., Hvolbñk, B., Abild-Pedersen, F., Chorkendorff, L., and Christensend, C.H. (2008) *Chem. Soc. Rev.*, **37**, 2163.

43 Politano, A., Formoso, V., and Chiarello, G. (2008) *J. Chem. Phys.*, **129**, 164703.

44 Jenkins, S. and King, D.A. (2000) *J. Am. Chem. Soc.*, **122**, 10610–10614.

45 McCrea, K., Parker, J.S, Chen, P.L., and Somorjai, G. (2001) *Surf. Sci.*, **494**, 238–250.

46 Vannice, M.A. and Garten, R.L. (1979) *J. Catal.*, **56**, 236–248.

47 Vannice, M. (1997) *Top. Catal.*, **4**, 241–248.

48 Shetty, S., van Santen, R.A., Stevens, P.A., and Raman, S. (2010) *J. Mol. Catal. A – Chem.*, **330**, 73–87.

49 Shriver, D. and Sailor, M.J. (1988) *Acc. Chem. Res.*, **21**, 374–379.

50 Chisholm, M.H., Johnston, V.J., Streib, W.E., and Huffman, J.C. (1992) *J. Am. Chem. Soc.*, **114**, 7056–7065.

51 Miller, A.J.M., Labinger, J.A., and Bercaw, J.E. (2010) *Organometallics*, **29**, 4499–4516.

52 Maitlis, P.M. (2004) *J. Organomet. Chem.*, **689**, 4366–4374.

53 Ciobica, I.M., Kramer, G.J., Ge, Q., Neurock, M., and van Santen, R.A. (2002) *J. Catal.*, **212**, 136.

54 Davis, B.H. (2009) *Catal. Today*, **141**, 25–33.

55 de Smit, E. and Weckhuysen, B.M. (2008) *Chem. Soc. Rev.*, **37**, 2758–2781.

56 Hahn, C. (2004) *Chem. Eur. J.*, **10**, 5888.

57 Maitlis, P.M. and Zanotii, V. (2009) *Chem. Commun.*, 1619–1634.

58 Storch, H.H., Golumbic, N., and Anderson, R.B. (1951) *The Fischer–Tropsch and Related Syntheses*, John Wiley & Sons, Inc., New York.

59 Saib, A.M., Borgna, A., van de Loosdrecht, J., van Berge, P.J., Geus, J.W., and Niemantsverdriet, J.W. (2006) *J. Catal.*, **239**, 326–339.

60 Madon, R.J., Braden, D., Kandoi, S., Nagel, P., Mavrikakis, M., and Dumesic, J.A. (2011) *J. Catal.*, **281**, 1–11.

61 Botes, F.G. (2007) *Energy Fuels*, **21**, 1379–1389.

62 Schulz, H., Erich, E., Gorre, H., and Steen, E.V. (1990) *Catal. Lett.*, **7**, 153–167.

63 Bartholomew, C.H. (2001) *Appl. Catal. A – Gen.*, **212**, 17–60.

64 Kolasinski, K. (2004) *Surface Science*, John Wiley & Sons, Inc., New York.

65 Torres Galvis, H.M., Bitter, J.H., Khare, C.B., Ruitenbeek, M., Dugulan, A.J., and de Jong, K.P. (2012) *Science*, **335**, 835.

Part Four
Environmental Aspects

Greener Fischer-Tropsch Processes for Fuels and Feedstocks, First Edition. Edited by Peter M. Maitlis and Arno de Klerk.

13
Fischer–Tropsch Catalyst Life Cycle

Julius Pretorius and Arno de Klerk

Synopsis

The term "catalyst life cycle" defines the manufacture and the disposal of the catalyst as well as the manner in which it is used in the synthesis reactor.

13.1
Introduction

An important aspect of the commercial application of the Fischer–Tropsch (FT) process is the catalyst life cycle. This includes not only what happens to the catalyst in the synthesis reactor but also its manufacture and the disposal of the spent catalyst. As the part of the catalyst life cycle under synthesis conditions is fully discussed elsewhere in this book, this chapter deals only with the catalyst preparation and the fate of the spent catalysts.

At present, iron and cobalt are the only two Fischer–Tropsch active metals employed industrially; they may be grouped into three distinct families: precipitated, supported, and fused catalysts. A number of issues related to the manufacture, consumption, and disposal of the FT catalysts need to be addressed. In catalyst manufacturing, the material requirements and waste streams generated during manufacture are important in determining the viability and cost of a catalyst. The amount of solid waste generated as spent catalyst is a direct consequence of the catalyst lifetime, which is a function of the catalyst's resistance to deactivation and its ability to be regenerated. Another aspect of importance is the fate of the spent catalyst. A number of options, ranging from recycling to disposal are available, and the one selected will depend on the interplay between regulatory requirements and economic considerations.

Much of the information related to these issues is proprietary and confidential, which limits our ability to delve too deeply into the specifics. However, sufficient information from the scientific and patent literatures is available to allow the generic discussion given here.

Greener Fischer-Tropsch Processes for Fuels and Feedstocks, First Edition. Edited by Peter M. Maitlis and Arno de Klerk.

13.2 Catalyst Manufacturing

The main FT catalyst families in commercial use today are (a) precipitated iron low-temperature FT (Fe-LTFT) catalysts, (b) supported cobalt low-temperature (Co-LTFT), catalysts and (c) fused iron high-temperature (Fe-HTFT) catalysts.

13.2.1 Precipitated Fe-LTFT Catalysts

The preparation of low-temperature iron-based FT (Fe-LTFT) catalysts may take place via a variety of routes, not all of which are in commercial use. All the routes utilize water as solvent and contain the same generic steps. These involve the preparation of aqueous solutions of the iron salt precursor and the required transition metal promoter(s) and the formation by precipitation of an iron oxyhydroxide that contains the appropriate promoters at the required level. This precipitate is then washed to remove unwanted residual cations and anions; alkali metal promoters and a structural promoter (if required) are then added. The final material is then ready for calcination and reduction to metal.

Ferric nitrate, $Fe(NO_3)_3$) is most often used as the iron source. It may be prepared by dissolution of metallic iron in concentrated nitric acid or from the commercially available hydrated salt, $Fe(NO_3)_3{\cdot}9H_2O$ [1, 2]. Other iron salts have been used, but the resulting catalysts did not perform satisfactorily [3]. This was ascribed to anion poisoning, neglecting the influence of the anion on iron oxide phase formation and also on the precursor surface area and pore volume of the final catalyst [4, 5]. The importance of the precursor oxide phase in determining catalyst performance has been illustrated recently [1, 6]. Dissolution of metallic iron in nitric acid also gives rise to the formation of nitrogen oxides (NO_x), which need to be controlled. Apart from adverse environmental effects, the formation of NO_x also leads to the loss of nitric acid. Benham *et al.* [2] mitigated acid loss by purging the solution with O_2 to oxidize NO to NO_2, which is more soluble in water than NO and dissociates in water to form nitric acid and NO.

Precipitation is effected by neutralizing the acidic metal nitrate solution with a base containing a Group 1 or Group 2 metal or NH_4^+ as cation, resulting in the formation of an iron (hydr)oxide phase that also contains the required promoter(s). The precipitation removes more than 99% of the iron and transition metal promoters from solution, while the mother liquor has a high concentration of Group 1 or 2 metal or NH_4^+ cations and nitrate. These salts adversely affect catalyst performance and must be removed by extensive washing of the iron (hydr)oxide precipitate. A significant volume of water containing highly soluble nitrate salts at concentrations that preclude discharge into the environment is produced. For example, if Na_2CO_3 is used as precipitation agent, the mother liquor may contain 1.4 kg Na^+ and 3.4 kg NO_3^- per kg of iron precipitated.

During the calcination step, there is a possibility of NO_x emissions, depending on the efficiency of the washing step. The reduction step does not produce a waste stream, however, if it is performed outside the reactor, reduction gases will have to be dealt with.

13.2.2 Supported Co-LTFT Catalysts

The patent literature reveals that commercial supported cobalt FT catalysts are prepared by impregnating an oxidic support, such as alumina [7, 13, 14, 20, 25], silica [15, 20, 21, 25], titania [16, 20], or mixtures of these oxides [8–12, 20, 22, 25, 26], with solutions of the catalytically active metallic compounds by incipient wetness or slurry impregnation using water as solvent [27]. Cobalt and precious metal promoters are typically added as nitrate salts, for example, $Co(NO_3)_2{\cdot}6H_2O$ and $[Pt(NH_3)_4](NO_3)_2$ [9]. Organic solvents may also be employed, but the handling of large volumes of organic solvents adds to the complexity and cost of the catalyst preparation facility [18]. Upon completion of the impregnation process, excess solvent is removed; the resulting solvent stream is practically free of metal salts, thus, limited liquid effluent streams may be expected to arise.

After drying, the catalyst precursor is heated to decompose the metal nitrate salts to the oxides; this calcination will result in the liberation of gaseous NO_x [28]. A typical supported cobalt catalyst contains approximately 300 g Co per 1000 g support. Assuming a 100% impregnation efficiency, achieving this level of cobalt will require 5 mol $Co(NO_3)_2{\cdot}6H_2O$; thus, 10 mol of nitrogen as nitrous oxides per kg support may be expected during calcination. The environmental impacts of NO_x emissions are well known [29, 30] and typical NO_x emission control technologies such as chemical scrubbing, adsorption, and catalytic reduction need to be put in place [31, 32]. Depending on the NO_x emission control technology being used, waste streams consisting of aqueous solutions of nitrate or sulfate salts or spent catalyst will have to be managed.

13.2.3 Fused Fe-HTFT Catalysts

The manufacture of commercial iron-based HTFT catalysts has been discussed by Dry [1] and Steynberg *et al.* [33]. The process consists of the electrical fusion of mill scale with the required promoters. The main chemical promoter is K_2O and the structural promoters are MgO or Al_2O_3. The molten product from the fusion furnace is poured into ingots and allowed to cool. The rate of cooling determines the optimal distribution of promoters through the catalyst. The resulting ingots are crushed to the required particle size and reduced in hydrogen. A number of parameters are of importance.

First, the particle size of the crushed material needs to be accurately controlled over a narrow range to maximize fluidization efficiency and minimize catalyst loss from the reactor. In this respect, the particle size distribution for fixed fluidized bed operation is more sensitive than that for circulating fluidized bed operation. It has been suggested that for a fixed fluidized bed, an average particle size range of

30–50 μm can be maintained by continuously removing catalyst and adding larger catalyst particles, typically around 80 μm [34]. As the catalyst ages, carbon builds up resulting in a decrease in bed density, and ultimately leading to the entrainment of small Fe-FT catalyst particles. Catalyst particles smaller than 30 μm should therefore be limited [34]. For fixed fluidized bed operation, a cutoff diameter of 22 μm is industrially employed to define catalyst particles as fines [33].

Second, the interplay between the mechanical strength of the catalyst and its surface area is of importance. Higher surface area results in higher activity but lower mechanical strength, resulting in fines formation and catalyst loss. An important control parameter is the molar ratio $Fe^{3+}/2Fe^{2+}$ in the final prereduced catalyst. Fusion conditions determining the mole ratio include the furnace temperature, residence time, and reducing agent consumption.

Third, the distribution of promoters is heterogeneous. Most iron ores have some silica (SiO_2) contamination and the alkali (K_2O) promoter forms an alkali silicate phase, which does not enter into a solid solution with the magnetite. On cooling, the alkali silicate phases are present as occlusions distributed through the catalyst ingot. When the ingot is crushed, the finer catalyst particles often lack the alkali promoter [35], so the catalyst size is important even from a catalysis perspective apart from a hydrodynamic perspective. Loss of alkali promoter can be compensated for by co-feeding some alkali separately and the manner in which it is added (during catalyst synthesis or operation) has little effect on performance [1].

Not much has been reported about the waste streams produced during the manufacture of fused FT catalysts. It was noted that the manufacturing of the fused iron FT catalyst used industrially by Sasol (and PetroSA) is similar to that of an ammonia synthesis catalyst [35]. Undersized material from ammonia catalyst preparation is usually recycled to the electric arc furnace [36], thereby avoiding solid waste from the catalyst manufacturing operation. The same is possible during fused Fe-FT catalyst manufacturing. The nature of the impurities in the iron oxide raw material is important, with sulfur being especially deleterious. It was reported that some mill scales employed for catalyst manufacturing had to be roasted to remove associated sulfur before being used for catalyst preparation [35]. In such cases, some SO_x emissions will be associated with raw material preparation.

13.3 Catalyst Consumption

Economic rather than environmental considerations, determine catalyst consumption in industrial FT facilities. This has skewed the perception of the longevity of iron-based compared to cobalt-based catalysts. The price ratio of Fe and Co is around 1: 1000 [37], making regeneration and metal reclamation of Co-based catalysts imperative. The same is not true of Fe-based catalysts. Furthermore, it has been reported that Fe-based catalysts can be safely disposed of by land filling, whereas Co-based catalysts are potentially harmful to the environment and cannot be disposed in the same way [38]. Although land filling of spent Fe-FT catalysts has been an accepted practice in certain jurisdictions, this may not be the case elsewhere or in the future. Without proper trace metal composition and leachability analysis of spent Fe-FT catalysts, the "benign" nature of spent Fe-FT catalysts is unconfirmed at present. The low cost of iron resulted industrially in more wasteful practices than for cobalt, but this does not have to continue, since some iron may be regenerated.

13.3.1
Catalyst Lifetime during Industrial Operation

FT catalysts deactivate with time on stream. The rate of catalyst deactivation depends on the nature of the catalyst and its operation. The nature of the deactivation will determine whether the catalyst can be rejuvenated *in situ*, regenerated *ex situ*, or whether it should be replaced. The deactivation of Fe-based and Co-based catalysts is described in Chapters 8 and 9 and in the literature [39, 40].

Iron-based catalysts are employed in various processes and with different reactor configurations. The reported catalyst lifetime of precipitated iron-based catalysts used industrially for low-temperature FT (LTFT) synthesis in fixed bed and slurry bubble column reactors is 70–100 days [41]. The iron-based catalyst used for medium-temperature FT (MTFT) synthesis in a slurry bubble column reactor has a similar lifetime, while the fused iron-based catalyst employed for high-temperature FT (HTFT) synthesis in fluidized bed reactors has an equivalent lifetime of around 40–45 days [41].

Generally speaking, cobalt-based catalysts, when properly designed and operated, have longer catalyst lifetimes than iron-based catalysts. The first-generation Co-LTFT catalyst for the Shell middle distillate synthesis (SMDS) process requires *in situ* rejuvenation only every 9–12 months and it has an overall lifetime around 5 years [42]. The new-generation catalyst for SMDS reportedly has an even longer lifetime [43]. Deactivation of the Sasol Co-LTFT catalyst for slurry bubble column operation is faster, with 50% activity loss reported after 50 days on stream [44]. This catalyst can be regenerated *ex situ* to fully restore activity [45], but the industrial cycle length and overall catalyst lifetime have not been reported for the Sasol Co-LTFT catalyst.

13.3.2
Fe-LTFT Catalyst Regeneration

Due to the low cost of Fe-LTFT catalysts, limited effort has been expended on catalyst regeneration. However, Rentech patented a process for the regeneration of Fe-LTFT catalysts [46, 47], where spent catalyst from a slurry bubble column reactor is dewaxed and then oxidized to its original preactivated state. The oxidized catalyst precursor is then reactivated. No information on the effectiveness of the regeneration procedure is provided, but it was stated that an issue with the regeneration regime is controlling temperature and preventing sintering [46]. The link between catalyst precursor properties and eventual catalyst performance [1, 6], together with the thermodynamically metastable nature of the precursor iron oxide phase [5], suggests that reoxidation followed by reactivation may not be a viable regeneration approach to fully restore catalyst activity.

13.3.3
Fe-HTFT Catalyst Regeneration

The low cost of the iron-based HTFT catalyst, together with the perceived environmentally innocuous nature of the spent catalyst, has led commercial operations to dispose of spent catalyst rather than regenerate it [38]. This practice is contingent on local environmental regulations regarding the disposal of solid wastes. Should regulations change, operators may be forced to dispose of spent catalyst in specially licensed hazardous waste disposal facilities. This may provide a cost-driven incentive to regenerate spent catalyst. Lanning [48] invented a process for the regeneration of "fluidized synthesis" iron catalysts. The process is based on the premise that the major deactivation mode is the deposition of carbonaceous material on the catalyst. Regeneration is effected by contacting deactivated catalyst with oxygen at a temperature sufficiently high to (a) oxidize surface carbonaceous deposits to carbon oxides and (b) melt the iron oxides formed during the process. The molten iron oxide droplets are cooled, reduced by hydrogen, and returned to the synthesis reactor. No evidence of the efficacy of the regeneration process was offered. As discussed in Section 13.2.3, parameters of importance in the preparation of a Fe-HTFT catalyst are fusion and melt cooling conditions, and it is not apparent to what extent such a regeneration process allows the control of these parameters. Since this regeneration procedure does not address the change in particle size distribution, a more practical approach would be to re-fuse the regenerated catalyst with promoter top-up, as was suggested by Dry [35].

In order to develop a Fe-HTFT catalyst regeneration strategy, one must be able to address the following key issues that require Fe-HTFT catalyst replacement, not all of which are related to catalyst activity decline.

13.3.3.1 Fouling by Carbon

This is the main cause of catalyst activity loss. Spent fused Fe-HTFT catalysts can be rejuvenated by reduction in H_2 at a temperature above 350 °C, which restores catalyst activity and selectivity close to the original values [1]. The success of this procedure is ascribed mainly to the removal of the carbon deposits and not due to reduction of the catalyst itself.

13.3.3.2 Loss of Alkali Promoter

As mentioned before, the alkali promoter can be added as a separate material, without affecting catalyst performance [1]. Loss of alkali promoter does not necessitate a rejuvenation procedure that requires the catalyst to be refused. Online addition of alkali promoter can take place directly to the FT reactor.

13.3.3.3 Mechanical Attrition

The particle size distribution of the Fe-HTFT catalyst is very important (Section 13.2.3). If the fines content becomes too high, the voidage of the fluidized bed increases significantly [33], resulting in decreased reactor productivity even when catalyst activity has not deteriorated. Furthermore, the load on the cyclones to separate the catalyst from the product gas increases, which may also lead to operational problems. Whatever regeneration procedure is employed to rejuvenate the catalyst, it must include catalyst screening to remove catalyst fines. The catalyst fines can be reused only by returning them to the electric arc furnace used to manufacture the catalyst.

13.3.3.4 Sulfur Poisoning

Sulfur cannot be removed during rejuvenation. In ammonia synthesis that employs a fused Fe catalyst very similar to that used in FT synthesis even reduction at 500 °C with pure gas cannot reverse the effect of sulfur poisoning [36]. Since low levels of sulfur are capable of substantially deactivating Fe-HTFT catalysts, the ability to rejuvenate the Fe-HTFT strongly depends on the contribution of sulfur poisoning to the overall catalyst deactivation. It also stands to reason that over time a purge stream of catalyst will have to be roasted to remove sulfur before being recycled to the electric arc furnace.

13.3.4
Co-LTFT Catalyst Regeneration

Not surprisingly, in addition to rejuvination practices, significantly more effort has been expended on the regeneration of cobalt catalysts than for iron catalysts. Patents on this topic have been filed by Exxon [49–53], Syntroleum [54], and Conoco [55]. Exxon utilize a continuous reactivation of catalyst in a separate reactor vessel, operated at the synthesis reactor pressure but at a lower temperature. Hydrogen is used as the regeneration agent. Syntroleum do not describe the regeneration conditions employed in their process, but the approach used is very similar to that described by Exxon in that regeneration takes place in one or more separate reactor vessels. Conoco developed a regeneration regime that takes place in the synthesis reactor and utilizes hydrogen-containing steam as regeneration agent. None of the process descriptions provide any hard evidence that allows an assessment of the effectiveness of the proposed procedures, and it is therefore difficult to judge what the effect of the regeneration procedures are on catalyst activity and life span.

Saib *et al.* [45] recently described a reduction–oxidation–reduction regeneration procedure that was capable of restoring activity to that of fresh catalyst. In this approach, the catalyst is removed from the synthesis reactor, dewaxed using heptane, oxidized in an air/N_2 mixture, and reduced in pure hydrogen. It is claimed that this regeneration approach reverses deactivation arising from sintering, carbon deposition, and surface reconstruction. These workers provide evidence that their regeneration method is successful in restoring catalyst activity. However, the rate of deactivation after the activation step was not discussed. According to Tsakoumis *et al.* [40], it appears that the deactivation rate after a regeneration step is higher than before the regeneration.

13.4 Catalyst Disposal

Catalyst disposal practices are determined by regulatory and economic considerations. Typically, jurisdictions regulate the disposal of solid wastes according to a waste classification system. This is a highly specialized area and a number of Acts and Regulations are usually applicable, the discussion of which falls outside the scope of this chapter. In general, regulators may require an assessment and classification of spent catalyst based on chemical composition and the leachability of its components. Properties such as combustibility will also be assessed. If the waste is considered hazardous, its disposal is strictly regulated and this may add significant cost, particularly if catalyst volumes are taken into account.

At older facilities, such as the original Sasol plant in South Africa, spent Fe-LTFT catalyst was codisposed with ash from the gasifiers [35]. The pyrophoric nature of the spent catalyst was a complicating matter in this practice. Because coal as feedstock for syngas generation was discontinued at the facility in 2004, the disposal practice was halted, and, currently, spent catalyst is fed into coal-fired boilers used for steam generation. This practice takes advantage of the spent catalyst heating value. Furthermore, it oxidizes the spent catalyst, thereby removing its pyrophoric properties. The passivated spent catalyst is then codisposed with the boiler ash.

The value of metals constituting the catalyst is significant in determining catalyst disposal practices. If it is economical, the catalyst components may be recovered for reuse. This is not a trivial matter in the case of supported catalysts, as the metallic, catalytically active material is dispersed throughout a metal oxide support. The problem of recovering valuable metals from spent catalyst is thus akin to extraction of metals from ore, requiring either a hydrometallurgical or a pyrometallurgical approach [56]. Matjie *et al.* [57] utilized a hydrometallurgical approach to recover aluminum, cobalt, and platinum from a spent GTL catalyst in a rather involved 13-step process. It employs a high-temperature, high-pressure caustic leach to solubilize aluminum, a nitric acid leach to solubilize cobalt, and an aqua regia leach to solubilize platinum. The metals are then separated by selective precipitation. Metal recovery efficiencies reportedly range from 91% for platinum to more than 99% for cobalt and aluminum. Brumby *et al.* [58] commented that the hydrometallurgical approach is fraught with difficulty due to its dependence on support type; they proposed a pyrometallurgical approach with fewer processing steps and a decreased dependence on support type.

Very limited evidence of efforts into the recovery of FT catalyst components is found in the literature [56, 57]. This may be due to companies not making public their activities in this area or it may simply indicate that it is not being pursued at all. However, comparing production rates of metals commonly used in FT catalysts with projected requirements based on future FT-based developments, the need for recycling of metallic components may become a necessity. For example, the global supply of the commonly used promoters platinum, ruthenium, and rhenium are 200, 20, and 50 t/a, respectively. Cobalt supply is in the region of 45 000 t/a. Assuming that a 100 000 bpd GTL plant requires approximately 500 tons of cobalt and 2.5 tons of the chosen promoter, it is evident that markets for promoter metals may be significantly impacted, adding to the impetus to recover and recycle catalyst components [58].

References

1 Dry, M.E. (1981) The Fischer–Tropsch synthesis, in *Catalysis Science and Technology* (eds J.R. Anderson and M. Boudart), Springer, pp. 159–255.

2 Benham, C.B., Bohn, M.S., and Yakobson, D.L. (1996) Process for the production of hydrocarbons, US Patent 5,504,118.

3 Hofer, L.J.E., Anderson, R.B., Peebles, W.C., and Stein, K.C. (1951) Chloride poisoning of iron–copper Fischer–Tropsch catalysts. *J. Phys. Chem.*, **55**, 1201–1206.

4 Cornell, R.M. and Schwertmann, U. (2003) *The Iron Oxides: Structure, Properties, Reactions, Occurrences and Uses*, 2nd edn, Wiley-VCH Verlag GmbH, Weinheim.

5 Schwertmann, U., Friedl, J., and Stanjek, H. (1999) From Fe(III) ions to ferrihydrite and then to hematite. *J. Colloid Interface Sci.*, **209**, 215–223.

6 Pretorius, P.J. (2011) On the preparation of low temperature iron Fischer–Tropsch catalysts: strategies that work and those that do not, in *ACS Symposium Series 1084: Synthetic Liquids Production and Refining* (eds A. de Klerk and D.L. King), American Chemical Society, Washington, DC.

7 van Berge, P.J., van de Loosdrecht, J., Caricato, E.A., Barradas, S., and Sigwebela, B.H. (2001) Impregnation process for catalysts, US Patent 2001/0051589 A1.

8 Van Berge, P.J., van de Loosdrecht, J., and Visagie, J.L. (2003) Cobalt catalysts, US Patent 2003/0211940.

9 Van Berge, P.J., van de Loosdrecht, J., and Visagie, J.L. (2008) Cobalt catalysts, US Patent 7,375,055.

10 Van Berge, P.J., van de Loosdrecht, J., and Visagie, J.L. (2004) Cobalt catalysts, US Patent 6,835,690.

11 Van Berge, P.J., van de Loosdrecht, J., and Visagie, J.L. (2005) Cobalt catalysts, US Patent 6,897,177.

12 Visagie, J.L., Botha, J.M., Koortzen, J.G., Datt, M.S., Bohmer, A., van de Loosdrecht, J., and Saib, A.M. (2010) Catalysts, US Patent 2010/0144520.

13 Eri, S., Kinnari, K.J., Schanke, D., and Hilmen, A.M. (2008) Fischer–Tropsch catalyst, preparation and use thereof, US Patent 7,351,679.

14 Rytter, E., Eri, S., and Schanke, D. (2004) Fischer–Tropsch catalysts, WO 2004/043596.

15 Arcuri, K.B., Agee, K.L., and Agee, M.A. (2001) Structured Fischer–Tropsch catalyst system and method, US Patent 6,262,131.

16 Rytter, E. (2008) Promoted Fischer–Tropsch catalysts, US Patent 2008/0255256.

17 Rytter, E., Skagseth, T.H., Wigum, H., and Sincadu, N. (2005) Fischer–Tropsch catalysts, WO 2005/072866.

18 Singleton, A.H., Oukaci, R., and Goodwin, J.G. (1999) Processes and catalysts for conducting Fischer–Tropsch synthesis in a slurry bubble column reactor, US Patent 5,939,350.

19 Rytter, E. (2006) Promoted Fischer–Tropsch catalysts, WO 2006/032907.

20 Zennaro, R., Gusso, A., Chaumette, P., and Roy, M. (2000) Catalytic composition suitable for the Fischer–Tropsch process, US Patent 6,075,062.

21 Espinoza, R.L., Jothimurugesan, K., Raje, A.P., Coy, K.L., and Srinivasan, N. (2004) Attrition resistant bulk metal catalysts and

methods of making and using same, US Patent 2004/0259960.

22 Espinoza, R.L., Jothimurugesan, K., Coy, K.L., Ortego, J.D., Srinivasan, N., and Ionkina, O.P. (2009) Silica–alumina catalyst support, catalysts made therefrom and methods of making and using same, US Patent 7,541,310.

23 Jin, Y. and Espinoza, R.L. (2004) High hydrothermal stability catalyst support, US Patent 2004/0127353.

24 Raje, A.P. and Espinoza, R.L. (2005) Novel method for improved Fischer–Tropsch catalyst stability and higher stable syngas conversion, US Patent 2005/0049317.

25 Minderhoud, J.K. and Post, M.F.M. (1985) Preparation of catalyst for producing middle distillates from syngas, US Patent 4,522,939.

26 Visagie, J.L. and Veltman, H.M. (2007) Producing supported cobalt catalysts for the Fisher–Tropsch synthesis, US Patent 2007/0287759.

27 Haber, J., Block, J.H., and Delmon, B. (1995) Manual of methods and procedures for catalyst characterization. *Pure Appl. Chem.*, **67**, 1257–1306.

28 van de Loosdrecht, J., Barradas, S., Caricato, E.A., Ngwenya, N.G., Nkwanyana, P.S., Rawat, M.A.S., Sigwebela, B.H., van Berge, P.J., and Visagie, J.L. (2003) Calcination of Co-based Fischer–Tropsch synthesis catalysts. *Top. Catal.*, **26**, 121–127.

29 Bailey, R.A., Clark, H.M., Ferris, J.P., Krause, S., and Strong, R.L. (2002) *Chemistry of the Environment*, 2nd edn, Academic Press, San Diego.

30 Manahan, S. (2000) *Environmental Chemistry*, Lewis Publishers, Boca Raton.

31 Thiemann, M., Scheibler, E., and Wiegand, K.W. (2000) Nitric acid, nitrous acid, and nitrogen oxides, in *Ullmann's Encyclopedia of Industrial Chemistry*, Wiley-VCH Verlag GmbH, Weinheim.

32 Berglund, R.L. (2000) Emission control, industrial, in *Kirk-Othmer Encyclopedia of Chemical Technology*, John Wiley & Sons, Inc., New York.

33 Steynberg, A.P., Espinoza, R.L., Jager, B., and Vosloo, A.C. (1999) High temperature Fischer–Tropsch synthesis in commercial practice. *Appl. Catal. A*, **186**, 41–54.

34 Storch, H.H. (1954) The Fischer–Tropsch process, in *The Chemistry of Petroleum Hydrocarbons* (eds B.T. Brooks *et al.*), Reinhold, New York, pp. 631–646.

35 Dry, M.E. (2004) FT catalysts, in *Fischer–Tropsch Technology* (eds A. Steynberg and M.E. Dry), Elsevier, Amsterdam, pp. 533–600.

36 Strelzoff, S. and Pan, L.C. (1964) Synthetic ammonia, in *Advances in Petroleum Chemistry and Refining* (eds K.A. Kobe and J.J. McKetta), John Wiley & Sons, Inc., New York, pp. 283–350.

37 Dry, M.E. (2004) Chemical concepts used for engineering purposes, in *Fischer–Tropsch Technology* (eds A. Steynberg and M.E. Dry), Elsevier, Amsterdam, pp. 196–257.

38 Vosloo, A.C., Dancuart, L.P., and Jager, B. (1998) Environmental aspects of Sasol Fischer–Tropsch technologies and products. in 11th World Clean Air and Environment Congress, International Union of Air Pollution Prevention Associations, Durban, South Africa, p. 6.

39 Weckhuysen, B.M. and de Smit, E. (2008) The renaissance of iron-based Fischer–Tropsch synthesis: on the multifaceted catalyst behaviour. *Chem. Soc. Rev.*, **37**, 2758–2781.

40 Tsakoumis, N.E., Rønning, M., Borg, Ø., Rytter, E., and Holmen, A. (2010) Deactivation of cobalt based Fischer–Tropsch catalysts: a review. *Catal. Today*, **154**, 162–182.

41 Mako, P.F. and Samuel, W.A. (1984) The Sasol approach to liquid fuels from coal via the Fischer–Tropsch reaction, in *Handbook of Synfuels Technology* (ed. R.A. Meyers), McGraw-Hill, New York, pp. 2.5–2.43.

42 Schrauwen, F.J.M. (2004) Shell middle distillate synthesis (SMDS) process, in *Handbook of Petroleum Refining Processes* (ed. R.A. Meyers), McGraw-Hill, New York, pp. 15.25–15.40.

43 Overtoom, R., Fabricius, N., and Leenhouts, W. (2009) Shell GTL, from bench-scale to world-scale, in *1st Annual Gas Processing Symposium*, Elsevier, Amsterdam.

44 Saib, A.M., Borgna, A., van de Loosdrecht, J., van Berge, P.J., and

Niemantsverdriet, J.W. (2006) XANES study of the susceptibility of nano-sized cobalt crystallites to oxidation during realistic Fischer–Tropsch synthesis. *Appl. Catal. A*, **312**, 12–19.
45 Saib, A.M., Moodley, D.J., Ciobîcă, I.M., Hauman, M.M., Sigwebela, B.H., Weststrate, C.J., Niemantsverdriet, J.W., and van de Loosdrecht, J. (2010) Fundamental understanding of deactivation and regeneration of cobalt Fischer–Tropsch synthesis catalysts. *Catal. Today*, **154**, 271–282.
46 Demirel, B., Bohn, M.S., Benham, C.B., Siebarth, J.E., and Ibsen, M.D. (2005) Method and apparatus for regenerating an iron-based Fischer–Tropsch catalyst, US Patent 6,838,487 B1.
47 Demirel, B., Bohn, M.S., Benham, C.B., Siebarth, J.E., and Ibsen, M.D. (2007) Method and apparatus for regenerating an iron-based Fischer–Tropsch catalyst, US Patent 7,303,731 B2.
48 Lanning, W.C. (1953) Regeneration of a Fischer–Tropsch reduced iron catalyst, US Patent 2,661,338.
49 Hsia, S.J. (1993) External catalyst rejuvenation system for the hydrocarbon synthesis process, US Patent 5,260,239.
50 Pedrick, L.E., Mauldin, C.H., and Behrmann, W.C. (1993) Draft tube for catalyst rejuvenation and distribution, US Patent 5,268,344.
51 Behrmann, W.C. and Leviness, S.C. (1994) Temperature control in draft tubes for catalyst rejuvenation, US Patent 5,288,673.
52 Leviness, S.C. and Mitchell, W.N. (1998) Catalyst rejuvenation in hydrocarbon synthesis slurry with reduced slurry recontamination, US Patent 5,811,363.
53 Chang, M., Coulaloglou, C.A., Hsia, S.J., and Mart, C.J. (1998) Slurry hydrocarbon synthesis process with multistage catalyst rejuvenation, US Patent 5,821,270.
54 Beer, G.L. (2001) Process and apparatus for regenerating a particulate catalyst, US Patent 6,201,030 B1.
55 Wright, H.A. (2002) Regeneration procedure for Fischer–Tropsch catalyst, US Patent 6,486,220 B1.
56 Brumby, A., Verhelst, M., and Cheret, D. (2005) Recycling GTL catalysts: a new challenge. *Catal. Today*, **106**, 166–169.
57 Matjie, R.H., de Wet, E.W., and Mdleni, M.M. (2004) Selective recovery of aluminium, cobalt and platinum values from a spent catalyst composition, US Patent 2004/0219082.
58 Brumby, A., Verhelst, M., and Cheret, D. (2005) Recycling GTL catalysts: a new challenge. *Catal. Today*, **106**, 166–169.

14
Fischer–Tropsch Syncrude: To Refine or to Upgrade?

Vincenzo Calemma and Arno de Klerk

Synopsis

The Fischer–Tropsch (FT) synthesis largely produces mixtures of linear hydrocarbons with some similarity to crude oil, that is known as *syncrude*. However, it also contains significant amounts of other organics, such as oxygenates and waxes. A rather different refining procedure than for crude oil is therefore needed. For many purposes, the syncrude can simply be upgraded and sold to oil refineries. However, selling products directly to the end-user requires them to have specific qualities that can only be produced by refining. Various cracking, isomerization, hydrotreating, and oligomerization procedures used in refining are described and their environmental impacts evaluated.

14.1
Introduction

The question to be considered is whether FT syncrude should be refined or upgraded? Since both approaches have been proposed this is not a trivial question. What are the options and why?

a) **Syncrude as product.** If syncrude is the final product from FT synthesis, the business model is analogous to that of crude oil production, where the producers of crude oil are not necessarily the refiners and the crude oil is traded as a commodity. Trading syncrude as a commodity and thereby avoiding the cost of downstream processing may have some business appeal, but there are technical difficulties that limit the practicality of this approach. In order to trade the syncrude, it must be a liquid. Syncrude is not produced as a single liquid phase, but as a mixture containing three to four different phases at ambient conditions (Chapter 4). The patent literature contains some proposals for converting part of the syncrude into a single "crude oil" product [1], but at least some upgrading is implied. If it is important to produce a commodity product, syngas-to-methanol production should be considered instead of FT synthesis, since it does not require upgrading before it can be shipped.

Greener Fischer-Tropsch Processes for Fuels and Feedstocks, First Edition. Edited by Peter M. Maitlis and Arno de Klerk.

b) **Upgrading.** The purpose of upgrading is to produce a higher quality oil that can then be marketed as a synthetic crude oil to refiners. This approach is extensively employed in the Canadian oil sands business, where the bitumen recovered from the oil sands is upgraded to enable pipeline transport. Ultimately, this bitumen is refined in conventional crude oil refineries that are capable of processing this type of heavy oil. Upgraded syncrude from FT synthesis can in principle also be sold in this way, but there is a caveat. The refining of FT syncrude is quite different from crude oil refining [2], since oxygenates present in the syncrude could wreak havoc in a refinery that has not been modified to process such material. Upgrading will likely require extensive deoxygenation of the syncrude in addition to addressing fluidity issues [1]. Nevertheless, the justification of this approach is to simplify product logistics, operation, and reduce the overall cost of the facility.

c) **Partial refining.** In partial refining, part of the syncrude is refined to blending materials or final products, while the remainder is upgraded to intermediates for off-site refining. This business model has been adopted for some of the new industrial FT-based gas-to-liquids (GTL) facilities. For example, the products from the Oryx GTL facility in Qatar are mainly naphtha and distillate, which are sold as intermediates for thermal cracking (i.e., off-site refining) and as blending material for diesel fuel, respectively. The justification for this approach is based on economy of scale. A FT refinery is usually much smaller than a present-day crude oil refinery, since it is argued that it is more efficient to refine the small volumes of difficult-to-refine syncrude fractions off-site in a larger facility than it is to invest in on-site refining capacity. To put this into perspective, the average capacity of a US crude oil refinery is around 150 000 b/day (or 7 500 000 t/a; 1000 m^3/h) [3], whereas this is close to the maximum capacity of only the largest of the industrial FT facilities (Shell Pearl GTL and Sasol Synfuels) [2]. Some of the larger crude oil refineries have two to three times this capacity.

d) **Refining.** Historically, on-site refining of syncrude was the norm in the design of FT facilities [2]. Combining FT synthesis with refining allows integration of the stepwise cooling and recovery of syncrude procedures into the refinery design, which improves refining efficiency. This is one of the important environmental benefits that integrated FT refining has over crude oil refining [4]. Furthermore, the capital cost of an associated FT refinery is only 10–15% of the total cost of the FT facility (Chapter 7); however, this is where most value addition takes place. One can also turn the economic justification for partial refining around by asking: If you cannot justify spending 10% more to refine syncrude to final products, how can you justify spending nine times as much to make the syncrude in the first place?

14.1.1
To Refine or to Upgrade?

The arguments that are usually presented to justify upgrading, partial refining, or full refining are mainly based on financial and business considerations. Does this picture change when we consider the environmental impact of each decision? The answer is likely to be "yes." The moment carbon is considered valuable and we calculate a carbon-based *E-factor* (defined as kilogram waste made per kilogram of desired product) [5] or *intensity factors* (material use, energy use, or effluent per amount of value addition), there is a strong argument in favor of refining that is integrated with FT synthesis.

Put in simple terms, FT synthesis to produce syncrude for off-site refining has the same drawback as conventional crude oil production, but with a potentially larger environmental impact (Box 14.1). The environmental impact associated with the decision to upgrade syncrude, instead of refining it on-site, is related to the justification for practicing FT synthesis in the first place. Indirect liquefaction based on FT synthesis enables us to convert a carbon-based raw material, which is not a useful carbon carrier in its native state at a specific location, into a carbon carrier that is useful and transportable. Syncrude is just an intermediate in this transformation into a useful carbon carrier. For an economic and environmental comparison, one should therefore compare FT synthesis and associated upgrading with a production well for crude oil delivery. One may argue that the syncrude from FT synthesis is cleaner than a typical crude oil, but syncrude production requires more energy and has a larger carbon loss (as CO_2) compared to crude oil production. Upgrading only does not make sense if one considers FT synthesis as alternative to crude oil.

It is important to realize that the final products from syncrude refining may in some applications be energy carriers and carbon carriers (e.g., transportation fuels). However, the value lies in the nature of the carrier, not in its energy value *per se*, as the carbon-based raw material has a much higher energy value before its

Box 14.1 Gaseous by-products

Crude oil production is often associated with the production of some gas. Some of the gas can be reinjected in the oil well to maintain pressure and thereby extend its production life. When the production location has access to a pipeline infrastructure, the gas can be sold as natural gas and has value as a product. If the excess gas production is in a remote location, the gas has little value and may be flared. This creates an opportunity for gas-to-liquids applications, since syngas-to-methanol or FT synthesis can convert the gas into a liquid product that is transportable. However, FT synthesis suffers from the same drawback as crude oil production, because syncrude contains a substantial gas fraction not suitable for transportation. If it were, the gas would be more valuable and there would be no environmental incentive for GTL conversion. Furthermore, as the gas from FT synthesis is clean and is produced through an energy-intensive process, using it as fuel gas would imply that the CO_2 footprint of the heating fuel is two to three times higher than necessary. Not having an associated refinery (and appropriate gas loop) that can convert most of the gas fraction from FT synthesis into liquid products undermines the potential environmental benefit of GTL conversion.

Box 14.2 Justification for FT synthesis

The justification for indirect liquefaction and FT synthesis, as well as for the critical importance of refining, can be illustrated by analogy. Wheat is a staple food in many countries. Wheat can be compared with the raw material used as feed for indirect liquefaction. It has all the energy value, but it is not a preferred foodstuff in its original form. The wheat can be ground to produce flour, which is a very useful intermediate in the preparation of desirable foodstuffs, such as bread. The syncrude produced by FT synthesis is a useful intermediate. Like flour, the syncrude must first be converted, here into transportation fuels, lubricants, or chemicals, before it becomes a desirable final product.

conversion by indirect liquefaction. If the main objective is only the energy, there is no reason for indirect liquefaction or FT synthesis (Box 14.2). The need to refine syncrude and the burden on the refinery to convert the syncrude with the highest possible efficiency is therefore clear. Despite the importance of thermal efficiency in energy applications and its impact on the environment, the real environmental impact of indirect liquefaction and FT synthesis is related to the overall carbon efficiency.

When refining is combined with syncrude production by FT synthesis, the refinery offsets some of the carbon cost associated with the production of syncrude compared to the use of crude oil. FT refining is more efficient than crude oil refining [4] as can be seen from a comparison of the fuels refinery yields for the production of on-specification transportation fuels (Table 14.1) [3, 6]. The yields for the FT

Table 14.1 Comparison of transportation fuel yields from modern crude oil refining and Fischer–Tropsch refining.

Description	Transportation fuel yield (%)			
	US crude oil Syncrude as an intermediate refinery average in 2003[a)]	Fe-HTFT motor gasoline refinery[b,c)]	Fe-LTFT jet fuel refinery[b,d)]	Fe- LTFT diesel fuel refinery[b,d)]
Motor gasoline	46.9	58.3	21.7	32.6
Jet fuel	9.5	18.4	62.3	28.2
Diesel fuel/distillate	23.7	8.0	0	24.6
Total	80.1	84.7	84.1	85.4

a) The source did not clearly specify the yield units for crude oil; it is likely a volumetric yield, that is, m^3 fuel per m^3 oil.
b) The yield is gravimetric, that is, kg fuel per kg syncrude; the volumetric yield is difficult to define because syncrude is multiphase.
c) Gas loop design includes cryogenic separation and methane recycle.
d) Gas loop design includes C_3 and heavier recovery, but no cryogenic separation.

refineries are not for optimized designs, but for designs to maximize a specific fuel type. Yet, it is clear that when a FT refinery is combined with an appropriate gas loop and syncrude recovery design, the transportation fuel yield that can be obtained is higher than that obtained with crude oil refining. However, this advantage is lost when a crude oil refinery design is used for FT syncrude refining.

14.1.2
Refining of Fischer–Tropsch Syncrude

The selection of refining catalysts, the technology, and the design of FT refineries for transportation fuel, lubricant, and petrochemical production are described in the literature [1, 2, 6, 7]. As an example, let us look at a FT fuels refinery (Figure 14.1) that produces mainly distillate for diesel fuel blending, as well as on-specification motor gasoline and jet fuel, as the conversion processes found here are common to many FT refinery designs:

a) **Hydrocracking and hydroisomerization.** Heavy syncrude fractions, such as waxes that are solids at ambient conditions, can be converted into lighter liquid

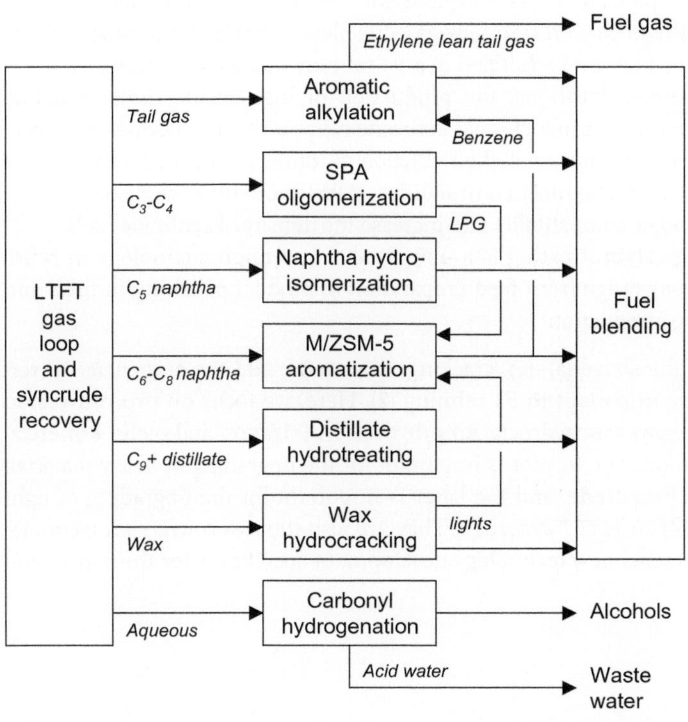

Figure 14.1 Fuels refinery design employing Fe-LTFT syncrude as feed material for the production of motor gasoline, jet fuel, and distillate for diesel fuel blending. The transportation fuel yield from this refinery design is 88%, and excludes the mixed alcohols (3% yield) that are coproduced.

products. Wax hydrocracking and wax hydroisomerization are useful for the production of blending materials for transportation fuel applications, as well as for lubricant base oils. Under milder conditions, hydroisomerization leads to better product properties without significant chain degradation. In addition to wax conversion, the hydroisomerization of butane and light naphtha fractions is used to produce blending components for motor gasoline.

b) **Olefin dimerization and oligomerization.** One of the primary functions of a syncrude refinery is to convert normally gaseous olefins into liquid products; this can also be useful in the production of specific products with desirable properties after hydrogenation, such as high-octane paraffinic motor gasoline (alkylate), synthetic jet fuel blendstock (isoparaffinic kerosene), polyalpha-olefin lubricant base oil, and linear α-olefins.
c) **Aromatization and naphtha reforming.** Aromatics are required for transportation fuels and petrochemical applications. The production of mononuclear aromatics from gaseous and naphtha range paraffins improves the usefulness of these fractions. In the case of gaseous paraffins, it also provides a refining pathway for the production of liquid products.
d) **Aromatic alkylation.** This is a useful technology for transportation fuel and petrochemical applications, as it exploits the olefin-rich syncrude and provides another pathway to convert normally gaseous olefins into liquid products. Some of the functions that can be fulfilled in a FT refinery include the beneficial use of ethylene in remote locations, the production of high-octane motor gasoline blending material, the production of kerosene range aromatics necessary to meet synthetic jet fuel specifications, the reduction of refinery benzene levels, and the production of petrochemical commodities. Alkyl aromatics are also useful to improve elastomer compatibility and increase the density of synthetic fuels.
e) **Hydrotreating.** Hydrotreating is a ubiquitous conversion technology in refining. Applications range from feed preparation to product polishing in fuels and petrochemicals production.

In addition to the above (a)–(e), cracking, in general, and the oxygenate conversion processes are important in FT refining [2]. Here, we focus on two conversion technologies, namely, wax hydrocracking/hydroisomerization and olefin dimerization/oligomerization. The former is important for the upgrading of heavy material, especially in LTFT syncrude, and the latter is important for the upgrading of light material, especially in HTFT syncrude. They are also the two conversion technologies that have attracted new technology development specifically for the conversion of FT syncrude.

14.2
Wax Hydrocracking and Hydroisomerization

An important consequence of the chain growth mechanism for the formation of aliphatic hydrocarbons in the FT reaction is the impossibility to synthesize a product with a narrow range of chain lengths. Regardless of the catalyst and operating conditions, the FT synthesis gives rise to a wide carbon number range, from methane to high molecular weight paraffins (Chapter 4). LTFT synthesis, either on iron- or cobalt-based catalysts, gives a mixture of hydrocarbons, strongly shifted toward high molecular weight products. The need, for economic reasons, to decrease light gas selectivity as much as possible led to the development of LTFT processes with high α-values. A large fraction of these FT products have boiling points >370 °C, while the yield in the middle distillate range is rather limited. With the current LTFT technologies, typical α-values are close to 0.90 and the fraction of heavier than C_{22} material is around 40–45%.

The product from LTFT synthesis is a mixture of linear paraffins (>85–90%) together with smaller amounts of olefins and oxygenates. The latter are mainly primary alcohols plus minor amounts of carboxylic acids, esters, and ketones [8]. As it is mainly made up of linear (*n*-) paraffins, the middle distillate fraction exhibits very high cetane numbers, but poor cold flow properties [9, 10]. The impossibility of producing high yields of middle distillate with acceptable cold flow properties directly from FT synthesis leads to the necessity of converting the FT wax to maximize middle distillate yield and improve the automotive fuel properties; this is most effectively done by wax hydrocracking technology, where two main reactions occur in parallel: hydroisomerization and hydrocracking. The former leads to a marked improvement of cold flow properties, while the latter is responsible for the increase of middle distillate yield. Theoretically, ideal wax hydrocracking in combination with FT synthesis will result in a monotonic increase in distillate yield with increasing α-value (Figure 14.2) [11]. The challenge is to develop a hydrocracking catalyst and an associated technology that displays ideal hydrocracking behavior.

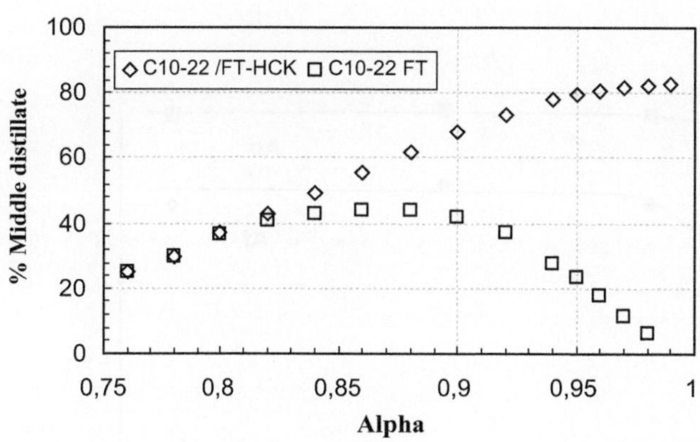

Figure 14.2 Middle distillate yield from ideal hydrocracking of Fischer–Tropsch wax. The product yield from FTS was calculated based on the Anderson–Shultz–Flory distribution, while in the hydrocracking stage it was assumed that each paraffin gives rise to a equimolar production ratio for all the cracking products except those resulting from breakage of terminal α, β, and γ C—C bonds whose probability of breakage was set to zero. Moreover, it was assumed that that lighter than C_{22} material is not further cracked.

14.2.1 Hydrocracking and Hydroisomerization Catalysts

A wide range of catalysts have been developed for specific applications depending on the desired products and the characteristics of the feedstock [1, 12, 13]. Bifunctional catalysts are used for hydroconversion, characterized by the presence of *acidic sites* that provide the isomerization/cracking function and *metal sites* with hydrogenation/dehydrogenation function. Typical acidic supports are amorphous metal oxides or mixtures of metal oxides, for example, HF-treated Al_2O_3, SiO_2–Al_2O_3, ZrO_2/SO_4^{2-}, zeolites (Y, beta, mordenite, and ZSM-5), and silicoalumino-phosphates (SAPO-11, SAPO-31, and SAPO-41). The metals most commonly used are Pt, Pd, and sulfided bimetallic systems, Co/Mo, Ni/W, or Ni/Mo. The latter are mainly used for the hydrocracking of residue fractions containing sulfur in crude oil refining.

The balance between the acidity of the support, as determined by the concentration and strength of acidic sites, and the hydrogenation/dehydrogenation activity of the metal function is of primary importance in determining the activity for and selectivity to hydroisomerization and distribution of the cracking products [14, 15]. Giannetto *et al.* [15] observed a rapid increase of activity up to 0.5% of platinum loading followed by a leveling of activity at higher concentrations (Figure 14.3). The initial increase shows that the increasing hydrogenation/dehydrogenation function enables hydrocracking by the faster reaction pathway through the formation of olefinic intermediate, while the leveling out of activity suggests that the formation of olefinic intermediates is no longer rate-limiting. In addition to the increase of activity, higher levels of platinum result in a remarkable increase of selectivity for hydroisomerization, while another beneficial effect is the improvement of catalyst stability [14].

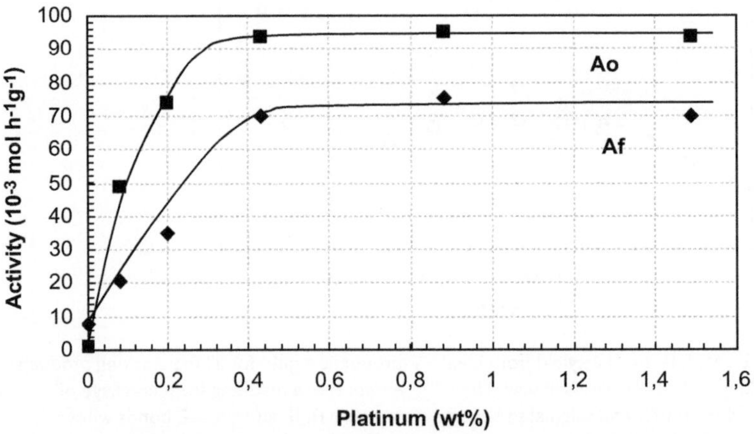

Figure 14.3 Hydrocracking of *n*-heptane showing the change of initial activity (A_o) and final activity (A_f) as a function of platinum loading.

The presence of a strong hydrogenation/dehydrogenation function on a support with medium–low strength acid sites generally fosters the formation of isomerized products with carbon distributions more favorable for middle distillate production. Catalysts loaded with Pt or Pd show a considerably higher selectivity to hydroisomerization and a better distribution of cracking products compared to those using sulfided base metals such as Ni, Co, Mo, and W [16–19]. Sulfided base metal hydrocracking catalysts for FT waxes lead to poorer selectivity and a decrease in middle distillate yield at higher conversion compared to noble metal hydrocracking catalysts [20].

The effect of the hydrogenation/dehydrogenation function is shown in Figure 14.4 [19]. The sulfided CoMo catalyst leads to a cracking products distribution shifted toward the lighter compounds, while the Pt catalyst characterized by a stronger hydrogenation/dehydrogenation activity leads to a symmetric molar distribution of cracked products, indicative of a situation where the further conversion of primary products is negligible.

In addition to the influence of the hydrogenation/dehydrogenation power of the catalyst, the nature of the acidic support employed for the catalyst also has a strong influence on product distribution and isomerization selectivity. Although the behavior of each particular support can vary over a wide range, zeolite-based catalysts are generally more active than amorphous silica–alumina catalysts. However, the latter catalysts are better suited to maximize middle distillate yields, whereas zeolites lead to product distributions shifted more toward the gasoline range [12]. High isomerization selectivity has been reported for amorphous silica–alumina [21] and amorphous silica–alumina phosphate [22]. However, zeolitic materials can also exhibit shape-selective behavior leading to high isomerization selectivity, as in case

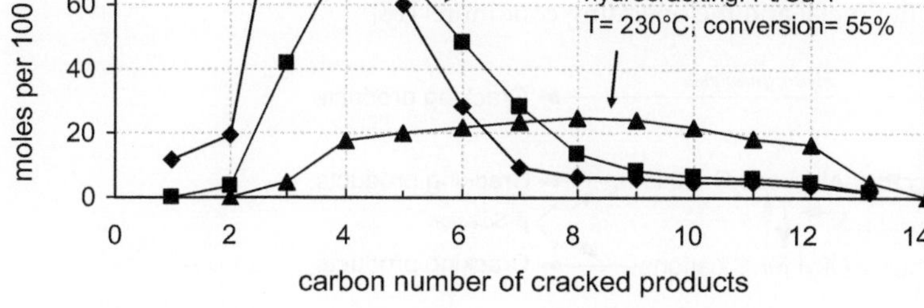

Figure 14.4 Molar carbon number distribution of cracked products from the cracking of *n*-hexadecane over different catalysts at 50% conversion. The hydrogenation/dehydrogenating power of the catalysts increase in the order SiO_2–Al_2O_3–ZrO_2 < sulfided CoMo/SiO_2–Al_2O_3 < Pt/Ca–Y-zeolite.

of the medium pore zeolite ZSM-22 [23] and the silicoaluminophosphate SAPO-11 [24].

14.2.2
Mechanism of Hydrocracking and Hydroisomerization

The mechanism of hydroconversion of linear paraffins over bifunctional catalysts began to be actively investigated at the beginning of the 1960s. Based on the previous work by Mills *et al.* [25] and Weisz [26], a mechanism proceeding through carbenium intermediates with hydrogenation/dehydrogenation and isomerization steps was proposed [27]. Macroscopically, hydrocracking and hydroisomerization take place as a series of consecutive reactions where the linear paraffins are converted first into monobranched paraffins, then into dibranched paraffins, and thus into isomers with progressively higher degrees of branching. In parallel with the isomerization reaction, reactant molecules are cracked. The rate of cracking increases as the isomerization reaction progresses to produce more branched isomers.

Subsequent investigations of model compounds offered a clearer picture of the mechanism, the roles of the acidic and of the hydrogenation/dehydrogenation sites, and their interaction [15, 28, 29]. Currently, the most widely accepted scheme for paraffin conversion on a bifunctional catalyst is by the sequence illustrated in Figure 14.5:

a) The paraffin adsorbs on the catalyst surface.
b) The paraffin is dehydrogenated at a metal site.
c) The adsorbed olefin forms a carbenium ion at a Brønsted acid site.
d) The secondary carbenium ion rearranges to a tertiary carbenium ion.
e) The carbenium ion can either desorb to give the corresponding iso-olefin or undergo β-scission to give smaller carbenium ions and *n*/*iso*-olefins.
f) The *n*/*iso*-olefins are subsequently hydrogenated at a metal site with formation of the corresponding *n*/*iso*-paraffins.

In addition to the carbenium-based mechanism, Figure 14.5 shows that hydrogenolysis occurs on metal-promoted catalysts, the degree to which it takes place depend on the metal and the operating conditions [18].

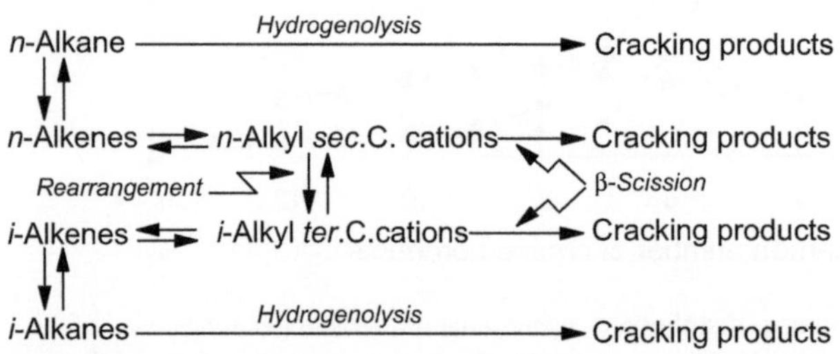

Figure 14.5 Reaction scheme for hydroisomerization and hydrocracking of paraffins on a bifunctional catalyst.

If there is an optimal balance between acid and hydrogenation/dehydrogenation functions, the rearrangement from the secondary to the tertiary carbenium ion is considered to be rate-determining. Both isomerization and hydrocracking reactions occur through the same type of carbenium ion intermediate. The rearrangement most likely occurs via protonated cyclopropane (PCP) intermediates as originally proposed by Condon [30] and Brouwer and Hogeveen [31]. In agreement with the postulated bifunctional mechanism, the activation energies of hydroisomerization and hydrocracking are similar [32–34], while the reaction rate shows a negative order with respect to H_2 pressure for pressures above a threshold value [32, 33, 35–37]. The negative order with respect to H_2 pressure is strong evidence in favor of the bifunctional mechanism since it indicates that (ideally) all the steps before the rate-determining step can be considered in quasi-equilibrium. Hence, an increase of H_2 pressure will shift the dehydrogenation equilibrium toward hydrogenation, thereby lowering the olefin concentration and decreasing the equilibrium concentrations of *n*-alkyl secondary carbenium ions, which in turn leads to a decrease in the rate-determining step. It is consequently advantageous from a kinetic viewpoint to conduct hydrocracking and hydroisomerization at the lowest practical pressure that still maintains an ideal behavior.

With bifunctional catalysts characterized by a balance of, and a proximity between, acid and metal sites, the formation of the olefin intermediates is fast enough not to be rate-limiting. When olefin formation is not rate-limiting, the acid function determines the kinetics of the system [15, 29]. Under these conditions and in the absence of diffusional limitations, the reactivity under mild hydrocracking conditions strongly depends on the chain length of the hydrocarbon reactant [28, 38]. The underlying reasons for this can be twofold:

a) An increase of secondary carbons would lead to an increase in the probability of formation of carbenium intermediates. The reactivity is expected to be proportional to the number of secondary carbon atoms per molecule. Based on the PCP intermediate, Sie [39] proposed that reactivity should be proportional to C_{n-6} and to C_{n-4} for hydrocracking and hydroisomerization, respectively.
b) It is known that the Henry adsorption constant of normal and isoparaffins increases exponentially with carbon number at least up to C_{16} [39, 40]. Consequently [41, 42], the higher reactivity of heavier paraffins can also be ascribed to their stronger physisorption, leading to a higher density on the catalyst surface and consequently to higher reaction rates.

Hydrocracking reactivity also depends on the nature of branching of the reactant molecule. As shown in Figure 14.6, β-scission of type A is by far the fastest reaction, but it requires a tribranched carbenium structure with an α,γ,γ-configuration of branching groups. Cracking via mode B1 is much slower, implying the presence of gem-type structures; it occurs through a less energetically favorable route in comparison to β-scission of type A. The other modes of cracking are in decreasing order: type B2 for dibranched paraffins with isolated branching groups, C for monobranched paraffins, and D for linear paraffins.

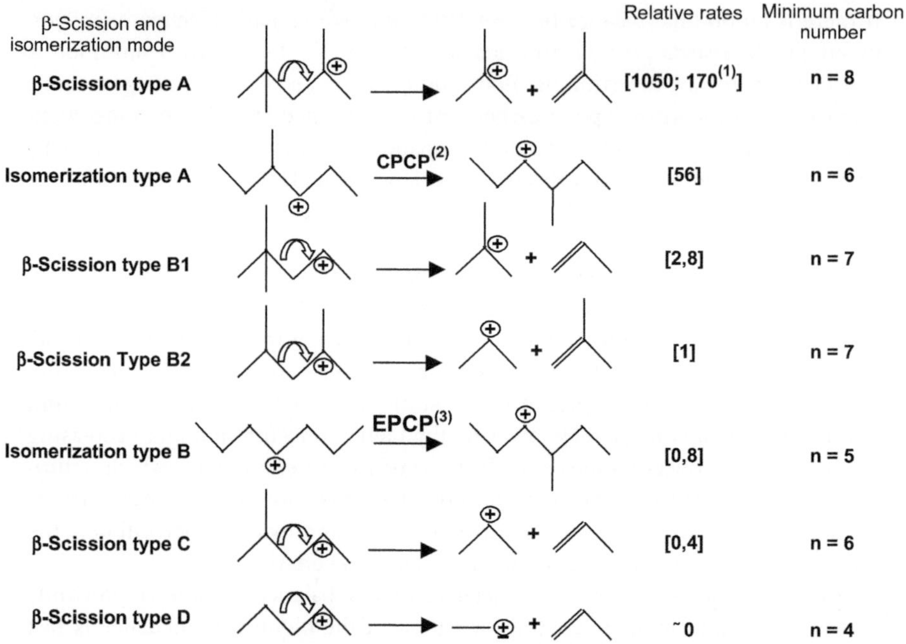

Figure 14.6 Different β-scission and skeletal rearrangement reactions and their relative rates that occur during paraffin hydrocracking and hydroisomerization [43]. Values in (1) from Ref. [44]; (2): CPCP, corner protonated cyclopropane; (3): EPCP, edge protonated cyclopropane.

Two types of isomerization reactions are possible: A and B. Type A isomerization is faster and does not involve a change of branching degree. It occurs through a corner protonated cyclopropane (CPCP), or π-complex, and it is responsible for the alkyl shift along the aliphatic chain. The type B isomerization occurs through an edge protonated cyclopropane (EPCP). This leads to a change of branching degree and is significantly slower than a type A isomerization. The most abundant branching formed during the isomerization involves the methyl group, followed (in decreasing order) by ethyl and propyl [45, 46]. It is widely accepted that the formation of methyl branching occurs via protonated cyclopropane intermediates, while the formation of ethyl and propyl isomers, by analogy with the PCP mechanism, may result by the formation of protonated cyclobutanes (PCB) and protonated cyclopentane intermediates, respectively [45, 47]. The lower formation of ethyl branching and higher homologs is most likely due to the much less energetically favored route for the formation of these intermediates. It has been reported that protonated cyclobutane is 130 kJ/mol less stable than PCP [48].

In the light of the relative reaction rates of the various β-scission and isomerization modes shown in Figure 14.6, it can easily be understood why the concentration of tribranched isomers never attains high values during hydrocracking and hydroisomerization of paraffins: once the tribranched isomers attain a suitable configuration, they rapidly undergo β-scission.

14.2.3
Products from Hydrocracking Conversion

In ideal hydrocracking and at medium to low conversion where subsequent conversion of primary products is negligible, cracking products exhibit an approximate wide bell-shape distribution centered on chain lengths with carbon numbers half of the original molecules (Figure 14.7) [19, 28].

While the fragments corresponding to the breaking of the internal ($>C_3$) C—C bonds are produced in almost equimolar amounts, there is a virtual absence of methane, ethane, and the corresponding heavier fragments $C_{n-1}$and C_{n-2} (Figure 14.7); propane and C_{n-3} are formed in intermediate amounts. Such a carbon number distribution can be explained on the classical β-scission route by assuming that the reactivities of inner bonds are equal, whereas those of the "outer" bonds that give rise to the C_1, C_2, and C_3 fragments are lower, due to the formation of fewer of the less-stable carbenium ion intermediates there. The carbon number distribution can also be explained in terms of the PCP intermediate for isomerization and cracking. At higher conversions, the distributions of the products become less symmetrical and, owing to the further reactions of the primary products, the carbon number distribution is progressively skewed toward lighter fragments (Figure 14.8) [49].

Thus far, hydrocracking behavior was illustrated with data obtained by hydrocracking shorter chain (C_{16} or less) paraffins (Figures 14.7 and 14.8). There have also been studies on the hydrocracking of real FT waxes. In the 1970s, Sasol carried out a study on wax hydrocracking to produce transportation fuels [50]. In a project sponsored by the US Department of Energy, UOP investigated the hydrocracking of Arge Fe-LTFT wax using two types of catalysts (probably on silica–alumina and zeolite supports). In recycling mode, the maximum yield of the 149–371 °C cut material with acceptable cold flow properties was around 80% [51]. The recently renewed interest in the FT synthesis as a means to convert natural gas into distillate fuels has led to increasing studies on hydrocracking of FT waxes by many oil companies [1, 20, 46, 52–54].

Leckel [53] studied a commercial $NiMo/SiO_2–Al_2O_3$ catalyst for the hydrocracking of Fe-LTFT waxes. Middle distillate ($C_{10}–C_{22}$) yields up to 65–70% were achieved with a wax containing 90% of C_{23} and heavier material. Selectivity to the $C_{10}–C_{22}$ fraction was found to be significantly affected by the operating conditions and decreased with the increase in conversion. In two subsequent studies [20, 55], LTFT waxes were hydrocracked using silica–alumina-supported sulfided metal and platinum catalysts. Both sulfided and noble metal-based catalysts were able to produce diesel fuel with high cetane number (>70) and good cold flow properties (cloud point < -15 °C). However, the data showed that the performance of the best noble metal-based catalyst was significantly better than the sulfided base metal-based catalyst, both in terms of selectivity to middle distillate and isomerization of products.

Hydrocracking of a LTFT wax (61% $C_{10}–C_{22}$, 39% $\geq C_{23}$) using a platinum on amorphous silica–alumina catalyst was reported by Calemma *et al.* [56]. The middle

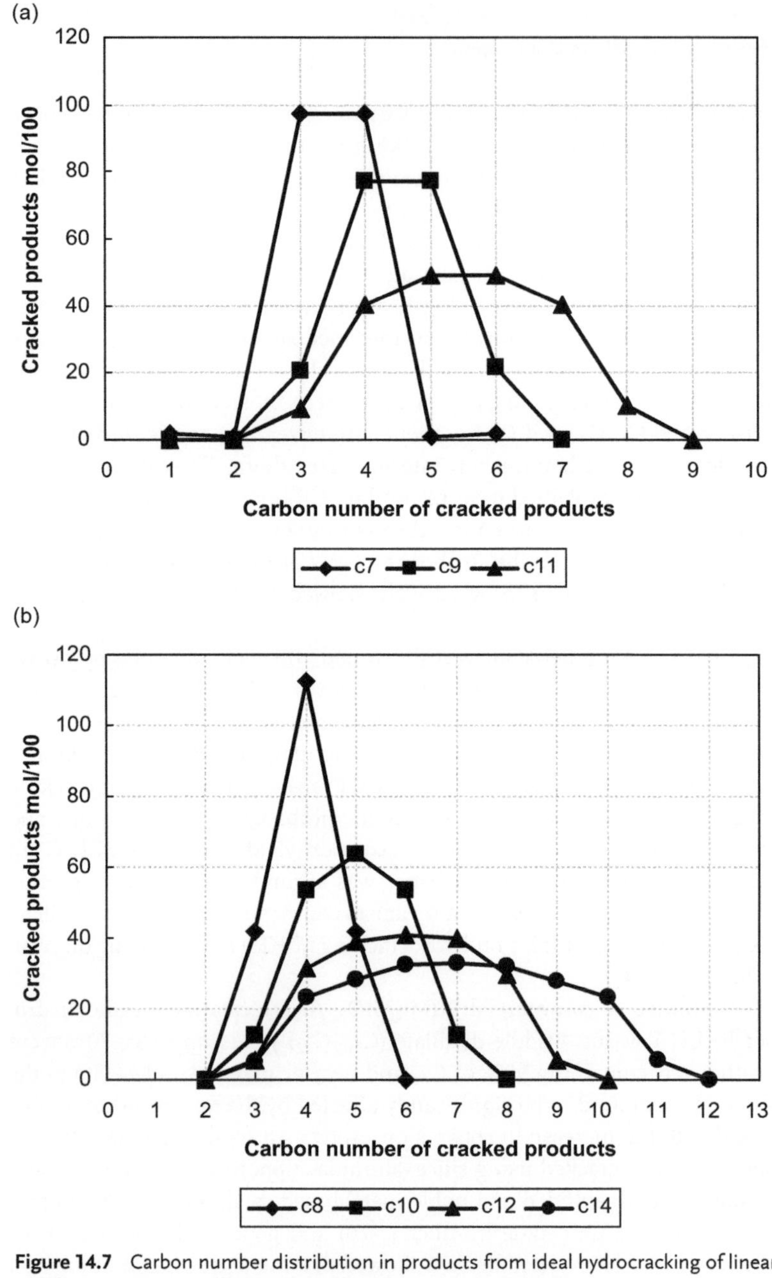

Figure 14.7 Carbon number distribution in products from ideal hydrocracking of linear paraffins of different carbon numbers in the range of C_7–C_{14} as indicated.

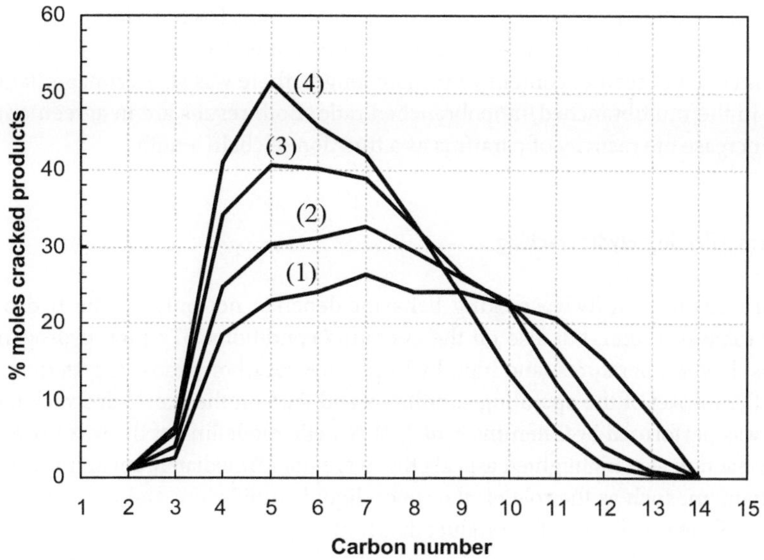

Figure 14.8 Molar carbon number distribution of products from *n*-hexadecane hydrocracking over a Pt/USY catalyst at different levels of cracking conversion – (1): 70%; (2): 83%; (3): 93%; and (4): 97%.

distillate yields increased with conversion up to 85–90% and thereafter decreased owing to further reactions of the first formed products. The maximum yield achieved was 82–87%. A strong increase of isomer content at higher conversion was observed and as a consequence the freezing point and pour point decreased, reaching values of −50 and −30 °C, respectively. The isomer content of the C_{10}–C_{14} fraction ranged between 20% and 65%, while that of the C_{15}–C_{22} fraction changed from 50% to 92%. The higher degree of isomerization of the C_{15}–C_{22} in comparison to the lighter C_{10}–C_{14} fraction was explained in terms of both higher reactivity of longer chains and partial evaporation of the feed leading to a liquid phase rich in heavier components. The results of this study suggested that vapor–liquid equilibrium (VLE) plays a significant role in determining both isomer content and selectivity to middle distillate.

In a subsequent study [46], a light Co-LTFT wax was subjected to hydrocracking in the temperature range of 319–351 °C and at H_2 pressure between 35 and 60 atm. The results in terms of selectivity to middle distillate, maximum yields achievable, and cold flow properties were similar to those obtained previously and the maximum middle distillate yield achieved was 85%. A detailed analysis of the C_{10}–C_{22} fraction in the product mixture indicated that wax hydrocracking took place analogous to that in shorter paraffins. Wax conversion occurred through consecutive isomerization reactions and concomitant cracking, as depicted in Figure 14.5. At low conversion, isomers are mainly made up of monobranched paraffins, while multibranched isomers are more abundant at the highest conversion level. Furthermore, the degree of isomerization always increases with the chain length. The isomer concentration in a test at 84% of conversion of $>C_{22}$ fraction changed from 11% to 99% as the carbon number in the product increased from C_{11} to C_{21}. Along with the increase of isomer content with chain length, there was also a concomitant increase in the multibranched/monobranched ratio. Both results are in agreement with an increase of reactivity of paraffins as a function of chain length.

14.2.4 Parameters Affecting Hydrocracking

The approach to ideal hydrocracking behavior depends not only on the hydrocracking catalyst design, but also on the operating conditions [57]: high hydrogen pressures, low temperatures, and high hydrogen/hydrocarbon ratios. A systematic study of the impact of the operating conditions and feed on the wax hydrocracking reaction was performed by Calemma *et al.* [52]. Kinetic modelling of the wax hydrocracking reaction also highlighted aspects that were not immediately apparent from the mechanism, such as the role of the vapor–liquid equilibrium and the effect of the nature of the wax feed on the product distribution [58–60].

14.2.4.1 **Effect of Temperature**

Conversion of FT wax is strongly affected by the temperature [52, 55, 58, 61]. An increase in reaction temperature from 351 °C to 367 °C increases the conversion of the $>C_{22}$ fraction from 20% to 81% (Figure 14.9) [58].

The energy of activation reported in literature for the isomerization and hydrocracking of paraffins ranges between 145 and 170 kJ/mol [34, 36]. The apparent energy of activation for $>C_{22}$ conversion, assuming a first-order reaction, was found to be 338 and 377 kJ/mol for two FT waxes with 70% and 48% $>C_{22}$, respectively (Calemma, V., unpublished data.). A change in reaction temperature affects the selectivity to middle distillate, isomerization degree of products, and their

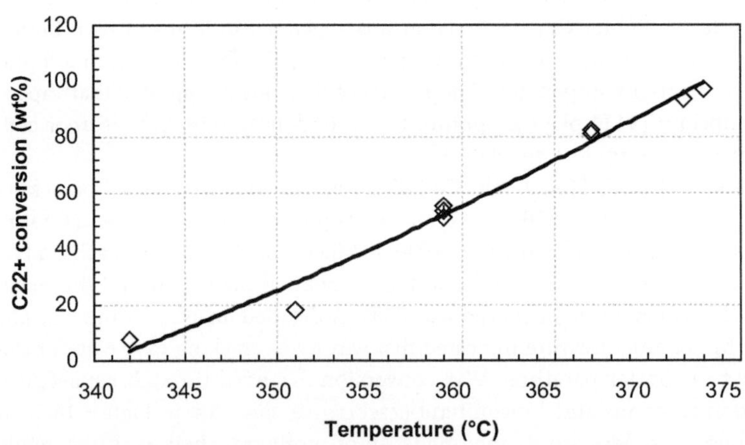

Figure 14.9 Effect of reaction temperature on the conversion of LTFT wax hydrocracked at 4.7 MPa, 2 h^{-1} WHSV, and 0.105 wt/wt H_2/wax ratio.

Table 14.2 Effect of temperature on hydrocracking conversion, product distribution, and isomer content of the distillate fraction.

Description	Hydrocracking temperature		
	350 °C	360 °C	365 °C
Conversion of heavier than C_{22} material (mass%)	17	69	86
Naphtha to distillate ratio	0.11	0.28	0.33
Isoparaffin content in distillate fraction (%)	77	82	84

distribution. These changes arise from differences in degrees of conversion, rather than from reaction temperature [56]. As shown in Table 14.2 [61], higher temperatures result in higher conversion, higher naphtha to distillate ratio, and higher isomerization conversion of the products. The increase in naphtha to distillate ratio results from the consecutive cracking of primary products. Since the middle distillate fraction in the feedstock is very low (about 4%), the high isomer content in the diesel fraction, even at low conversion, is consistent with a high degree of isomerization of the cracking products.

14.2.4.2 **Effect of Pressure**

The conversion of paraffins exhibits a negative order of -1 to -0.85 with respect to the hydrogen pressure. In agreement with the results obtained on model compounds, it is generally found that an increase in pressure leads to a decrease in hydrocracking conversion (Figure 14.10) [61].

Hydrocracking tests of coal-derived FT waxes over silica–alumina-supported sulfided NiW/NiMo and noble metal catalyst showed that in the range of 2.5 and 7.0 MPa, conversion of the $>C_{22}$ fraction is inversely proportional to the hydrogen pressure [55]. For the sulfided NiW/NiMo, an increase of pressure from 2.5 to 4.0 MPa resulted in a decrease of conversion from 97% to 85%, while for the noble

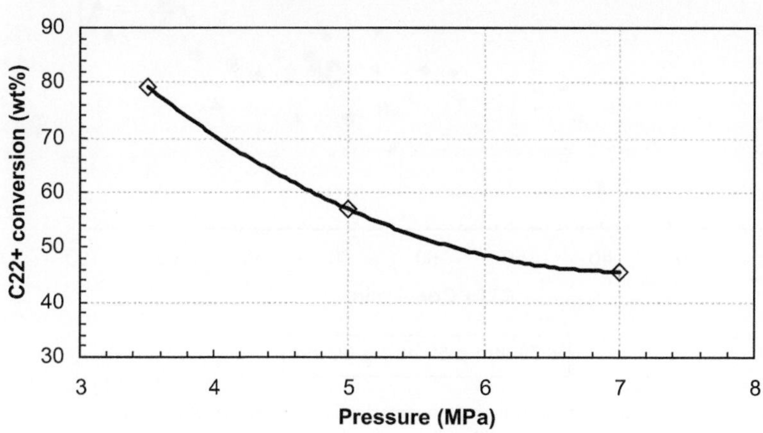

Figure 14.10 Effect of reaction pressure on the cracking conversion of LTFT wax.

metal-based catalyst an increase of pressure from 3.5 to 7.0 MPa led to a decrease from 90% to 65%.

Higher pressures result in lower conversions and, consequently, it has been found that the products are characterized by lower isoparaffin content and higher distillate to naphtha ratios. However, these changes seem to be merely the consequence of the lower extent of the reaction. When the selectivities are compared at the same level of conversion, the values at different pressures are very similar [55, 61]. However, there is an effect on the degree of isomerization. (Figure 14.11) [56]; thus, the increase of isomer content at higher pressures could be due to an increase in the fraction of C_{10}–C_{14} material present in the liquid phase in direct contact with the catalyst. It can also be explained by the modified hydrocracking mechanism proposed by De Klerk [2].

14.2.4.3 **Effect of H_2/Wax Ratio**

Vapor–liquid calculations show that in the range of operating conditions used for the hydrocracking of FT wax, the feed is present in both the liquid and the gas phases. The liquid phase becomes rich in heavier components, while the opposite happens for the gas phase. It has been reported that the most evident effects of an increase in H_2/wax ratio on hydrocracking of LTFT wax are a decrease in the isomer concentration in the middle distillate (Figure 14.11) and an increase of apparent conversion rate of the $>C_{22}$ fraction (Figure 14.12) [52]. An increase in the selectivity to middle distillate, as indicated by a lower naphtha to distillate ratio, has also been observed.

At first sight the increase of conversion with an increase in H_2/wax ratio seems surprising, because it is not related to any of the variables defining the reaction

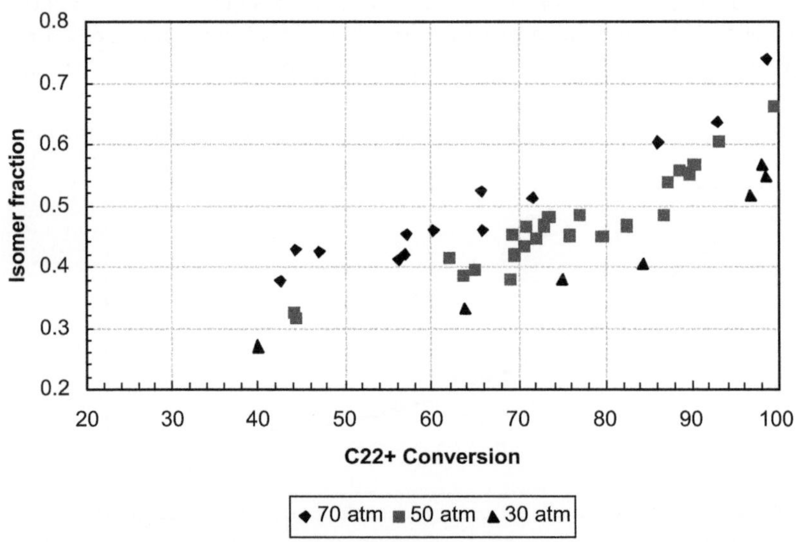

Figure 14.11 Effect of reaction pressure on the isomerization conversion of LTFT wax as indicated by the isoparaffin fraction in the C_{10}–C_{14} material obtained by hydrocracking over a Pt/SiO_2–Al_2O_3 catalyst at 340 °C, 0.5–4 h^{-1} WHSV, and 0.125 wt/wt H_2/wax ratio.

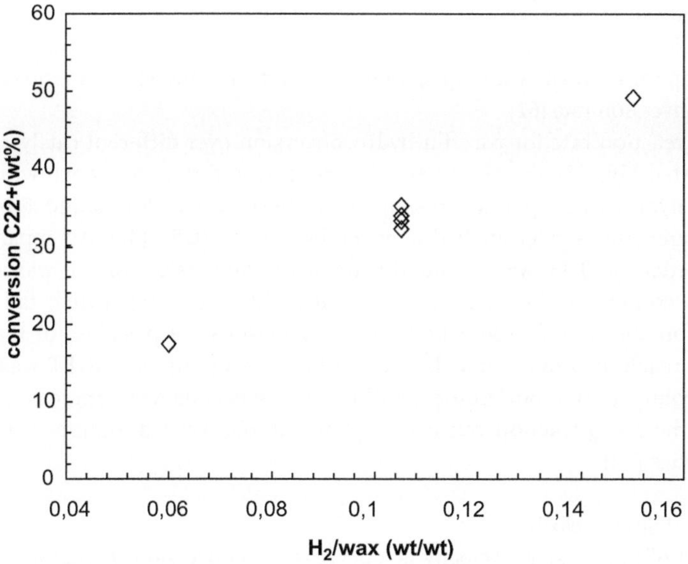

Figure 14.12 Conversion of the heavier than C_{22} fraction as a function of H_2/wax ratio at 4.75 MPa and 2 h^{-1} WHSV.

rates of paraffins. In fact, a decrease would be expected from the kinetic expression, because higher H_2/wax ratios would lead to lower partial pressures of alkanes and higher effective hydrogen pressures. A possible explanation for the observed effects may be due to changes in the vapor–liquid equilibrium [56]. As shown in Table 14.3 [62], higher H_2/wax ratios lead to higher vapor/liquid ratios of the reacting mixture and a liquid phase rich in $>C_{22}$ fraction. In these circumstances, while the overall WHSV remains constant, an increase in the H_2/wax ratio leads to a lower space velocity of the liquid phase in direct contact with the catalyst surface

Table 14.3 Effect of pressure, temperature, and H_2/wax ratio on the vapor–liquid equilibrium.

Operating conditions			Vapor to feed ratio on H_2-free basis (wt/wt)			
Pressure (MPa)	Temperature (°C)	H_2/wax (wt/wt)	C_{10}–C_{14}	C_{15}–C_{22}	$>C_{22}$	Total feed
3.5	324	0.06	0.785	0.402	0.047	0.360
3.5	324	0.15	0.918	0.627	0.120	0.528
3.5	360	0.06	0.888	0.616	0.132	0.476
3.5	360	0.15	0.965	0.843	0.294	0.657
7.0	324	0.06	0.604	0.242	0.024	0.271
7.0	324	0.15	0.811	0.460	0.061	0.434
7.0	360	0.06	0.742	0.412	0.069	0.387
7.0	360	0.15	0.900	0.672	0.164	0.547

and, consequently, to an increase in the conversion of $\geq C_{22}$ materials. The changes in the actual weight hourly space velocity of the liquid phase due to the changes in the degree of vaporization allow an explanation of the effect of the H_2/wax ratio on the apparent conversion rate [62].

A first-order reaction rate for paraffin hydroconversion over different catalysts has been reported [36, 63–65]. However, a nearly zero-order dependence with respect to the hydrocarbon partial pressure was observed by Steijns and Froment for *n*-decane and *n*-dodecane hydroconversion over Pt/USY [32]. An apparent reaction order of 0.33 was found during hydroconversion of FT waxes having a $>C_{22}$ content of 48% (Calemma, V., unpublished data), which falls between the aforementioned observations. Various reasons are possible for the differences in reaction order found between model compounds and FT wax; however, the changes in vapor–liquid equilibrium associated with increasing conversion of the $>C_{22}$ fraction can also explain the observed deviation from first-order kinetics [62].

14.2.4.4 **Effect of Space Velocity**

The main effect of higher space velocity is a decrease in conversion. A change in LHSV from 0.5 to 1 h^{-1} during wax hydrocracking led to a decrease in $>C_{22}$ conversion from 57% to 30% [55]. In another study, a change of WHSV from 1 to 3 h^{-1} resulted in decrease of the $>C_{22}$ conversion from 88% to 21% [58]. Similarly, the changes in degree of isomerization and selectivities are essentially due to the differences in the progress of the reaction. At higher conversion, the products are more isomerized, while a decrease of selectivity to middle distillate is caused by a further cracking of the initial products.

14.2.4.5 **Effect of Oxygenates**

The wax fraction produced from LTFT synthesis also contains primary alcohols and minor amounts of carboxylic acids, esters, and ketones [8]. Oxygenates have a significant impact on FT refining catalysis [1] and on hydrocracking. Removal of oxygenated compounds by hydrotreating was found to improve hydrocracking conversion of $>C_{22}$ and a 15 °C increase in operating temperature was needed to achieve the same conversion in unhydrogenated waxes as in hydrogenated waxes [53]. A study of the effect of oxygenated compounds and of the addition of tetradecanol and lauric acid on the hydrocracking of FT wax over both sulfided NiMo and noble metal catalysts [66] concluded that the presence of oxygenated compounds modifies the metal–acid balance of the catalyst, which in turn leads to a change in catalyst behavior. The addition of 5% of either the alcohol or the carboxylic acid to an otherwise oxygenate-free wax feed resulted in a loss of activity of the noble metal-based catalyst. There was a strong decrease of conversion in both cases, but more pronounced for the alcohol. Comparison at the same level of conversion showed that the presence of tetradecanol resulted in a slight increase in diesel selectivity, a similar decrease in naphtha selectivity, and less isomerization of products. A later study [54] found that addition of 5% of 1-decanol to a previously hydrotreated commercial FT wax led to a significant loss of activity. Activity could be recovered by a 5 °C increase in reaction temperature. The middle distillate selectivity and the isomerization degree were not affected.

14.2.5 Comparative Environmental Impact

Published results show that hydrocracking of FT waxes aimed at maximizing the middle distillate cut is a complex reaction. The selectivity to the desired product strongly depends on the type of catalyst used, operating conditions, feed distribution, and the presence of oxygenates. The absence of sulfur allows the use of noble metals, which, if loaded on suitable acidic supports, should lead to optimal selectivity to middle distillates and their degree of isomerization. Noble metal hydrocracking of FT waxes also has a number of other advantages over conventional crude oil hydrocracking that decreases the environmental impact:

a) High conversion can be achieved under moderate conditions (<350 °C, <50 atm), which reduce the energy requirements.
b) The product is sulfur free and requires no further hydrotreating.
c) The selectivity to middle distillate is better, with less feed material converted into C_5–C_9 naphtha and C_3–C_4 LPG.

14.3 Olefin Dimerization and Oligomerization

Olefin dimerization involves the coupling of two olefins, whereas olefin oligomerization refers to the coupling of more (usually up to five) olefins. Together with cracking, reforming, and hydrotreating, it has long been practiced, originally as a thermal refining process. In the 1930s, solid phosphoric acid (SPA) was introduced as catalyst [67], which was then widely adopted. Historically, the main purpose of this technology in the refinery was to convert the olefinic gases into liquid products.

However, the value of light olefins has increased with the growth of the petrochemical industry. Whenever possible, the direct sale of some of the light olefins as petrochemical feedstocks is preferred to converting them into fuels, thus propylene has become a valuable commodity. Although many crude oil refineries still employ olefin dimerization and oligomerization the overall use of such reactions has decreased. Olefin dimerization and oligomerization remain key FT refining technologies [2], since syncrude is comparatively rich in olefins that are primary products during FT synthesis.

14.3.1 Dimerization and Oligomerization Catalysts

Oligomerization catalyst selection is not an operating parameter that can be independently changed to improve efficiency, because the catalyst determines

Table 14.4 Olefin oligomerization catalyst selection in relation to the production objectives for Fischer–Tropsch syncrude refining.

Product	Catalyst	Feed requirements	
		Carbon range	Oxygenate tolerance
Olefinic gasoline (polymer)	SPA	C_2–C_5 (C_3–C_4 preferred)	Limited (~300 μg/g)[a]
	H-ZSM-22/57	C_3–C_4 preferred	Not reported
	Homogeneous, Ni	C_2 (C_3–C_4 also possible)	Sensitive (<10 μg/g)
Paraffinic gasoline (alkylate)	SPA	*n*-Butenes and/or isobutene	Limited (~300 μg/g)[a]
	Acidic resin	Isobutene	Limited (~1%)[a]
Jet fuel	SPA	C_2–C_5 (C_3 preferred)	Limited (~300 μg/g)[a]
	H-ZSM-5	C_2–C_{10}	Percentage levels[b]
	ASA[c]	C_2–C_{10}	Percentage levels[b]
	H-ZSM-22/57	C_3–C_4 Preferred	Not reported
Diesel fuel	H-ZSM-5	C_2–C_{10}	Percentage levels[b]
	Thermal	C_2–C_{10}	Percentage levels
Lubricant base oil	Homogeneous	C_8–C_{12} linear α-olefins preferred	Sensitive (<10 μg/g)

Petrochemical applications are numerous and are not included here.

a) Oxygenates/water used for selectivity control, but excess inhibits catalysis.

b) Deoxygenated over catalyst, but some side reactions retain oxygenate functionality; of specific importance is carbonyl to acid conversion.

c) Amorphous silica–alumina.

the type of product that is produced, and many catalyst types have been reviewed [1]. Based on product type and quality, some recommendations are presented in Table 14.4.

We will discuss only the catalyst types that are employed industrially in conjunction with FT synthesis, namely, SPA and H-ZSM-5. These two are the most likely for application in future FT refineries and the choice depends on the refining objective. Motor gasoline and jet fuel applications favor the use of SPA with light olefin feeds (C_2–C_5). Distillate applications favor the use of H-ZSM-5, which can effectively convert a broader range of olefins (C_2–C_{10}) and is less sensitive to oxygenates in the feed.

14.3.2
Mechanisms of Dimerization and Oligomerization

The difference in products (and their properties) that can be obtained with the same olefinic feed over different dimerization and oligomerization catalysts arises

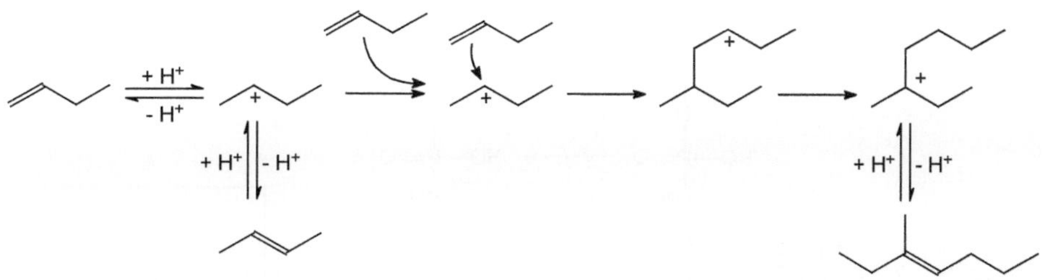

Figure 14.13 Olefin dimerization by a Whitmore-type carbocation mechanism as illustrated by the dimerization of 1-butene.

from differences in the catalytic mechanism. The mechanisms can broadly be grouped into four main categories:

a) **The Whitmore-type carbocation mechanism.** This mechanism (Figure 14.13) [1, 68, 69] is the Brønsted acid-catalyzed mechanism by which H-ZSM-5 catalysts operate. It involves the protonation of an olefin by the acid site of the catalyst giving a carbocation intermediate that initiates the reaction. The carbocation is an electrophile and can attack the double bond of another olefin to form an adduct that is also a carbocation. This type of propagation can be repeated, growing the chain length of the molecule. The carbocation intermediates are the same that were encountered during hydrocracking and hydroisomerization (Section 14.2.2). As the chain length increases, the configuration becomes more favorable for cracking (Figure 14.6). Oligomerization and cracking are opposite reactions and as the temperature increases, the rate of cracking increases relative to that of oligomerization. The strength of the acid sites determines at what temperature cracking becomes significant; at that point, a further increase in temperature leads to the change in the reaction from kinetic to "equilibrium" control. Only a partial equilibrium is attained and only the carbon number distribution is equilibrated. One important consequence of operating under "equilibrium" control is that the carbon number distribution of the product becomes insensitive to the carbon number distribution of the feed [70]. The reaction network is terminated when the proton is returned to the catalyst and an olefin is regenerated.
b) **Ester-based mechanism.** This explains how solid phosphoric acid catalysts operate [1, 71]. The most important difference between the Whitmore-type carbocation mechanism and the ester-based mechanism is how the initiation creates the propagating electrophile (Figure 14.14). There are two important consequences of this type of mechanism. The first is that esters of butene bonded through a terminal carbon can behave like C_5 species, because the α-carbon is no longer a primary carbon, but a secondary carbon stabilizing intermediates that would otherwise have required the formation of a primary carbocation. The second is that the stability of the ester drives rearrangement reactions, including low-temperature skeletal isomerization [72], which is enabled because the α-carbon is a secondary carbon in the ester. An important consequence of this is that dimerization of *n*-butenes over SPA can produce trimethylpentenes in high yield.

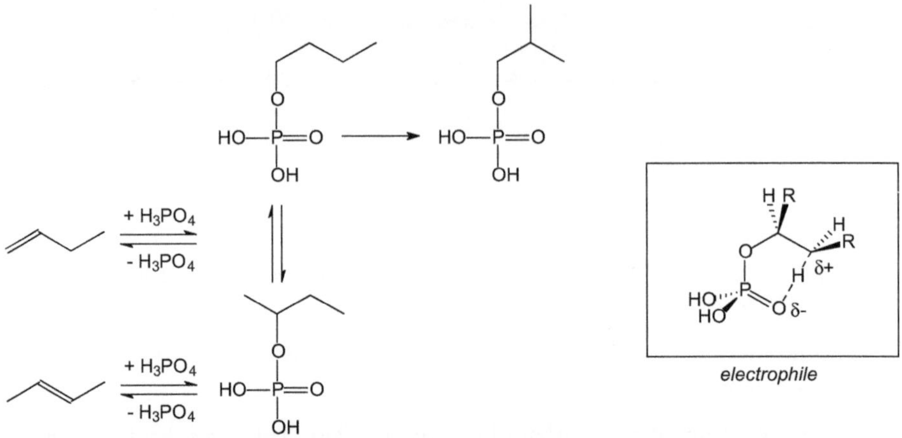

Figure 14.14 Ester-based mechanism to create the electrophile that takes part in olefin addition as illustrated by *n*-butene conversion over phosphoric acid (R = hydrogen or alkyl).

c) **Organometallic olefin insertion and β-hydride elimination mechanism.** This is quite different from (a) and (b) above. Dimerization and oligomerization by metal-based catalysts (notably Ni) take place by this mechanism [1, 73].
d) **Free radical mechanism.** When homolytic bond dissociation takes place, a free radical species is formed. Olefin addition can easily propagate by free radical addition [1, 2].

Geometric constraints imposed by the catalyst may limit the possible reaction intermediates and hence the products that can be formed, as is the case with H-ZSM-5 conversion. Although SPA does not have geometric constraints, it is a supported liquid-phase catalyst, imposing solubility constraints on the intermediates, which likewise affect the nature of the products [71].

14.3.3
Products from Solid Phosphoric Acid and H-ZSM-5 Conversion

The products from SPA and H-ZSM-5 conversions are different (Table 14.4), even using the same feed material, since the catalysis is affected by a number of variables, some of which can be manipulated to tune the product quality (Section 14.3.4).

Solid phosphoric acid can convert C_2–C_5 range olefins with ease, although conversion of ethylene tends to be lower than that of the heavier olefins, for which very high conversions (>95%) are obtained. The octane number of the olefinic motor gasoline is high and almost constant, irrespective of the feed composition. Typical values for Research Octane Number (RON) are 95–97 and for Motor Octane Number (MON) are 81–82 [1]. The product is highly isomerized and can be hydrogenated to produce high-quality isoparaffinic kerosene (IPK) for jet fuel production. The freezing point of the kerosene is less than −47 °C, which is required for jet

Box 14.3 Aliphatic alkylation reactions

Alkylate production in refineries is normally performed at low temperatures using either H_2SO_4 or HF as catalysts. Aliphatic alkylation is essentially a dimerization reaction, but it employs isobutane as source for the production of the electrophile and an olefin in the C_3–C_5 range (preferably C_4) as the nucleophile. These direct alkylation processes produce a highly branched paraffinic product, with a RON in the range of 89–98 and MON in the range of 87–95. Side product formation and the need for spent acid disposal increases the environmental footprint of H_2SO_4 and HF aliphatic alkylation compared to indirect alkylation by butene dimerization over SPA.

fuel. The distillate can be employed as diesel fuel, but it has a low cetane number (~30) on account of the high degree of branching. Fortunately, the product from SPA-catalyzed olefin oligomerization is almost exclusively in the motor gasoline and kerosene boiling range.

It is also possible to employ solid phosphoric acid to selectively dimerizate butenes. An ester-based mechanism operates that results in skeletal isomerization of the *n*-butenes during low-temperature dimerization (140–160 °C) to produce a product that is rich in trimethylpentenes. When that product is hydrogenated, it yields an alkylate-equivalent motor gasoline with RON and MON in the range of 86–88 [1], remarkably considering that the *n*-butene to isobutene ratio in a typical FT syncrude is ~10 : 1. This gives SPA oligomerization of FT butenes a considerable environmental benefit over alternative refining pathways (Box 14.3) [74].

A wide range of olefins can be converted over H-ZSM-5, with C_2–C_{10} olefins yielding a very similar carbon number distribution when the catalyst is operated at a temperature high enough for thermodynamic control [70]. In H-ZSM-5 catalysts, this threshold is around 200–230 °C. The octane number of the naphtha fraction is sensitive to the feed, with typical FT olefins yielding olefinic motor gasoline with 81–85 RON and 74–75 MON [75]. The pore-constrained geometry of H-ZSM-5 limits branching and the products are more linear than that from an unconstrained dimerization/oligomerization following a Whitmore-type carbocation mechanism. Although this property is less good for motor gasoline, it enables the production of good quality distillates for jet fuel and diesel fuel. A cetane number of 51–54 is typical for the distillate from FT olefins. The distillate selectivity is around 65%, but under industrial practice with a naphtha recycle, an overall distillate yield of around 84% can be obtained [2].

14.3.4
Parameters Affecting Solid Phosphoric Acid and H-ZSM-5 Conversion

To achieve the best quality product, one must not only select the most appropriate catalyst, but should also operate appropriately. The reactor and heat management must be properly designed because the olefin addition reaction is very exothermic (85–105 kJ/mol).

14.3.4.1 Effect of Temperature

The effect of temperature is rather different for SPA and H-ZSM-5. The low-temperature skeletal isomerization pathway requires the phosphoric acid ester intermediate to be sufficiently stable to allow the rearrangement. At temperatures above 160 °C, the ability to skeletally isomerize *n*-butenes before dimerization declines rapidly. As the temperature is increased beyond 205 °C, the rate of catalyst deactivation also becomes significantly higher and this sets an upper temperature limit on the use of SPA [71]. The converse is true of H-ZSM-5. At temperatures below the transition from kinetic to thermodynamic control, the cracking rate over H-ZSM-5 is too low to prevent the formation of heavy oligomers. These heavy oligomers are difficult to remove from the catalyst resulting in deactivation. At temperatures above 200–230 °C, the carbon number distribution of the product is regulated by the combination of temperature and pressure. The operating temperature also governs side reactions and at 280 °C there is a transition from high to low acid production with FT naphtha [1].

14.3.4.2 Effect of Olefinic Composition

When olefinic motor gasoline or distillates are produced, both SPA and H-ZSM-5 are insensitive to the olefinic composition of the feed. A wide carbon number range can be converted over H-ZSM-5, whereas conversion over SPA is limited to C_2–C_5 olefins since the reaction rate for heavier olefins is much lower. However, when SPA is employed to produce a motor gasoline component for hydrogenation, the quality of the hydrogenated motor gasoline becomes extremely sensitive to the feed composition [2, 74, 76]. In fact, high-quality alkylate-equivalent motor gasoline can only be produced where the olefinic feed is almost exclusively C_4.

14.3.4.3 Effect of Oxygenates

SPA is very sensitive to the concentration of oxygenates and water in the feed material as the acid strength of the SPA catalyst is determined by the combination of water partial pressure and temperature [71]. In addition, the different classes of oxygenates promote different reactions over SPA, causing inhibition and often leading to the production of water, which directly affects the catalyst and catalysis [1]. The amounts of oxygenates and of water must therefore be controlled in relation to the operating temperature. The impact of oxygenates on H-ZSM-5 is less severe and the conversion of olefins to distillate (COD) process was specifically developed to process HTFT naphtha that contains significant levels of oxygenates. Although some inhibition occurs, the most important industrial impact is the production of carboxylic acids during oxygenate conversion [1, 2].

14.3.5 Comparative Environmental Impact

The proportion of olefins as primary products in FT refineries gives a feedstock advantage over crude oil refineries. It is this feedstock advantage that makes dimerization/oligomerization a key FT refining technology, whereas limited olefin availability restricts its application in crude oil refining. Nevertheless, a fair number of crude oil refineries operate olefin oligomerization units (mainly employing SPA catalysts) to produce olefinic "polymer" gasoline. In this respect, there is little advantage in FT applications, except with respect to the potential volume as a fraction of the total refinery production. The main benefit of SPA-catalyzed dimerization is that it provides an indirect alkylation alternative to aliphatic alkylation (Box 14.3) and, more specifically, employs *n*-butenes directly without the need for skeletal isomerization [74]. Of all the conversion units in a conventional crude oil refinery, aliphatic alkylation is the most energy intensive [77]. The ability to replace an aliphatic alkylation unit in a FT refinery with indirect alkylation by SPA dimerization therefore constitutes an energy-saving reduction in environmental impact.

The most often cited disadvantage of SPA is that the catalyst lifetime is rather limited, around 600–900 kg product per kg catalyst [71]. As long as the kieselguhr support structure is not damaged, the catalyst can be regenerated; however, owing to its low cost, the standard refinery practice is to replace SPA, not regenerate it. Although regular catalyst replacement creates the impression that the solid waste from SPA oligomerization increases its environmental footprint, a life cycle analysis of the SPA catalyst indicates that it has a remarkably small impact on the environment.

The SPA catalyst is made by mixing phosphoric acid with kieselguhr (diatomaceous earth), which is then extruded and calcined [78]. The production process requires energy, as does the production of the phosphoric acid, but the process is not wasteful. The kieselguhr is a natural silica source and no templating molecules or organic solvents are required during catalyst production. The spent catalyst can be neutralized with ammonia to produce an ammonium phosphate fertilizer for agricultural use [79]. In this way the spent catalyst is not returned to a landfill, but beneficially employed as a product that would otherwise have been produced directly from phosphoric acid and ammonia. The spent SPA catalyst from the FT refinery in Secunda, South Africa, is disposed of in this fashion.

The catalyst lifetime of H-ZSM-5 when employed for olefin oligomerization is measured in years, but it requires periodic oxidative regeneration.

References

1. De Klerk, A. and Furimsky, E. (2010) *Catalysis in the Refining of Fischer–Tropsch Syncrude*, Royal Society of Chemistry, Cambridge, UK.
2. De Klerk, A. (2011) *Fischer–Tropsch Refining*, Wiley-VCH Verlag GmbH, Weinheim.
3. Gary, J.H., Handwerk, G.E., and Kaiser, M.J. (2007) *Petroleum Refining. Technology and Economics*, 5th edn, CRC Press, Boca Raton, FL.
4. De Klerk, A. (2007) *Green Chem.*, **9**, 560–565.
5. Sheldon, R.A. (2007) *Green Chem.*, **9**, 1273–1283.
6. De Klerk, A. (2011) *Energy Environ. Sci.*, **4**, 1177–1205.
7. De Klerk, A. (2008) *Green Chem.*, **10**, 1249–1279.
8. Bertoncini, F., Marion, M.C., Brodusch, N., and Esnault, S. (2009) *Oil Gas Sci. Technol.*, **64** (1), 79.
9. Lynch, T.R. (2008) *Process Chemistry of Lubricant Base Stocks*, CRC Press, Boca Raton, pp. 21–42.

10 Murphy, M.J., Taylor, J.D., and McCormick, R.L. (2004) *Compendium of Experimental Cetane Number Data*, NREL/SR-540-36805.
11 Sie, S.T., Senden, M.M.G., and Van Wechem, H.M.H. (1991) *Catal. Today*, **8**, 371–394.
12 Scherzer, J. and Gruia, A.J. (1996) *Hydrocracking Science and Technology*, Marcel Dekker, New York, pp. 13–39, 96–111.
13 Sequeira, A., Jr. (1994) *Lubricant Base Oil and Wax Processing*, Marcel Dekker, New York, pp. 119–152, 194–224.
14 Alvarez, F., Ribeiro, F.R., Perot, G., Thomazeau, C., and Guisnet, M. (1996) *J. Catal.*, **162**, 179–189.
15 Giannetto, G.E., Perot, G.R., and Guisnet, M. (1986) *Ind. Eng. Chem. Prod. Res. Dev.*, **25** (3), 481–490.
16 Archibald, R.C., Greensfelder, B.S., Holzman, G., and Rowe, D.H. (1960) *Ind. Eng. Chem.*, **52**, 745–750.
17 Gibson, J.W., Good, G.M., and Holzman, G. (1960) *Ind. Eng. Chem.*, **52**, 113–116.
18 Weitkamp, J. and Ernst, S. (1990) *Guidelines for Mastering the Properties of Molecular Sieves* (eds D. Barthomeuf, E.G. Derouane, and W. Hölderich), Plenum Press, New York, pp. 343–363.
19 Weitkamp, J. (1978) *Erdoel Erdgas Kohle*, **31**, 13–22.
20 Leckel, D. (2007) *Ind. Eng. Chem. Res.*, **46**, 3505–3512.
21 Corma, A., Martinez, A., Pergher, S., Peratello, S., Perego, C., and Bellusi, G. (1997) *Appl. Catal. A*, **152**, 107–125.
22 Calemma, V., Flego, C., Carluccio, L.C., Parker, W., Giardino, R., and Faraci, G. (2009) US Patent 7,534,340 (Eni SpA and Intitut Français du Pétrole).
23 Souverijns, W., Martens, J., Froment, G.F., and Jacobs, P.A. (1998) *J. Catal.*, **174**, 177–184.
24 Miller, S.J. (1994) *Microporous Mater.*, **2** (5), 439–449.
25 Mills, G.A., Henemann, H., Milliken, T.H., and Oblad, A.G. (1953) *Ind. Eng. Chem.*, **45**, 134–137.
26 Weisz, P.B. (1962) *Adv. Catal.*, **13**, 137–190.
27 Coonradt, H.L. and Garwood, W.E. (1964) *Ind. Eng. Chem. Proc. Dev.*, **3** (1), 38–45.
28 Weitkamp, J. (1975) *ACS Symp. Ser.*, **20**, 1–27.
29 Guisnet, M., Alvarez, F., Giannetto, G., and Perot, G. (1987) *Catal. Today*, **1**, 415–433.
30 Condon, F.E. (1958) Catalysis, in *Alkylation, Isomerization, Polymerization, Cracking and Hydroreforming*, vol. **6** (ed. P.H. Emmet), Reinhold, New York, pp. 43–189.
31 Brouwer, D.M. and Hogeveen, H. (1972) *Prog. Phys. Org. Chem.*, **9**, 179–240.
32 Steijns, M. and Froment, G.F. (1981) *Ind. Eng. Chem. Prod. Res. Dev.*, **20**, 660–668.
33 Ribeiro, F., Marcilly, C., and Guisnet, M. (1982) *J. Catal.*, **78**, 267–274.
34 Debrabandere, B. and Froment, G.F. (1997) *Stud. Surf. Sci. Catal.*, **106**, 379–389.
35 Weitkamp, J., Jacobs, P.A., and Martens, J.A. (1983) *Appl. Catal.*, **8**, 123–141.
36 Calemma, V., Peratello, S., and Perego, C. (2000) *Appl. Catal. A*, **190**, 207–218.
37 Roussel, M., Norsic, S., Lemberton, J.-L., Guisnet, M., Cseri, T., and Benazzi, E. (2005) *Appl. Catal. A*, **279**, 53–58.
38 Eilers, J., Posthuma, S.A., and Sie, S.T. (1990) *Catal. Lett.*, **7**, 253–270.
39 Sie, S.T. (1993) *Ind. Eng. Chem. Res.*, **32**, 403–408.
40 Denayer, J.F., Baron, G.V., Martens, J.A., and Jacobs, P.A. (1998) *J. Phys. Chem. B*, **102**, 3077–3081.
41 Denayer, J.F., Souverijns, W., Jacobs, P.A., Martens, J.A., and Baron, G.V. (1998) *J. Phys. Chem. B*, **102**, 4588–4597.
42 Denayer, J.F., Ocakoglu, R.A., Huybrechts, W., Dejonckheere, B., Jacobs, P., Calero, S., Krishna, R., Smit, B., Baron, G.V., and Martens, J.A. (2003) *J. Catal.*, **220**, 66–73.
43 Marcilly, C. (2003) *Catalyse Acido-Basique: Application au Raffinage et à la Pétrochimie*, vol. **1**, Editions Technip, p. 217.
44 Martens, J.A., Tielen, M., Jacobs, P.A. (1987) *Catal. Today*, **1**, 435.
45 Weitkamp, J. (1982) *Ind. Eng. Chem. Prod. Res. Dev.*, **21** (4), 550–558.
46 Calemma, V., Gambaro, C., Parker, W.O., Jr., Carbone, R., Giardino, R., and Scorletti, P. (2010) *Catal. Today*, **149**, 40–46.
47 Martens, J.A. and Jacobs, P.A. (2008) *Handbook of Heterogeneous Catalysis*, vol. **3** (eds G. Ertl, H. Knözinger, and J. Weitkamp), Wiley-VCH Verlag GmbH, pp. 1137–1149.

48 Fiaux, A, Smith, D.L., and Futrell, J.H. (1977) *J. Mass. Spectrom. Ion Phys.*, **25** (3), 281–294.

49 Martens, J.A., Weitkamp, J., and Jacobs, P. A. (1985) *Stud. Surf. Sci. Catal.*, **20**, 427–436.

50 Dancuart, L.P., De Haan, R., and De Klerk, A. (2004) *Stud. Surf. Sci. Catal.*, **152**, 482–582.

51 Shah, P.P., Sturtevant, G.C., Gregor, J.H., Humbach, M.J., Padrta, F.G., and Steigleder, K.Z. (1988) *Fischer–Tropsch Wax Characterization and Upgrading*, Final Report, DOE Contract No.: AC22-85PC80017, June 6.

52 Calemma, V., Correra, S., Perego, C., Pollesel, P., and Pellegrini, L. (2005) *Catal. Today*, **106**, 282–287.

53 Leckel, D. (2005) *Energy Fuels*, **19**, 1795–1803.

54 Bouchy, C., Hastoy, G., Guillon, E., and Martens, J.A. (2009) *Oil Gas Sci. Technol.*, **64** (1), 91–112.

55 Leckel, D. (2007) *Energy Fuels*, **21**, 1425–1431.

56 Calemma, V., Peratello, S., Pavoni, S., Clerici, G., and Perego, C. (2001) *Stud. Surf. Sci. Catal.*, **136**, 307–312.

57 Thybaut, J.W., Laxmi Narasimhan, C.S., Denayer, J.F., Baron, G.V., Jacobs, P.A., Martens, J.A., and Marin, G.B. (2005) *Ind. Eng. Chem. Res.*, **44**, 5159–5169.

58 Gambaro, C., Calemma, V., Molinari, D., and Denayer, J. (2010) *AIChE J.*, **57**, 711–723.

59 Calemma, V. and Gambaro, C. (2011) *ACS Symp. Ser.*, **1084**, 239–253.

60 Gambaro, C. and Calemma, V. (2011) *ACS Symp. Ser.*, **1084**, 254–277.

61 Leckel, D. and Liwanga-Ehumbu, M. (2006) *Energy Fuels*, **20**, 2330–2336.

62 Correra, S., Calemma, V., Pellegrini, L., and Bonomi, S. (2005) *Chemical Engineering Transactions, Vol. 6: 7th Italian Conference on Chemical and Process Engineering* (ed. S. Pierucci), AIDIC Servizi S.r.l., p. 849.

63 Stangeland, B.E and Kittrel, J.R. (1972) *Ind. Eng. Chem. Proc. Des. Dev.*, **11**, 15–20.

64 Jaffe, S.B. (1976) *Ind. Eng. Chem. Chem. Proc. Des. Dev.*, **15**, 410–416.

65 Girgis, M.J. and Tsao, Y.P. (1996) *Ind. Eng. Chem. Res.*, **35**, 386–396.

66 Leckel, D. (2007) *Energy Fuels*, **21**, 662–667.

67 Ipatieff, V.N. (1936) *Catalytic Reactions at High Temperatures and Pressures*, Macmillan, New York.

68 O'Connor, C.T. (1997) *Handbook of Heterogeneous Catalysis* (eds G. Ertl, H. Knözinger, and J. Weitkamp), Wiley-VCH Verlag GmbH, Weinheim, pp. 2380–2387.

69 Sanati, M, Hörnell, C., and Järås, S.G. (1999) *Catalysis*, **14**, 236–287.

70 Garwood, W.E. (1983) *ACS Symp. Ser.*, **218**, 383–396.

71 De Klerk, A. (2011) *Catalysis*, **23**, 1–49.

72 De Klerk, A. (2004) *Ind. Eng. Chem. Res.*, **43**, 6325–6330.

73 Chauvin, Y., Hennico, A., Leger, G., and Nocca, J.L. (1990) *Erdoel Erdgas Kohle*, **106**, 309–315.

74 De Klerk, A. and De Vaal, P.L. (2008) *Ind. Eng. Chem. Res.*, **47**, 6870–6877.

75 De Klerk, A. (2007) *Energy Fuels*, **21**, 3084–3089.

76 De Klerk, A., Engelbrecht, D.J., and Boikanyo, H. (2004) *Ind. Eng. Chem. Res.*, **43**, 7449–7455.

77 Sittig, M. (1978) *Petroleum Refining Industry: Energy Saving and Environmental Control*, Noyes, Park Ridge, NJ.

78 Coetzee, J.H., Mashapa, T.N., Prinsloo, N. M., and Rademan, J.D. (2006) *Appl. Catal. A*, **308**, 204–209.

79 Van der Merwe, W. (2010) *Environ. Sci. Technol.*, **44**, 1806–1812.

15
Environmental Sustainability

Roberta Miglio, Roberto Zennaro, and Arno de Klerk

Synopsis

A key aim when designing a Fischer–Tropsch (FT) plant is for it to produce as little waste as possible. Designs for the first industrial FT facilities were not primarily concerned with the environmental impact of the operations but this is no longer the case. Following the principles of green and sustainable development in chemical processing and manufacture, all processes (whether they are making fuels or specialty chemicals) should aim to be as efficient as possible, with lowest energy requirements and fewest waste products. For example, effluent treatment can be avoided if the effluent is not created in the first place in the plant. The principles underlying safer and cleaner process designs are similar in many respects. If it is not possible to eliminate effluent production, the process designs that will produce the smallest amounts of the most innocuous effluents are sought. Large amounts of water (H_2O) and of carbon dioxide (CO_2) are produced in FT plants and their amounts are central to any discussion of the environmental sustainability of FT processes. The generation of heat from the reaction and how it is used in the plant are very important too. Feedstock availability is a key aspect of the upstream impact of FT facilities, while the carbon efficiency determines the downstream efficiency. Although biomass is a renewable resource, it should be remembered that its production requires the expenditure of much energy in addition to sunlight. Natural gas is likely to remain the cleanest of the feedstocks overall. Other key aspects that must be considered in the design of an FT plant today are the water management, the water treatment technologies, solid waste and air quality management, and the energy footprint of the refinery.

15.1
Introduction

When the first industrial FT facilities were built, the designs were not overly concerned with the environmental impact of the plants. For example, solid waste had been allowed to accumulate in waste heaps, liquid effluents had been contained in

Greener Fischer-Tropsch Processes for Fuels and Feedstocks, First Edition. Edited by Peter M. Maitlis and Arno de Klerk.

Box 15.1 Early practices

In 1955, when Sasol 1 was commissioned, it was a coal-to-liquids facility employing German and American FT technology. The raw syngas from Lurgi dry ash moving bed coal gasification was purified in a Rectisol cold methanol absorption process to remove the sulfur compounds and carbon dioxide. The dilute effluent from this process (containing 1.5% H_2S in a CO_2 stream) was discharged to the atmosphere at a rate of 54 000 m^3/h (standard conditions; that is, around 1 t/h H_2S) [1]. In 1973, a Stretford unit was constructed to convert the H_2S into elemental sulfur, but it was not fully commissioned [2]. A later attempt to deal with the H_2S emissions progressed to the piloting stage [3], but the problem of H_2S in the air was only resolved in 2004 when Sasol 1 was converted into a GTL facility, with natural gas feedstock.

dams, and gaseous effluents had been emitted as stack gases. These practices were not due to carelessness, but were in line with accepted industrial practice at that time. Since then, much has changed. Environmental legislation has become stricter and there is awareness of the importance of environmental sustainability.

Older FT facilities that are still operating have over time been modified to mitigate the effects of older design practices and the need to address pollution, despite the design solutions not always being easy or successful (Box 15.1) [1–3]. Process design to achieve the smallest environmental footprint must be started during the conceptual design phase. Effluent treatment can be avoided if the effluent is not created in the first place. The principles underlying *safer* and *cleaner* process designs are similar in many respects. As it is unfortunately not always possible to eliminate effluent production from process design, one must strive for a design that will produce the least amount of effluent and the most innocuous effluents practicable.

The literature dealing with synthetic fuels processing, including Fischer–Tropsch synthesis, indicated considerable interest in health and environmental issues in the late 1970s and early 1980s [4–7], much of the work being related to the coal conversion programs started in response to the 1973 Oil Crisis. The abundance of heteroatoms and trace elements in coal made it a challenging carbon source as raw material. Since the 1990s, all new industrial FT facilities have been designed using natural gas as raw material; this avoids the complexity of solid carbon feedstocks such as coal, biomass, shale oil, or organic waste. Although natural gas is an attractive raw material for GTL, in many locations it is already a very desirable energy carrier (e.g., for heating) and gas-to-liquids conversion is unnecessary and may even be wasteful. The environmental sustainability of FT-based processes also needs to be considered for potential carbon sources other than gas.

The production of water (H_2O) and carbon dioxide (CO_2) are central to the discussion of the environmental sustainability of FT-based processes. As they are stable compounds, they strongly influence the thermodynamics of the carbon conversion process. In order to generate the synthesis gas that is needed as feed for an FT process, a carbon source and water are required (Chapter 2). FT synthesis ($CO + 2\,H_2 \rightarrow -CH_2- + H_2O$) can be regarded as much about water as it is about the synthesis of hydrocarbons (HC). In addition to water consumed and produced by the process, water is also a key utility for cooling and for steam generation. Significant quantities of heat are transferred during both synthesis gas generation and FT synthesis. Water is therefore a key aspect of any FT facility and water and energy supply are fundamentally connected [8].

In the same way, carbon dioxide is involved both in the process and as a utility effluent. The production of CO_2 is a necessary consequence of the H : C balance and the energy requirements of the process, with about two-thirds of the carbon ending up as CO_2 [9]. Even for an ideal process, the second law of thermodynamics requires energy use in excess of work performed. Carbon-based energy is derived from combustion ($C + O_2 \rightarrow CO_2 + \text{energy}$) and the energy cost of a FT-based process can therefore be expressed in terms of an equivalent CO_2 emission. The stoichiometry of the FT reaction also requires the use of carbon in a 1:4 ratio with hydrogen.

A vital aspect of sustainability is not only the environmental impact but also the ability to continue the practice. To paraphrase the United Nations report on sustainable development [10], *environmentally sustainable FT synthesis is a process that meets the needs of the present, without compromising the ability of future generations to meet their own needs.* Many past practices were not sustainable (Box 15.1). How will future generations judge our present designs and their impact on the environment? Present and future practitioners of FT must defend their use of both carbon and water.

15.2
Impact of FT Facilities on the Environment

The life cycle assessment of any process must consider both the upstream and downstream impacts [11], and Chapter 13. The upstream impact assessment considers the resources that are required to produce the raw feed needed by the process. The amount of resources consumed per unit of product made is an important metric and is equivalent to a system-wide *E*-factor (kg waste/kg product) [12]. An upstream impact assessment can give valuable insights into the sustainability (or unsustainability) of a process by highlighting any hidden environmental costs (Box 15.2). The downstream impact assessment deals with the more immediate impact on the local environment arising from process emissions. It is the emission-based assessment that is more commonly found in the discussion of FT facilities.

15.2.1
Upstream Impact Assessment

When a mass and energy balance is performed over an FT facility, the key inputs can be identified. These inputs need to be traced to their origin in order to assess the upstream impact. There are two aspects: the raw material itself and the

Box 15.2 Biofuels

Biomass-derived fuels (biofuels) as sustainable/renewable fuel sources are increasingly under scrutiny since much of the production arises from agriculture and agricultural waste. An upstream assessment of biofuel production shows that although biomass is a renewable resource, that is not true of its production. Agriculture requires fertilizer that is carbon and energy intensive to produce, while planting, tending, watering, and harvesting rely on crude oil-derived fuels (carbon-intensive). Furthermore, the fertilizer can eutrophy rivers and has probably helped to bring about the 'Dead Zone' in the Gulf of Mexico. In addition, biofuel production competes with food production for land use.

processes that make the raw material available. As Hoffman [13] has put it: "The existence as well as the prosperity of a civilization is bound to the consumption and availability of resources. The economy of that civilization relate to the means by which those resources are utilized and the rewards disseminated."

Since fossil fuels are nonrenewable resources, an important upstream impact of FT facilities is to reduce the future availability of such raw materials. Even renewable and semirenewable raw materials, such as biomass and waste, are not indefinitely sustainable (Box 15.3). In fact, many of the environmental issues related to biomass as raw material are upstream impacts [14]. The inevitable impact of carbon-based resource consumption has to be justified by the usefulness of the products that are produced by FT.

Good access to water is of critical importance for synthetic fuel production, for example, to make hydrogen, since additional hydrogen is often needed to satisfy the stoichiometry of FT synthesis. However, most of the water consumed by a FT facility is needed for utilities. The original Sasol 1 CTL facility (Chapter 5) consumed more than $10\,m^3$ water/m^3 syncrude [15]. This was reduced in the design of the Sasol 2 and 3 CTL facilities (Chapter 5) to around $8\,m^3$ water/m^3 syncrude [16]. Most of the water is required for evaporative cooling, although some is needed to satisfy the stoichiometry of CTL conversion. Based on mass balance considerations only, a water consumption rate of $2\text{–}3\,m^3/m^3$ syncrude is achievable in a CTL facility [4]. The water consumption rate for gas-to-liquids conversion is lower, because methane has a high H : C ratio. With natural gas as a feed for syngas production, the FT conversion stoichiometrically becomes a net producer of water. This water can be used to offset the evaporative losses in the utility system (Section 15.3).

Box 15.3 Sustainable biomass production

The estimated annual fixation of CO_2 by photosynthesis is $150\text{–}200 \times 10^9$ tons globally [13]. If we assume that 1% of this biomass can be harvested annually without disrupting the ecosystem (i.e., sustainable), then we have around $1.5\text{–}2.0 \times 10^9$ tons available as raw material (current crude oil production is around 5×10^9 t/a).

Table 15.1 Coal losses during coal cleaning as an example of the upstream impact of raw material production that is not reflected by the mass balance of a coal-to-liquids FT facility.

Country	Cleaning plant output (kg/kg feed)			*E*-factor (kg waste/kg product)
	Cleaned coal	Clean middlings	Discard	
Australia	0.72	0.01	0.27	0.36
Canada	0.74	—	0.26	0.35
France	0.52	0.07	0.41	0.69
Germany	0.51	0.06	0.43	0.74
India	0.70	0.23	0.07	0.07
Japan	0.57	0.14	0.29	0.42
Poland	0.64	0.09	0.27	0.38
Roumania	0.53	0.26	0.21	0.27
South Africa[a)]	0.73	—	0.23	0.32
United Kingdom	0.70	—	0.30	0.43
United States	0.73	—	0.27	0.37

The differences in recovery are due to technology and the coal properties in each country.
a) Mass balance not closed in source reference.

Other input streams, such as catalysts, chemicals, and neutralizing agents, are small in comparison to the carbon-based raw material and water requirements for FT-based facilities, but must be taken into consideration when a full environmental impact assessment is performed.

The technology for producing the raw material, whether it is the mining of coal, biomass by agriculture, or natural gas by exploration and drilling, affects the environmental cost associated with such production. There are also inevitable losses associated with the production of raw materials that are related to the nature and technology of production (Table 15.1) [17]. One should also consider whether the raw materials consumed by an FT facility give "good value." This should be location specific, since the point of origin of the raw material rarely coincides with the point of use of the final product. For example, if natural gas can be supplied by pipeline for indoor heating, direct use may be better than its use in generating electricity or in FT conversion to produce a liquid fuel that in turn provides heating. However, we must always consider the use to which the energy is put; thus, there is no point in piping natural gas for transportation requirements if the end users, the drivers of cars and buses, need to run their engines on diesel or gasoline.

15.2.2
Downstream Impact Assessment

The carbon efficiency of indirect liquefaction is quite low, with only around a third of the feed carbon ending up in useful products [9]. The effluents from a FT-based facility are consequently substantial. Their downstream impact depends on the pollution control measures in the facility and on the nature of the effluents that leave the facility. In addition to those that are generic to FT-based facilities, the raw material employed as feed for syngas generation has a substantial impact on the effluents generated. The downstream impact is also affected by the extent of syncrude upgrading, as well as by the specifics of the FT refinery design [18].

The generic effluents are those that relate directly to an FT process and the utility requirements for syngas generation. The main effluents are wastewater (Section 15.3) and CO_2 (Section 15.5). Both are produced as a consequence of the stoichiometry and energy requirements of syngas production and FT synthesis. The spent FT catalyst (Section 15.4) is likewise an effluent that is generic to FT facilities. Some of the less obvious downstream impacts that should also be considered when a more exhaustive environmental impact assessment is performed are noise, waste heat (both heat pollution and the creation of microclimates), and losses of amenities.

The effluents formed also depend on the raw materials employed. All the non-CHO elements, present in the raw material must be removed before the FT synthesis, and this should take place during syngas generation and cleaning. The main classes are sulfur compounds, nitrogen compounds, trace elements (e.g., Hg and As), solid wastes (e.g., ash, feed rejects, and salts), and organic wastes (e.g., tar and pitch). Coal gives rise to the most effluents and dealing with the impurities and the effluents has been extensively discussed [4–7]. Some of these effluents may be beneficially employed, for example, as chemicals that can be recovered from the coal tar and ammoniacal liquor.

As described in Chapters 4 and 14, the conversion of FT syncrude into useful products requires refining. The nature and quantity of effluent generated depend on the design of the FT gas loop and particularly on the syncrude separation section, since the recovered syncrude is sent to the FT refinery (Chapter 5 and Section 15.6).

15.3
Water and Wastewater Management

FT facilities are water intensive, with large volumes of water recycled and consumed. FT facilities employing natural gas as feedstock can potentially contribute to the development of the surrounding area through the use of their coproducts: water and steam [19]. However, most FT facilities are net consumers of water. Water processing is significant for three reasons: Water is the main product of the FT reaction. It is an important reagent in syngas production since most H_2 from gasification of carbon-rich raw materials comes from water. It is also is employed as a utility, for example, to produce steam to remove the heat of reaction from FT synthesis.

To address the sustainability of water resources, the water sector is currently developing awareness in communities and businesses on how production chains affect water systems. The overall XTL processes (Section 2.2) are water intensive, with high water flow rates and many circulating flows and loops. Available tools, such as the "water footprint," have prompted the development of new standards. Apart from the book by Probstein and Gold [4], wastewater treatment received little attention in books dealing with FT technology [20]. This has been a major oversight in the FT literature given the amount of energy used to collect, treat, and move water.

All commercial FT facilities to date have been planned and built in the vicinity of the carbon feedstock. Unless water is not a scarce resource there, local and even national level planning is needed to make provision for the water use associated with a FT facility [21]. Cases in point are the GTL developments in the Middle East, where freshwater supply is limited, since this area is not only an arid region, but is also experiencing significant population growth and industrialization, placing additional stress on available water supplies. Most of the freshwater budget in the Middle East will come from desalination plants, for which the environmental effects must be considered. In general terms, the energy requirement to deliver clean water from seawater is two to three times higher than that needed for reuse of wastewater. It is nearly twice that for wastewater treatment for discharge, which is on the order of 0.5–1 kWh/m^3 (1 kWh = 3.6 MJ) just for the energy input [22]. Using existing water resources more efficiently or reducing water imports consequently has a beneficial impact on the carbon footprint.

Water management can take two forms during the design of an FT facility in a water-constrained environment. First, process cooling can be performed by other means than a traditional cooling water design. Air cooling and saltwater cooling are two alternatives to be considered. Second, the selection of raw water intake is key to the long-term success of the facility. Selecting a freshwater source is not always the most economical and some of the issues to consider are as follows:

a) *Supply security*: Tapping into a local resource must have the blessing of the local community over the longer term. Freshwater sources are also exposed to seasonal changes and one must be certain that this will not become a bottleneck to plant operability in dry seasons.
b) *Water quality*: The poorer the quality of the raw water, the more costly will be its cleaning for use. Freshwater sources are not necessarily less costly to clean than the desalinating seawater.
c) *Effluent disposal from water treatment*: There is an economic and environmental cost associated with the proper disposal of the effluent from water cleaning. This is usually more of an issue for inland facilities, but even at the coast the discharge from water cleaning cannot just be dumped into the ocean.
d) *Future rehabilitation and liabilities*: Tapping into a communal resource comes with current and future responsibilities.
e) *Distance from the water resource*.

15.3.1 Water Produced in FT Facilities

Water is the main product from FT synthesis in an equimolar ratio based on CO conversion. For every 1 ton of hydrocarbons that is produced, more than 1.2 tons of water are produced. This water is acidic and highly corrosive. The FT aqueous

Box 15.4 Water consumption rate (example of)

At the coal-to-liquids Sasol 2 site the net rate of water consumption is around 8 m^3 water/m^3 hydrocarbon product [21].

product can be refined to recover some of the dissolved oxygenates, but recovering the dissolved carboxylic acids is more challenging [18].

Syngas generation is another potential source of water, albeit not a net source of water. The amount and quality of water produced are technology and feedstock dependent. In a GTL scheme, water is produced in natural gas extraction processes; in a BTL scheme, water is the moisture of the raw biomasses; in the CTL case, water originates from mining or from the coal gasification step. Other more minor process water sources are associated with the downstream refining units; for example, any conversion process that involves oxygenate to hydrocarbon conversion coproduces water.

Despite the considerable amounts of water produced in the FT process, even larger quantities of water are required for cooling purposes. FT facilities can therefore be designed to have zero liquid effluent discharge (ZLED), but will still require water for evaporative cooling that is in excess of the water produced by the process (Box 15.4) [16].

15.3.2
Quantities and Quality of Water

The most important process-derived water stream is the FT aqueous product (reaction water). The quantity and quality of this stream depend on the FT technology and the design of the syncrude separation system in the gas loop. After separation of hydrocarbons, there is about 1.2–1.4 m^3 reaction water generated for each m^3 of liquid hydrocarbons. At this point, the aqueous product still contains dissolved organic compounds, mainly light alcohols, aldehydes, ketones, esters, and carboxylic acids (Table 15.2) [20]. The quantity of these compounds, within each group, decreases with increasing molecular weight. This water can also contain traces of metals coming from the reactor. Although this water has a low salt content, it has a high chemical oxygen demand (COD) level. A peculiarity of the

Table 15.2 Generic composition of aqueous products from different FT technologies.

Compound class[a)]	Fe-HTFT	Fe-LTFT	Co-LTFT
Nonacid oxygenates	4.47	3.57	1.00
Carboxylic acids	1.41	0.71	0.09
Hydrocarbons	0.02	0.02	0.02
Water	94.11	95.70	98.89

a) Inorganic compounds are $<0.005\%$.

reaction water is that the composition is known and relatively constant, quite different from a typical wastewater that has wide and fluctuating levels of contamination.

Another important process stream is water from condensation after syngas cooling. Expressed on the same basis, this stream is around 0.8–1 m^3/m^3 HC when natural gas is the feed. In the case of methane reforming, the water produced is of a high quality. Solid feed gasification technologies vary considerably in their water/steam requirements and consequently in the volume of water from condensation after syngas cooling [23]. There could also be a caustic scrubbing effluent from the syngas cleaning section or other specific aqueous effluents in a BTL or CTL configuration, with contaminants such as ammonia, HCN, phenols, and toxic metals, which are not present when the feedstock is natural gas. The flow of these aqueous streams is on the order of 0.3–0.6 m^3/m^3 HC.

The cooling tower blowdown, where climatic conditions require this unit operation, can be an important aqueous output, on the order of 0.3–0.7 m^3/m^3 HC. This water may be contaminated with antiscale corrosion inhibitors and chlorine. Moreover, in a XTL complex, brine water has to be managed if it comes out as by-product of desalination or from demineralization in the utilities section of the plant that is needed for steam production. Suggested approaches for this salt-laden stream include evaporation with subsequent solids disposal, or in the case of a CTL facility, chemically binding the salts with the ash from gasification [24]. There are also discontinuous streams originating from civil uses, sanitary water, or oily water from ground or tank washing, on the order of 0.2–0.5 m^3/m^3 HC.

In addition to the above-mentioned water products, the facility also requires a significant import of freshwater necessary for utilities, mainly in cooling towers and for syngas cooling. The quality requirements of this water include the absence of suspended solids and scaling components. Most of the water intake in the facility is for the production of boiler feed water, which needs complete removal of suspended solids, inorganics, and ions. There should also be provision for potable water and firewater on the plant site.

The overall water export and import balance in a conventional GTL facility can be closed with a net water import, on the order of at least 0.02 m^3/m^3 HC. For CTL, the net water import is much higher. A nearly neutral balance, with a net water import of less than 0.01 m^3/m^3 HC, can be reached when a GTL plant design includes treatment units for reaction water recycling as process water.

15.3.3 Water Management Approaches

The water treatment system is integral to the process design of an FT facility. It enables the water within the facility to be reused, for example, for cooling in the synthesis section. It enables disposal of water outside the facility and conditions the water for other uses, such as irrigation or drinking water. Different BATANEC (best available technology at nonexcessive cost) solutions can be identified as a function of the project-specific requirements. There is not a single best solution, and technology selection must be adapted to the geographical location and the

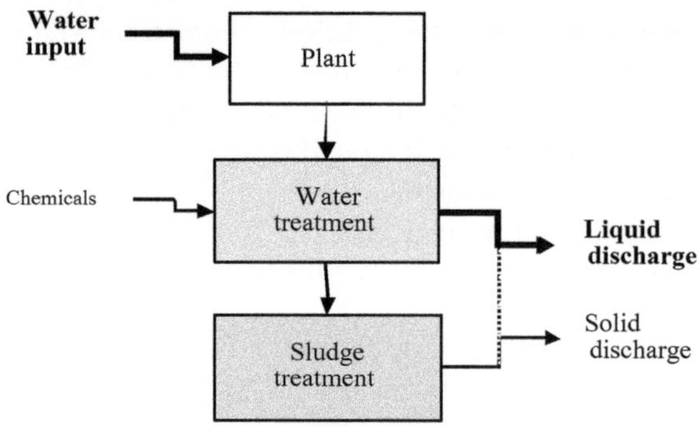

Figure 15.1 Water management by the "water discharge" approach.

industrialization degree of the area. For instance, in a cold climate, cooling water needs are reduced and cooling towers may not be necessary. Moreover, even two identically designed and constructed plants can have different efficiencies and emission profiles due to differences in local conditions and business focus. Where applicable, synergies with existing plants and other industrial facilities have to be exploited.

Water management practices have developed approaches, ranging from "water discharge" to "zero liquid effluent discharge" and recently "water valorization."

In the "water discharge" approach (Figure 15.1), the reaction water and the other aqueous streams are treated in order to satisfy the legislated criteria for surface water discharge or irrigation water discharge. The Sasol 1 facility is an example of this type of design. The treatment normally uses chemicals and produces sludge that has to be disposed. In a BTL scheme, the production of biomass needs energy, as well as sunlight and an abundance of water. In this situation, water for irrigation is a valuable product.

In the "zero liquid effluent discharge" approach (Figure 15.2), the water is treated in order to be recycled. By doing so, the treated water can be used to partially balance the water requirements of the utility systems in the plant,

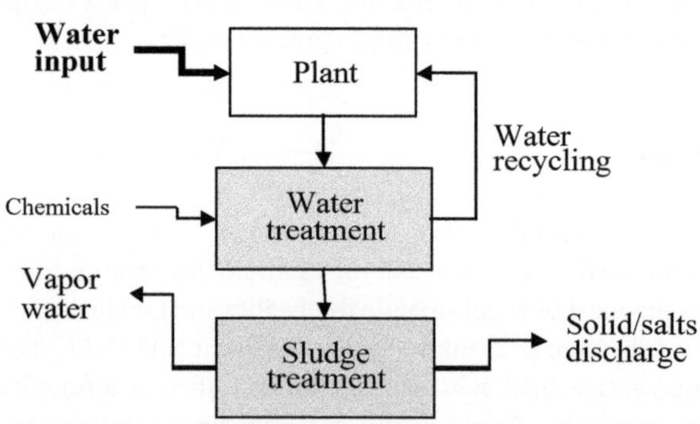

Figure 15.2 Water management by the "zero liquid effluent discharge" approach.

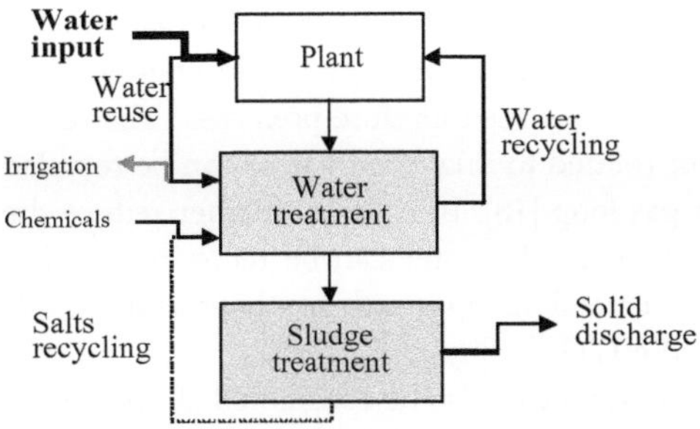

Figure 15.3 Water management by the "water valorization" approach.

which are large consumers of freshwater. Organic contaminants and salts are concentrated in a stream that undergoes evaporation and crystallization steps. Water is partially lost as vapor. This approach is the preferred solution where allowed emissions are very low. This approach has been used in the design of the Sasol 2 and 3 (Sasol Synfuels) CTL facility, which employs ZLED [16]. A ZLED design has also been implemented at Pearl GTL (Section 15.3.5).

The "water valorization" approach (Figure 15.3) combines elements of water discharge and ZLED with the objective of achieving a more sustainable overall water strategy. Part of the water is treated, just enough to be recycled in the process, and it is partially reused. Some of the water is used outside the plant after appropriate purification steps. Overall the strategy is to reduce the demand for new water intake, but in such a way that the intensity of water treatment is kept at the lowest practical limit. There is consequently a trade-off involved. In this more complex practice, additional costly technology can be required, but solid discharge and chemical imports are minimized.

15.3.4
Water Treatment Technologies

The water treatment system normally includes a biological step, which can be preceded by a stripping/distillation step, to separate the more volatile compounds, those boiling points less than that of water. The organic compounds remaining in the water are mainly carboxylic acids and this water is also referred to as *acid water*. The water after biological treatment is then normally subjected to further purifying steps to remove solids and residual salts.

Most of the COD of the integral water stream is due to alcohols, but this water can be successfully treated under anaerobic conditions [25], and a combination of anaerobic and aerobic processes can be suitable for the treatment. When the water is treated by a biological process, the organic compounds contained therein are degraded to CH_4, CO_2, and H_2O and the added chemicals lead to the production of a sludge. The anaerobic digestion of both the alcohol-rich wastewater [26] and the acid-rich wastewater [27] have been discussed in literature.

The carbon loss of the whole cycle is increased by residual oxygenates in the aqueous product sent to water treatment. Removal of the nonacid chemicals by distillation/stripping before biological treatment is therefore preferred. The recovered nonacid chemicals can be further refined to final products or can be recycled in various ways to the FT refinery or gas loop [18]. In order to further reduce the carbon footprint, part of the carbon in the acid water can be recovered during anaerobic biological treatment as methane. This approach has been industrially implemented with success at the PetroSA GTL facility.

Recovery of the carboxylic acids in the water can also be considered as an alternative to aerobic or anaerobic biological treatment. A liquid–liquid extraction process using a light solvent in a packed bed extractor can be used to recover residual carboxylic acids and such technology was piloted, but was not commercialized, for practical and economic reasons [18]. It has also been suggested that residual carboxylic acids can be separated by other chemical methods. For example, it was possible to reduce the carboxylic acid concentration in FT acid water by electrodialysis, while achieving a current efficiency of 30% (energy cost of 4 MJ/kg) [22]. Nevertheless, devising efficient processes for recovering carboxylic acids from dilute aqueous solutions remains a challenge.

Optimal separation/purification technology is driven by the physicochemical properties of the compounds. Several companies have developed and patented water treatment schemes, some of which have been reviewed [28]. Despite the variety of claims and approaches, most water treatment technologies include a separation step to separate the nonacid chemicals from the acid water, as well as suggested uses for the nonacid chemicals and a procedure to treat the acid water. The basic principles are illustrated by the Eni water treatment technology.

Eni developed water treatment schemes based on three main unit operations: a stripping column for separating volatile compounds (alcohols) from reaction water, a biological degradation treatment for residual compounds (carboxylic acids), and a saturator for reuse of water as process water and alcohol recovery for syngas conversion (Figure 15.4). The biological treatment produces water with surface water quality discharge or irrigation quality, by combining anaerobic and aerobic treatment as a function of target requirements.

Alternative schemes have been developed for complete water reuse in the plant (Figure 15.5). After anaerobic biological treatment in which microbiological activity converts dissolved organic material into biogas, a combination of membrane-type treatment units produces a high-quality permeate that can be reused as process water or it can be used in the steam circuit. Additional units, that is, of the electrodialysis type, can be used for recovery of chemicals for reuse and for brine or sludge minimization.

15.3.5
Benchmark Technology: Water Treatment at Pearl GTL

Pearl GTL, developed by Qatar Petroleum and Royal Dutch Shell, is the world's largest industrial GTL complex and the largest energy project in Qatar. The facility

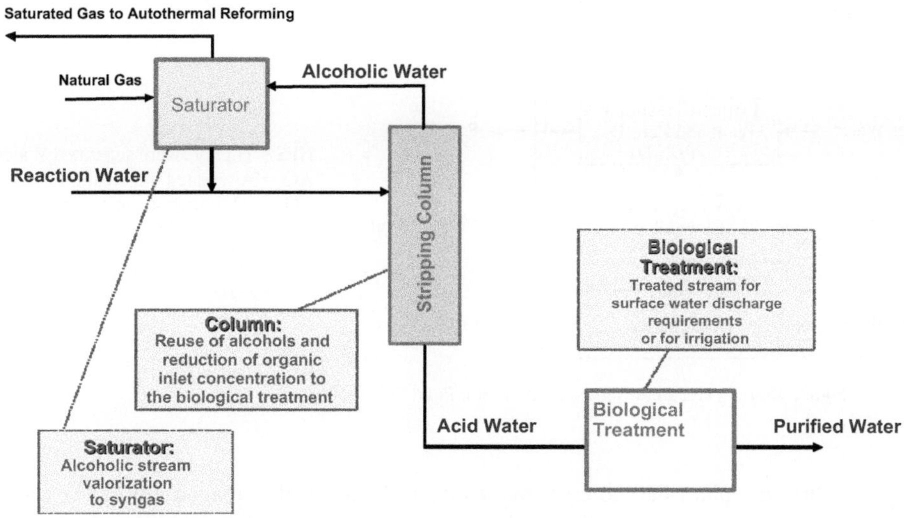

Figure 15.4 Eni GTL water treatment process with water discharge.

was commissioned in 2011 and at present represents the benchmark technology for water treatment associated with a FT based GTL facility.

Management of water in the Pearl GTL design is shown in Figure 15.6 [29]. The water is purified to a level where it can be mostly reused within the plant. As a

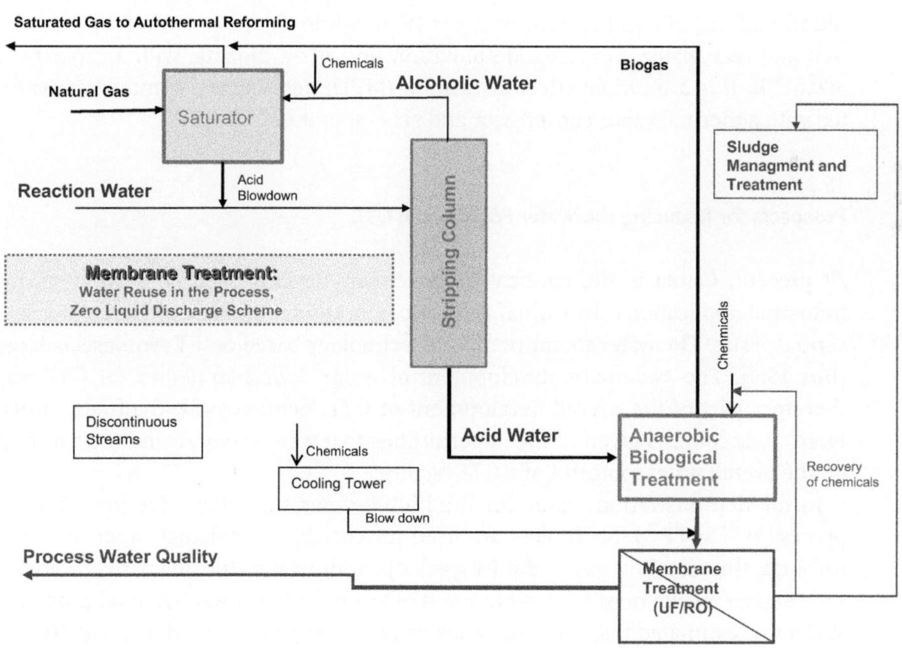

Figure 15.5 Eni GTL water treatment process for zero liquid effluent discharge.

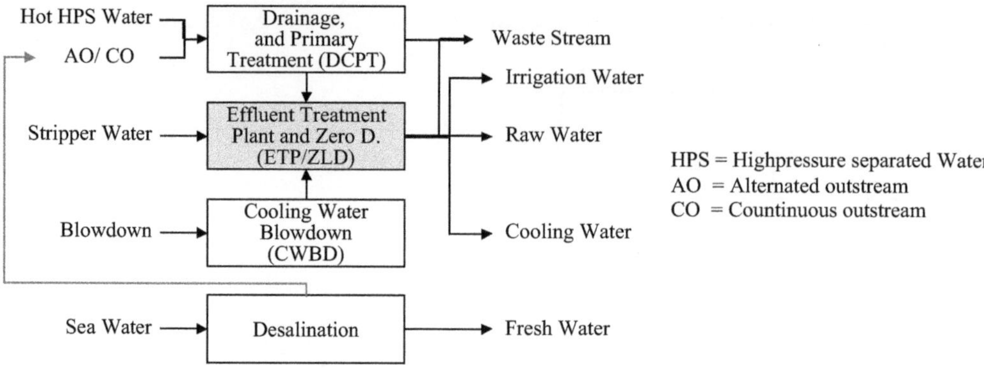

Figure 15.6 Water treatment scheme at the Pearl GTL facility.

result, the plant has no freshwater intake. Optimized management of the water cycle with an objective of zero liquid effluent discharge into the natural environment is a major aspect of the Pearl GTL project. Most of the water is used for cooling by evaporation and as boiler feed water for the steam systems. From this water, 8000 m^3/h is used as steam to run eight steam-driven air compressors on the air separation units (ASU) and other steam-driven equipment for power generation. Some water is also used for irrigation and site greening at the plant. The water treatment capacity is 1800 m^3/h (equivalent to 16 million t/a), which is comparable to a water treatment plant for a city of 140 000 people [30]. Wastewater is treated by ultrafiltration (UF) and reverse osmosis (RO), while sludge is treated by evaporation and recrystallization. One desalination unit is on the site, with a capacity of 300 m^3/h. It is a multiple effect distillation (MED) desalination plant, which is fitted with a thermal vapor compressor and six evaporation cells [31].

15.3.6
Prospects for Reducing the Water Footprint in CTL

At present, China is the country most actively developing CTL technology for industrial applications. In China, as in most locations, the lack of freshwater is a serious issue. The water footprint of CTL technology based on FT synthesis is large (Box 15.4). The systematic development of water saving strategies for CTL was therefore part of the overall development of CTL technology by Synfuels China. Here, an account is given of the opportunities that were explored and their impact on the overall water footprint of a CTL facility.

In the demonstration plants for the high-temperature slurry Fischer–Tropsch process (HTSFTP®), air coolers are used for cooling the exhaust vapor of steam turbines, the recycling gas in the FT gas loop, and most of the downstream nodes. This design results in water consumption of around 12 tons water/ton oil produced and it was estimated that the water consumption can be decreased to about 10 tons water/ton oil ($8\,m^3$ water/m^3 oil). This is similar to that of large-scale industrial CTL facilities such as Sasol Synfuels.

One strategy was to employ a closed-circuit water cooling system and air cooling for interstage cooling of the compressors in the air separation unit. It is estimated that about 2 tons water/ton oil can be saved by avoiding evaporative cooling of the water used for interstage cooling. However, adequate and reliable cooling is critical to the performance of the ASU and this modification of the cooling design needs to be well understood and accepted by suppliers of air separation technology before it can be adopted.

To further reduce the water consumption in CTL plants, Synfuels China have proposed a strategy that uses a separate unit in order to reduce the temperature of the cooling water streams to 30 °C. It uses an adsorption cooling system driven by low-pressure steam produced in the CTL processes. The exhaust heat from the adsorption unit is dissipated by air cooling. This closed-loop cooling water system avoids evaporative loss of cooling water and thus greatly reduces water consumption. According to the mass and energy balance analyses for several large-scale CTL projects, the low-grade heat (steam at 200–500 kPa) that can be recovered from the process can supply sufficient energy for closed-loop cooling water production to replace 50–70% of the evaporative cooling capacity of a standard CTL facility. Closed-loop cooling water production is more expensive, and it was estimated that changing from evaporative cooling to 50–70% closed-loop cooling would increase the overall capital cost by 2%. Yet, by doing so, the water consumption of the CTL facility is reduced to 3–4 tons water/ton oil (2–3 m^3 water/m^3 oil). In addition to the water saving, there are other operational benefits, such as reduced maintenance costs associated with heat exchangers due to reduced fouling and corrosion.

15.4 Solid Waste Management

The mass balance of an FT-based facility indicates various sources of solid waste, many of which are not unique to an FT facility. For example, spent refinery catalysts are common to any refining operation and procedures for reclamation and disposal are well documented [32]. Solid wastes generated from wastewater treatment are likewise common to wastewater treatment in general (Section 15.3).

An important source of solid waste not unique to FT facilities, but that is associated with syngas generation, is the solid residue from gasification. This form of solid waste is not present when natural gas is employed as feed material, but is formed when solid feed materials such as biomass or coal are used. The nature of the ash from gasification depends on both the raw material composition and the gasification technology. Thus, the ash from a dry ash gasifier and that from a slagging gasifier will have different properties even when the same feed material is employed. In addition to the solid waste from gasification, there are solid wastes from feed preparation before gasification (Table 15.1).

The type of solid waste that is unique to FT synthesis is the spent FT catalyst. There are basically two metals that are industrially employed to catalyze FT synthesis: iron and cobalt (see Chapters 8, 9, and 13).

a) Iron FT catalysts after exposure to the environment resemble iron ore in many respects and the spent catalysts are supposedly not hazardous. The unloaded, spent FT catalysts are covered with heavy syncrude and are pyrophoric [20]. Care must therefore be taken in handling these materials. In the case of LTFT synthesis, the spent catalyst is covered with wax; in the case of HTFT synthesis, the catalyst pores are filled with a waxy aromatic oil. Due to the syncrude covering the catalysts, they have heating value and catalyst disposal through coal-fed boilers has two advantages: it enables the beneficial use of the associated hydrocarbons and it dilutes the spent iron catalyst material with coal ash. Whatever the most practical way of dealing with this material, disposal does not pose major environmental concern and landfill is an option in some jurisdictions [33]. Since the starting material for the preparation of iron FT catalysts is iron oxide, recycling of the material is in principle possible, but it is not considered cost-effective. Iron oxide has a low value, making disposal economically more attractive than recycling.
b) Cobalt FT catalysts are potentially harmful to the environment [33], and unlike iron-FT catalysts, landfill is not an option. Metal recovery is not only necessary environmentally, but is also economically desirable. Cobalt is more expensive than iron and the value of a spent supported 20% Co-LTFT catalyst is around \$10/kg [34]. Furthermore, some industrial Co-LTFT catalysts contain 0.05% platinum, which more than doubles the value of the spent catalyst. One must also consider the sustainability impact on metals supply. For a 5 million t/a (100 000 bbl/day) equivalent Co-LTFT production, around 500 tons Co are required. This is a substantial proportion of the total annual global production of around 45 000 tons Co [34]. Recycling Co is consequently important.

15.5 Air Quality Management

Gaseous waste products from an FT facility can broadly be classified into three groups. First, there are non-WGSR products associated with syngas production and cleaning. These are mainly sulfur- and nitrogen-containing gases and unconverted hydrocarbons. The nature and concentration of the contaminants are determined by both the raw material used for syngas generation and the syngas generation technology. There is considerable literature on gas cleaning [35]. Second, there are water gas shift reaction (WGSR) products (H_2, CO, CO_2, and H_2O) associated with syngas production and conditioning in the FT gas loop. Of these, only CO_2 is a gaseous waste product (see Sections 15.5.1 and 15.5.2). Third, there are the gaseous waste products (including CO_2) from refining. FT refineries are different in some important respects from crude oil refineries and it has been claimed that FT refining is more environment-friendly than the crude oil refining (see Section 15.6) [36].

15.5.1
The CO_2 Footprint of FT Facilities

The CO_2 footprint of a FT facility is calculated from its carbon efficiency (*not* energy efficiency). For a CTL facility, it is about 6–7 kg CO_2/kg hydrocarbon product; for a GTL facility, it is about 3–4 kg CO_2/kg hydrocarbon product. The difference in CO_2 footprint between CTL and GTL is not related to the FT technology, but to the hydrogen content of the raw material.

The impression is sometimes erroneously created that cobalt-based FT technology has a smaller CO_2 footprint than iron-based FT technology, because it generates little CO_2 during FT synthesis. Cobalt has very low water gas shift activity and therefore it does not generate much CO_2, but by the same token it is not capable of consuming much CO_2 either. When performing a materials balance over the whole process, including both the syngas generation and the FT gas loop, the CO_2 footprints for all FT technologies are very similar. From an emissions perspective, it is immaterial where in the process the CO_2 is produced and potentially consumed (Section 15.5.2). The CO_2 produced during syngas generation, syngas conditioning, and FT synthesis all adds up and balances out.

The CO_2 footprint of FT-based facilities is intrinsic to indirect liquefaction technology. It is governed by the mass and energy balance over the process (see Section 15.1):

a) Mass balance requirements dictate that the reaction stoichiometry of the FT reaction must be satisfied. This implies that for ideal FT hydrocarbon synthesis, syngas will be consumed at a H_2 : CO ratio of 2 : 1. This is equivalent to a H : C atom ratio of 4 : 1. With the exception of methane (lean natural gas), most other carbon-based raw materials have lower H : C atom ratio (Table 15.3). The shortfall in hydrogen can be met by either adding hydrogen or by removing carbon. In practice, this is achieved by the water gas shift reaction (Section 2.3); some of the useful carbon (as CO) is converted into CO_2 and H_2 is produced from added water. It is important to note that it is not the nominal H : C ratio, but the effective H : C ratio of the feed that determines the CO_2 footprint of the feed [9]. Heteroatoms, such as oxygen, nitrogen, and sulfur, are hydrogenated and eliminated, which reduces the effective H : C ratio available for syngas generation. Biomass, which has a high nominal H : C ratio, has a very low effective H : C ratio (Table 15.3). Hence, the CO_2 footprints increase roughly in the order GTL < WTL < CTL < BTL.
b) Energy balance requirements can be classified into direct energy needs (first law of thermodynamics) and energy that must be used to perform useful work (second law of thermodynamics). Gasification and gas reforming are endothermic processes and requires energy input. This energy requirement can be expressed in terms of an amount of feed that must be combusted to provide the required energy. Feed combustion generates CO_2 and part of the CO_2 footprint associated with indirect liquefaction is due to the energy balance requirements. Again, the feed plays a key role in determining the CO_2 footprint. In feed

Table 15.3 CO_2 footprint due to the raw material selected as feed for an FT facility based purely on the stoichiometric requirements of FT hydrocarbon synthesis.

Feed material	Ultimate analysis, mineral ash-free basis (mass%)					H : C ratio		CO_2 footprint due to feed[a)] (kg CO_2/kg HC)
	C	H	N	S	O	Nominal	Effective	
Natural gas[b)] (lean)	74.9	25.1	0	<0.1	0	4.00	4.00	0.00
Natural gas[b)] (heavy)	75.2	24.8	0	<0.1	0	3.94	3.94	0.03
Oil sands bitumen (Athabasca)	83.1	10.6	0.5	4.9	0.9	1.52	1.44	2.34
Waste (municipal refuse)	53.3	8.2	4.9	0.1	33.5	1.83	0.65	3.96
Coal, bituminous (Deseret)	78.4	6.3	1.6	0.7	13.1	0.95	0.64	3.99
Coal, bituminous (Pocahontas #3)	91.2	4.4	1.3	0.5	2.5	0.58	0.50	4.40
Coal, subbituminous (Wyodak)	75.0	5.4	1.1	0.5	18.0	0.85	0.45	4.56
Coal, bituminous (Illinois #6)	77.7	5.0	1.4	2.4	13.5	0.77	0.43	4.60
Coal, lignite (Beulah-Zap)	73.0	4.8	1.2	0.7	20.3	0.79	0.32	4.97
Coal, lignite (Bienfait)	68.2	5.0	1.2	0.9	24.6	0.87	0.27	5.14
Coal, subbituminous (Coal Valley)	72.0	4.6	1.1	0.9	21.4	0.76	0.26	5.18
Coal, lignite (North Dakota)	69.9	4.6	1.0	1.4	23.2	0.78	0.23	5.30
Biomass, cellulosic (Switch grass)	50.1	6.1	0.9	0.1	42.8	1.45	0.12	5.75

a) Ideal Fischer–Tropsch hydrocarbon (HC) synthesis, that is, $CO + 2\,H_2 \rightarrow -(CH_2)- + H_2O$.
b) Dry nonassociated natural gas on an inert gas-free basis, with all C_3 and heavier hydrocarbons removed.

Table 15.4 Energy-related CO_2 footprint as function of the effective H : C ratio of the raw material.

Effective H : C ratio	CO_2 footprint for energy delivery (kg CO_2/GJ)[a]
4	55
2	74
1	89
0.5	99
0.25	105
0	112

a) Lower heating value calculated from the enthalpy of combustion of methane and carbon at 25 °C.

materials with a high effective H : C ratio, much of the energy can be derived from hydrogen combustion to yield water, thereby lowering the CO_2 footprint per energy unit (Table 15.4). FT synthesis is exothermic and can be used to offset the energy requirements somewhat, but there will always be a net energy requirement, since the facility performs work.

From the above discussion, it seems as if the FT technology itself has little impact on the CO_2 footprint. This is not entirely true. Compared to the mass and energy balance requirements, the impact of the FT technology is smaller, but should not be underestimated. Even though an exceptionally well-designed FT facility cannot reduce the CO_2 footprint beyond that dictated by the mass and energy balances, a poorly designed facility can increase the CO_2 footprint significantly. The syngas generation technology and gas loop design must be properly matched to the FT technology to minimize carbon loss in the gas loop [9]. Any syngas-derived products that do not end up as useful final products come at a significant CO_2 cost.

The preceding discussion also assumed "ideal" FT hydrocarbon synthesis as the basis for the calculation of the stoichiometric requirements. FT conversion does not lead to ideal hydrocarbon synthesis (Chapter 4). The FT technology determines the reaction stoichiometry and through the reaction stoichiometry it can influence the CO_2 footprint. By lowering the H : C ratio of the product, the CO_2 footprint is indirectly reduced. The product from HTFT synthesis has a lower H : C ratio than that from LTFT synthesis and the stoichiometric requirements are therefore different. This principle can be extended to the development of FT catalysts that produce a more aromatic product with an even lower H : C ratio than that from the HTFT synthesis. A number of approaches have been suggested, including the Kölbel–Engelhardt synthesis [37], zeolite–FT combinations [38, 39], and the incorporation of additional dehydrogenation functionality in the FT catalyst [40].

Other interventions to lower the CO_2 footprint of an FT facility are also possible. Noncarbon sources, such as nuclear energy, can be employed to provide energy to the process. When a noncarbon energy source is used, carbon does not have to be the change agent that is necessary for H_2 production, or for energy production.

The CO_2 footprint may also be reduced by strategies such as enhanced oil recovery (EOR), where the CO_2 produced by the process is captured and sequestered. Even though these strategies may be successfully implemented to reduce the CO_2 footprint of FT-based facilities, they do not alter the fundamentals of indirect liquefaction.

Strategies to reduce CO_2 that are based on noncarbon energy are independent of FT technology. These noncarbon energy strategies derive no synergy from their application with FT technology. Such strategies may equally be applied to reduce the CO_2 footprint of electric power generation or H_2 production by water electrolysis (Chapter 1).

15.5.2
Is CO_2 a Carbon Feed of the Future?

CO_2 conversion during FT synthesis is a recurring theme in the literature [41–46]. In addition to syngas-based conversion, there are also other catalytic pathways for direct CO_2 conversion [47]. Since Fe-based FT synthesis has the ability to convert CO_2-rich syngas, we must ask the question, is CO_2 a useful carbon feed of the future?

Although there may be cases where there is a location-specific advantage of employing CO_2-rich syngas, the use of CO_2 instead of CO as feed for FT does not change the overall mass balance requirements. As long as H_2 is produced by the WGSR, the reaction stoichiometry does not change. Even when the CO_2 is imported as an external carbon feed, H_2 must still be supplied and the energy requirement of the process dramatically increases. CO_2 is after all a very stable molecule and it does not contribute any reaction heat to the syngas production process. So, although Fe-based FT synthesis is capable of converting CO_2, it cannot be a CO_2 removal technology unless an external noncarbon-derived source of energy is employed.

While the best evidence to date indicates that CO_2 plays only a very minor role at best in the actual FT reaction, it is of course a key player in the WGSR (Section 2.3) and also in the heterogeneously catalyzed syngas to methanol synthesis (over Cu and ZnO) (Section 6.2.3).

15.6
Environmental Footprint of FT Refineries

Thus far, the discussion focused on the production of syncrude. Although no direct comparison has been made, the environmental footprint of FT syncrude production is likely to be considerably larger than that of crude oil production, which may be considered on a par with the raw material production for a FT facility (e.g., coal mining or natural gas production). One may therefore ask: what is the environmental or sustainability advantage related to an FT facility?

In an FT facility, the associated refinery contributes to the environmental footprint, but it makes only a small contribution to the emissions. This small environmental footprint of the refinery highlights one advantage of processing FT syncrude, namely, FT syncrude refining has a smaller environmental footprint than

Table 15.5 Externally supplied energy consumption data in conventional US West Coast crude oil refineries.

Process unit	Supplied energy use (GJ/m^3)[a,b]	Usage contribution of each unit (% of total)
Atmospheric distillation	1.29	40
Vacuum distillation	0.70	11
Fluid catalytic cracking	0.95	7[c]
Aliphatic alkylation (product basis)	4.66	5
Isomerization	1.53	2
Catalytic naphtha reforming	2.33	17
Hydrocracking	1.76	8
Delayed coking	0.73	6
Naphtha splitter	0.05	<1
Atmospheric gas oil splitter	0.08	<1
Naphtha desulfurization	0.55	3
Gas oil desulfurization	0.57	<1
Coker naphtha desulfurization	0.56	<1

a) Supplied energy consumption refers only to the external energy input and excludes energy derived from the process feed.
b) Energy equivalent of steam production was provided in source reference; the energy equivalent electricity consumption was calculated at a 45% conversion efficiency and was generally a minor component of the overall energy consumption.
c) A significant amount of energy is derived from coke combustion in the process, which is not reflected in this number.

conventional crude oil refining [36]. Thus, some of the environmental footprint associated with syncrude production can be offset by the gains made in refining.

FT refineries also have a sustainability advantage over crude oil refineries. Refineries are feed-specific and over time there has been a change in the composition of conventional crude oils, which have become heavier and more sulfur-rich. Procuring similar crude oils to those for which a refinery was designed is no longer as easy. Refineries can be modified, but such modifications affect the product slate. Crude oil as a feedstock has a direct impact on the product slate and the original product slate cannot be sustained when refining a heavier crude oil. This situation does not occur in FT facilities since the composition of the raw material feed is decoupled from the syncrude produced by FT synthesis. FT refineries can sustain their product slate even when there are substantial changes in the raw material feed composition.

15.6.1
Energy Footprint of Refining

The energy footprint of refining is directly related to the CO_2 footprint. Energy is usually supplied in four forms: steam, fuel, electricity, and process conversion. The first three are considered as supplied energy, because the origin of the energy source is external to the process. Table 15.5 provides a comparison of the energy

Table 15.6 Refining units found in on-specification FT fuel refineries with specific transportation fuel focus.

Refining unit	HTFT			LTFT		
	Motor gasoline	Jet fuel	Diesel fuel	Motor gasoline	Jet fuel	Diesel fuel
Atmospheric distillation[a)]	Yes	Yes	Yes	Yes	Yes	Yes
Fluid catalytic cracking	No	No	No	Yes	No	Yes
Isomerization	Yes	No	Yes	Yes	Yes	Yes
Catalytic naphtha reforming	Yes	Yes	Yes	Yes	Yes	Yes
Hydrocracking	Yes	Yes	Yes	No	Yes	Yes
Oligomerization	Yes	Yes	Yes	Yes	Yes	Yes
Hydrotreaters	Yes	Yes	Yes	Yes	Yes	Yes
Aromatic alkylation	Yes	No	Yes	Yes	No	No
Etherification	No	Yes	No	No	No	Yes
Alcohol dehydration	Yes	Yes	No	No	No	No

a) Distillation unit sees only a fraction of the total syncrude due to pre-separation in the gas loop or prerefining.

requirement and contribution of each refinery unit to the overall energy use in a typical crude oil refinery [48]. There are also process contributions and these are significant particularly in the case of fluid catalytic cracking and flexicoking (but not *delayed coking*) where coke is not used in the plant, but is burned to provide energy for the process. This energy that is derived from consumption of the process stream is not reflected in Table 15.5. To put this into perspective, in a modern refinery the CO_2 emission from a fluid catalytic cracking unit represents 40–45% of the total refinery CO_2 emissions [49].

There is no such thing as a generic FT or crude oil refinery design, but some comments can be offered to allow comparison. This assumes that the FT refinery has been designed for syncrude processing, with appropriate technology selection [18, 50]. A critique of industrial designs can be found in literature [51]. In order to make a direct comparison with crude oil refining, it is important to use only FT refinery designs that produce final on-specification transportation fuels. Table 15.6 shows a list of the refinery units needed for on-specification fuel production [52]. A few surprises emerge from this analysis:

a) In none of the FT fuels refinery designs, is there a need for a vacuum distillation unit, an aliphatic alkylation unit, or a delayed coker. Vacuum distillation is only required for petrochemical production from LTFT syncrude. "Alkylate," a mixture of branched chain paraffinic hydrocarbons (mostly isoheptane and isooctane) with a high octane rating, is produced differently from syncrude [53], eliminating the traditional HF- or H_2SO_4-based aliphatic alkylation unit.

The heavier products are of high quality, and since no carbon needs to be discarded, no delayed coker is required.

b) Although all designs require the equivalent of an atmospheric distillation unit (ADU), there is a significant difference. In the FT refinery designs, ADU is needed after refining or some pre-separation. It is not necessary as an entry point into the refinery as in crude oil refining. The ADUs in the FT refineries process only a fraction of the total syncrude and in some cases the cut point temperature for the bottoms is such that feed can be steam preheated and does not require a furnace. The energy requirement for the FT ADU is therefore much less than that for a crude oil ADU.
c) Fluid catalytic cracking is needed only for some of the LTFT refinery designs to supply olefins to the refinery in order to produce on-specification fuels.
d) Other units, which are less frequently encountered in crude oil refineries, are required for FT refining processes, namely, oligomerization, aromatic alkylation, etherification, and alcohol dehydration. Except for alcohol dehydration, all these operations are conducted at moderate temperatures ($<200\,^{\circ}C$) and are exothermic.
e) Hydrotreating in an FT refinery has a different objective than in a crude oil refinery and requires less severe operating conditions. Furthermore, the FT refinery as a whole is sulfur-free, which simplifies purification. In many of the other units that are common to crude oil refining, such as catalytic naphtha reforming and isomerization, the technology selection and/or mode of operation is different [50]. For example, the isomerization unit can be operated without a heater and with very little energy input by exploiting the heat of olefin hydrogenation [54].
f) One area where an FT refinery has a larger energy footprint than a crude oil refinery is for the processing of the aqueous product, as there is no equivalent stream in an oil refinery. However, as only small amounts of water-soluble products are generally formed in FT synthesis, this is not a greatly significant factor.

Although this is just a qualitative comparison, we conclude that the energy requirements in a FT refinery are less than those in a crude oil refinery and it is clear that a properly designed FT refinery will have smaller energy and CO_2 footprints than a similar sized crude oil refinery.

15.6.2
Emissions and Wastes in Refining

Emissions and wastes from refining are discussed along similar lines as for FT facilities as a whole: wastewater, solid waste, and air pollution. A more general discussion on refinery emissions can be found elsewhere [48, 55], here the focus is on aspects where FT refineries are substantially different from crude oil refineries.

a) **Wastewater:** The FT aqueous product refining section can be viewed as a wastewater cleanup section. The quality of the water leaving this section of the refinery is higher than that of the entering feed. The FT oil refining section produces water from the conversion of oxygenates present in the syncrude. The volume of wastewater produced in this way can be substantial; in HTFT refining, it is around 1.6% of the refined product [52]. The water is clean, but, depending on the refining technology, may contain dissolved light oxygenates. This water can either be sent directly to wastewater treatment (Section 15.3) or routed via the FT aqueous product refining section. Overall, the wastewater produced as consequence of refining is of comparable quality to that produced during crude oil refining.

b) **Solid waste:** Apart from spent refining catalysts, which are common to both FT and crude oil refineries, FT refineries have less carbon-based solid waste, since loss of carbon does not make sense in the FT context and has to be avoided. The situation is slightly different if pyrolysis products from gasification are corefined [18]. Pyrolysis products from gasification are usually less than 10% of the total syncrude production.

c) **Air pollution:** One of the main environmental advantages of FT syncrude is that it is sulfur- and nitrogen-free. Unless chemicals are introduced during refining, the gaseous process emissions from FT refining are limited to WGSR gases, light hydrocarbons, light oxygenates, and particulate matter. Gas cleaning is consequently much simpler than that in a crude oil refinery. Furthermore, the preferred refining technologies for FT syncrude in general do not require chemical additives [50]. It should be pointed out that some industrial FT facilities employ sulfided catalysts for hydrotreating and hydrocracking. This requires the addition of a sulfur additive (typically dimethyl disulfide) to the sulfur-free syncrude, in which case gas cleaning becomes analogous to that required for crude oil refining. Likewise, when gasification pyrolysis products are corefined with the syncrude, the off-gas contains sulfur and nitrogen compounds like that found with crude oil refining.

References

1 Hoogendoorn, J.C. and Salomon, J.M. (1957) Sasol: World's largest oil-from-coal plant. *Brit. Chem. Eng.*, 238–244.

2 Collings, J. (2002) *Mind over Matter. The Sasol Story: A Half-Century of Technological Innovation*, Sasol, Johannesburg.

3 Mashapa, T.N., Rademan, J.D., and Janse van Vuuren, M.J. (2007) Catalytic performance and deactivation of precipitated iron catalyst for selective oxidation of hydrogen sulfide to elemental sulfur in the waste gas streams from coal gasification. *Ind. Eng. Chem. Res.*, **46**, 6338–6344.

4 Probstein, R.F. and Gold, H. (1978) *Water in Synthetic Fuel Production: The Technology and Alternatives*, MIT Press, Cambridge, MA.

5 Cowser, K.E. and Richmond, C.R. (eds) (1980) *Synthetic Fossil Fuel Technology: Potential Health and Environmental Effects*, Ann Arbor Science Publishers, Ann Arbor.

6 Nowacki, P. (ed.) (1980) *Health Hazards and Pollution Control in Synthetic Liquid Fuel Conversion*, Noyes Data Corporation, Park Ridge, NJ.

7 Bentz, E.J., Jr. and Salmon, E.J. (1981) *Synthetic Fuels Technology: Overview with Health and Environmental Impacts*, Ann Arbor Science Publishers, Ann Arbor.

8 Scown, C.D., Horvath, A., and McKone, T.E. (2011) Water footprint of U.S.

transportation fuels. *Environ. Sci. Technol.*, **45**, 2541–2553.
9 De Klerk, A. (2011) Indirect liquefaction carbon efficiency. *ACS Symp. Ser.*, **1084**, 215–235.
10 Andrews, G.C. (2008) *Canadian Professional Engineering and Geoscience: Practice and Ethics*, 4th edn, Nelson Education, Toronto, pp. 359–386.
11 Ulgiati, S., Raugei, M., and Bargigli, S. (2006) Overcoming the inadequacy of single-criterion approaches to life cycle assessment. *Ecol. Model.*, **190**, 432–442.
12 Sheldon, R.A. (2007) *Green Chem.*, **9**, 1273–1283.
13 Hoffman, E.J. (1982) *Synfuels: The Problems and the Promise*, Energon, Laramie.
14 Braunstein, H.M., Kornegay, F.C., Roop, R.D., and Sharples, F.E. (1981) *Fuels from Biomass and Wastes*, Ann Arbor Science Publishers, Ann Arbor, pp. 463–504.
15 Mangold, E.C., Muradaz, M.A., Ouellette, R.P., Rarah, O.G., and Cheremisinoff, P.N. (1982) *Coal Liquefaction and Gasification Technologies*, Ann Arbor Science Publishers, Ann Arbor.
16 Mako, P.F. and Samuel, W.A. (1984) The Sasol approach to liquid fuels from coal via the Fischer–Tropsch reaction, in *Handbook of Synfuels Technology* (ed. R.A. Meyers), McGraw-Hill, New York, pp. 25–273.
17 Deurbrouck, A.W. and Hucko, R.E. (1981) *Chemistry of Coal Utilization, Second Supplementary Volume* (ed. M.A. Elliott), John Wiley & Sons, Inc., New York, pp. 571–607.
18 De Klerk, A. (2011) *Fischer–Tropsch Refining*, Wiley-VCH Verlag GmbH, Weinheim.
19 Valotti, E. and Zennaro, R. (2009) *Natural gas for sustainable development: the GTL approach.* Proceedings of the 8th International O&G Conference Petrotech, January 11–15, 2009, New Delhi, India.
20 Steynberg, A.P. and Dry, M.E. (eds) (2004) *FT Technology*, Elsevier, Amsterdam.
21 DWA (1986) *Management of the Water Resources of the Republic of South Africa*, Department of Water Affairs, Pretoria.
22 Vertova, A., Aricci, G., Rondinini, S., Miglio, R., Carnelli, L., and D'Olimpio, P. (2009) Electrodialytic recovery of light carboxylic acids from industrial aqueous wastes. *J. Appl. Electrochem.*, **39**, 2051–2059.
23 Higman, C. and van der Burgt, M. (2008) *Gasification*, 2nd edn, Elsevier, Amsterdam.
24 Dry, M.E. (1999) Fischer–Tropsch reactions and the environment. *Appl. Catal. A*, **189**, 185–190.
25 Majone, M., Aulenta, F., Dionisi, D., D'Addario, E.N., Sbardellati, R., Bolzonella, D., and Beccari, M. (2010) High-rate anaerobic treatment of FT wastewater in a packed-bed biofilm reactor. *Water Res.*, **44**, 2745–2752.
26 Lettinga, G., De Zeeuw, W., and Ouborg, E. (1981) Anaerobic treatment of wastes containing methanol and higher alcohols. *Water Res.*, **15**, 171–182.
27 Dinsdale, R.M., Hawkes, F.R., and Hawkes, D.L. (2000) Anaerobic digestion of short chain organic acids in an expanded granular bed reactor. *Water Res.*, **34**, 2433–2438.
28 De Klerk, A. and Furimsky, E. (2010) *Catalysis in the Refining of Fischer-Tropsch Syncrude*, Royal Society of Chemistry, Cambridge, UK, pp. 253–255.
29 Fabricius, N. (2008) Management of water resources & recycling for Pearl GTL. 7th GTLtec 2008, February 18–19, 2008, Doha.
30 www.shell.com/static/environment_society/downloads/environment/water/fresh_water_02042012.pdf (30 August 2012).
31 www.desalination.com/technologies/med-epc/veolia-water-solutions-technologies (30 August 2012).
32 Furimsky, E. (1996) Spent refinery catalysts: environment, safety and utilization. *Catal. Today*, **30**, 223–286.
33 Vosloo, A.C., Dancuart, L.P., and Jager, B. (1998) Environmental aspects of Sasol Fischer–Tropsch technologies and products. 11th World Clean Air and Environment Congress, Durban, South Africa, September 14–18, 1998, 6F-2.
34 Brumby, A., Verhelst, M., and Cheret, D. (2005) Recycling GTL catalysts: a new challenge. *Catal. Today*, **106**, 166–169.

35 Liu, K., Song, C. and Subramani, V. (eds) (2010) *Hydrogen and Syngas Production and Purification Technologies*, John Wiley & Sons, Inc., New York.

36 De Klerk, A. (2007) Environmentally friendly refining: Fischer–Tropsch versus crude oil. *Green Chem.*, **9**, 560–565.

37 Chaffee, A.L. and Loeh, H.J. (1985) Aromatic hydrocarbons from the Kölbel–Engelhardt reaction. *Appl. Catal.*, **19**, 419–422.

38 Chang, C.D., Lang, W.H., and Silvestri, A.J. (1979) Synthesis gas conversion to aromatic hydrocarbons. *J. Catal.*, **56**, 268–273.

39 Guan, N., Liu, Y., and Zhang, M. (1996) Development of catalysts for the production of aromatics from syngas. *Catal. Today*, **30**, 207–213.

40 Huffman, G.P. (2011) Incorporation of catalytic dehydrogenation into FT synthesis of liquid fuels from coal to minimize carbon dioxide emissions. *Fuel*, **90**, 2671–2676.

41 Puskas, I. (1997) Can carbon dioxide be reduced to high molecular weight Fischer–Tropsch products? *Prepr. Pap. Am. Chem. Soc. Div. Fuel Chem.*, **42** (2), 680–686.

42 Zhang, Y., Jacobs, G., Sparks, D.E., Dry, M.E., and Davis, B.H. (2002) CO and CO_2 hydrogenation study on supported cobalt FT synthesis catalysts. *Catal. Today*, **71**, 411–418.

43 Spadaro, L., Arena, F., Bonura, G., Di Blasi, O., and Frusteri, F. (2007) Activity and stability of iron based catalysts in advanced Fischer–Tropsch technology via CO_2-rich syngas conversion. *Stud. Surf. Sci. Catal.*, **167**, 49–54.

44 Lui, Y., Zhang, C.-H., Wang, Y., Li, Y., Hao, X., Bai, L., Xiang, H.-W., Xu, Y.-Y., Zhong, B., and Li, Y.-W. (2008) Effect of co-feeding carbon dioxide on FT synthesis over an iron–manganese catalyst in a spinning basket reactor. *Fuel Process. Technol.*, **89**, 234–241.

45 Srinivas, S., Malik, R.K., and Mahajani, S.M. (2008) Process alternatives for Fischer–-Tropsch synthesis of CO_2 rich syngas. *Prepr. Pap. Am. Chem. Soc. Div. Fuel Chem.*, **53** (1), 97–98.

46 James, O.O., Mesubi, A.M., Ako, T.C., and Maity, S. (2010) Increasing carbon utilization in Fischer–Tropsch synthesis using H_2-deficient or CO_2-rich syngas feeds. *Fuel Process. Technol.*, **91**, 136–144.

47 Ma, J., Sun, N., Zhang, X., Zhao, N., Xiao, F., Wei, W., and Sun, Y. (2009) A short review of catalysis for CO_2 conversion. *Catal. Today*, **148**, 221–231.

48 Sittig, M. (1978) *Petroleum Refining Industry: Energy Saving and Environmental Control*, Noyes, Park Ridge, NJ.

49 De Melloa, L.F., Pimentaa, R.D.M., Mourea, G.T., Pravia, O.R.C., Gearhart, L., Milios, P.B., and Melien, T. (2009) A technical and economical evaluation of CO_2 capture from FCC units. *Energy Procedia*, **1**, 117–124.

50 De Klerk, A. (2008) Fischer-Tropsch refining: technology selection to match molecules. *Green Chem.*, **10**, 1249–1279.

51 De Klerk, A. (2009) Refining Fischer–Tropsch syncrude: perspectives on lessons from the past, in *Advances in Fischer–Tropsch Synthesis, Catalysts, and Catalysis* (eds B.H. Davis and M.L. Occelli), Taylor & Francis, Boca Raton, pp. 331–364.

52 De Klerk, A. (2011) Fischer-Tropsch fuels refinery design. *Energy Environ. Sci.*, **4**, 1177–1205.

53 De Klerk, A. and De Vaal, P.L. (2008) Alkylate technology selection for Fischer-Tropsch syncrude refining. *Ind. Eng. Chem. Res.*, **47**, 6870–6877.

54 Lamprecht, D. and De Klerk, A. (2009) Hydroisomerization of 1-pentene to iso-pentane in a single reactor. *Chem. Eng. Commun.*, **196**, 1206–1216.

55 Gary, J.H., Handwerk, G.E., and Kaiser, M.J. (2007) *Petroleum Refining. Technology and Economics*, 5th edn, Taylor & Francis, Boca Raton, FL.

Part Five
Future Prospects

Greener Fischer-Tropsch Processes for Fuels and Feedstocks, First Edition. Edited by Peter M. Maitlis and Arno de Klerk.

16
New Directions, Challenges, and Opportunities

Peter M. Maitlis and Arno de Klerk

Synopsis

Our final chapter summarizes the arguments for building more FT facilities, as well as some of the problems and complications that may be encountered. At a time of uncertainty about the world energy situation and the rising costs of liquid fuels for transportation, it makes sense from an economic and strategic viewpoint to build new FT plants. We make the case for concentrating on smaller facilities that involve lower risk and lower capital costs instead of the more usually encountered approach that focuses on the economies of scale that large installations can bring. Opportunities for improving FT facilities are discussed, as well as some of the challenges that need to be overcome in the development of a globally significant FT-based synthetic liquids (XTL) industry. Increases in industry- and government-sponsored basic research will accelerate our greater understanding of and further improve the FT technology.

16.1
Introduction

There are several aspects of energy production that must be considered: energy efficiency is critical, but it is important to ensure that the overall process, from raw materials to final products, is as sustainable as possible. In other words, that the process does not consume an unreasonable proportion of the available resources, of which as many as possible are renewable, and that the amounts of by-products and of waste are minimized. "Green" factors are now important in the analysis of every chemical process and it is vital to ensure not only that the processes we develop and use are as environmentally benign as possible, but also that future generations will be able to use them in the same ways without detriment. This is known as *sustainable development,* which was defined by the Brundtland Commission as the use of natural resources to "meet present needs without compromising the ability of future generations to meet their own needs."

Greener Fischer-Tropsch Processes for Fuels and Feedstocks, First Edition. Edited by Peter M. Maitlis and Arno de Klerk.

In this context, it is also appropriate to define renewable energy as energy that is renewed naturally. It includes energy from biomass (biofuels), as well as hydroelectricity, wind, tidal, solar, and geothermal sources. It excludes raw materials that are depleted in use such as fossil fuels and nuclear power.

The wider significance of these terms also needs to be examined. For example, much play has been made with the concept of electric cars as the salvation of our transportation needs. Although it is true that widespread use of electric cars will reduce pollution in our cities, we must also consider from where the electricity to run them comes. The US Energy Information Agency has estimated that two-thirds of world electricity is generated from fossil fuels (coal 42%, natural gas 21%, and oil 4%), 14% from nuclear, and only 19% from renewables. Electricity generation is an energy transformation process and, just like other energy transformations such as the Fischer–Tropsch synthesis, it is subject to the second law of thermodynamics. The second law of thermodynamics requires that some energy must be used to perform work. A typical thermal efficiency from coal-fired electric power generation is 36–38% [1], which is not very different from the efficiency for FT coal-to-liquids conversion (Section 16.2.2). New technologies for electric power generation, such as integrated gasification combined cycle (IGCC), have the potential to improve the thermal efficiency to near 45% [1, 2], but this type of technology has not yet seen large-scale adoption. Electrically driven vehicles have a lower overall CO_2 footprint per distance traveled, but it has been estimated that the average CO_2 output for electric cars is 128 g/km, compared to an average of 105 g/km for hybrids such as the Toyota Prius, when the emissions from coal- and oil-fired electricity-generating stations are included. We must also remember that electric devices depend on batteries, the manufacture of which now requires considerable resources including some metals such as lithium.

Unfortunately, virtually none of the presently available and commercially effective sources of energy used by mankind is "clean" in the sense that it is completely nonpolluting and sustainable. There are serious difficulties with most of the currently popular technologies that can be used to supply energy for normal usage.

Sixty or seventy years ago, nuclear power was regarded as the panacea offering unlimited "clean" energy at low cost. Since then, many problems in the nuclear industry such as those highlighted by the disasters at Three Mile Island (the United States), Chernobyl (Russia/Ukraine), and Fukushima (Japan) have disillusioned many supporters.

The generation of electricity directly from wind-driven turbines is another very attractive source of "clean" energy. But the problem here is that wind is not steady or continuous and this form of power generation frequently does not match demand. Thus, supplementary energy from fossil fuels or from large storage facilities (e.g., batteries) must be supplied. The infrastructure for wind power is also expensive to build and to maintain. Similar concerns apply to solar energy.

Biomass as a "renewable source of energy" is currently widely espoused and some electricity-generating stations are now being run on wood chip fuels. Plant matter has long been used as the source of many chemicals. But growing and harvesting the biomass and then transporting it to its place of use is also labor-intensive, expensive, and energy demanding. Thus, although biomass is a renewable resource, its production is not. Some of the more visible impacts of intensified agriculture are the eutrophication of rivers and the development of "dead zones" (e.g., in the Gulf of Mexico) as a result of high fertilizer run-off leading to excessive growth of algae [3]. Furthermore, burning biomass also gives CO_2, which can potentially influence climate change. There are no silver bullets or other magic devices that will solve all our problems and, depending on the location, different solutions will be preferred.

Solar power that can be harnessed via photovoltaic cells, by solar furnaces, or other devices must be our best hope in the long term. Sunshine is plentiful, benign, and represents a source of energy that comes, at least in principle, at relatively low cost. As it can be used to generate electricity directly, it can therefore produce hydrogen and oxygen by electrolysis of water, which opens up a whole host of possibilities since such a hydrogen-based economy would avoid most of the problems arising from the use of fossil fuels. However, despite years of strenuous efforts by many scientists and engineers, we are still some way off an even moderately efficient conversion of sunlight into electricity or other form of energy on the scale needed and that can be used directly and easily for a variety of purposes including transportation. Photovoltaic cells on rooftops to generate electricity for domestic heating seem to be at the limit of commercial applications at present (2012), and the wider application of solar energy still looks some way off.

In the more immediate future, to produce energy on a larger scale, we need a more well-developed technology. In practice, to bridge the gap between the use of nonrenewable resources by unsustainable practices and future sustainable energy solutions requires using conversion technologies based on readily available fuels. An overview of the various options that we currently have to produce energy indicates that each one has some substantial drawbacks, thus we need to choose a technology that offers the most practicable methods at the least cost, both to ourselves today and to future generations. One such technology is the Fischer–Tropsch transformation of syngas into hydrocarbons. The FT process is already in use, and although not "green" in the strictest sense of the term, it does at least allow us to use proven technologies and to modify and improve them to reduce pollution.

16.2
Why Go Along the Fischer–Tropsch Route?

The extensive literature that already exists on FT synthesis as well as the industrial applications of FT technology shows the importance of FT today (Chapter 5), and we can analyze the strategic, economic, and environmental reasons for an expansion of interest in FT synthesis.

16.2.1
Strategic Justification

Conventional crude oil is at present the main source of transportation fuels, and demand for it increases year-on-year. As it is a finite resource, a point will come where the demand exceeds the supply and market forces will curb some demand and reduce some usage caused by price rises. Since a price-driven demand reduction is not selective, it will tend to exacerbate division along social and national lines. However, price alone does not govern availability, and resources may be allocated strategically and may not be subject to free market trading. The availability and trading of rare earth metals is a case in point [4, 5].

One simple answer to the immediate problem of diminishing crude oil supplies is to develop alternatives to crude oil: competition will increase choice and drive the price down. This strategic justification for the production of synthetic crude oil (syncrude) – to supplement conventional crude oil – will also foster a reasonable supply-and-demand balance globally. The production of syncrude by FT synthesis is a practical and well-understood way of achieving this.

The situation has been well summarized by James Boyd [6]: " . . . there has to be a major change in the patterns of usage of energy in order to maintain a healthy economic society. This is because the present delineated and known reserves of the basic energy raw materials exist in a far different ratio to each other than the current rate of usage. The issues we face then are those of taking what energy sources are available to use and creating a new supply system based more nearly on the ratio of availability." Bluntly put, it is in the strategic interest of any country to ensure energy diversification for the production of key energy carriers and for transportation fuels in particular.

16.2.2
Economic Justification

The raw material feed to a FT facility can be biomass, coal, natural gas, or organic waste. The price of these carbon-based feed materials is only weakly correlated with the price of crude oil. Conversely, the price of transportation fuels and petrochemicals are strongly correlated with the price of crude oil, since crude oil is the main input cost (see also Chapter 7). When there is a meaningful price difference between any of the alternative carbon sources, gas, and crude oil, there is a potential economic incentive for feed-to-liquids (XTL) conversion based on FT technology.

This is shown by the following simple example, using typical North American thermal coal, gas, and crude oil prices in 2011 (Table 16.1). The calculation takes into account the conversion efficiencies to LPG, gasoline, jet fuel, diesel fuel, and fuel oil products. For conventional crude oil, the conversion efficiency is quite high, about 89% [7]. Gas-to-liquids FT synthesis is somewhat less efficient, about 75% [8], but it is more efficient than CTL FT synthesis based on coal at about 33% [9]. The difference in carbon efficiency between GTL and CTL is due to the low H : C ratio of a typical bituminous coal, which requires significant carbon sacrifice in order to produce H_2 by the WGSR. Lignite contains even less carbon and hydrogen per mass than bituminous coal and the syngas production ratio by entrained flow gasification between lignite and bituminous coal is around 0.46 [2].

Table 16.1 Feed material cost comparison in 2011 between conventional crude oil and some alternative carbon sources.

Feed material	Feed material cost		Cost-efficiency[a)]	Product material cost[b)]
	$/bbl	$/GJ	%	$/bbl
Crude oil	100	18	89	113
Natural gas	22	4	75	30
Bituminous coal	17	3	33	51
Lignite (brown coal)	4	1	15[c)]	24

a) LPG, gasoline, jet, diesel, and fuel oil.
b) Effective feed-related cost; does not include operating and capital cost contributions.
c) Cost-efficiency calculated based on gasification efficiency of Illinois bituminous coal versus North Dakota lignite.

The price difference between any of the alternative carbon sources and crude oil on an oil equivalent barrel cost provides an indication of the economic incentive for investment in syngas generation and FT synthesis.

16.2.3 Environmental Justification

The environmental footprint of FT facilities was discussed in Chapter 15. FT facilities do not have a smaller environmental impact than those found elsewhere in the oil-and-gas industry, but for specific applications FT conversion has fewer drawbacks than the alternatives. For example, the exploitation of natural gas in a location that does not allow it to be used as pipeline gas (and which is therefore presently flared) must be compared with the reduced environmental impact of FT conversion. The environmental justification for implementing FT technology should always be made relative to that of alternative practices.

16.3 Considerations against Fischer–Tropsch Facilities

We have shown above why strong interest in FT synthesis continues. Why is it then that rather few industrial FT-based facilities have so far been built? The following are some of the chief factors that have constrained actions leading to alternative methods of converting carbon-containing raw materials into liquids (XTL):

a) *A FT facility is technically very complex.* This has been explained in Chapters 2, 3 and 5.
b) *The capital cost of XTL facilities is high.* In the early 2000s, the capital cost for GTL facilities was often quoted at $20 000–30 000 per daily barrel [10]. Much higher

Table 16.2 Estimated capital costs associated with recent Fischer–Tropsch-based GTL projects.

Facility	GTL capacity (bbl/day)	Total project cost (US$ billion)	GTL cost (US$/bbl/day)
Pearl GTL (Shell)	140 000	19	110 000[a)]
Escravos GTL (Sasol-Chevron)	34 000	6	180 000
Sasol 1 expansion (Sasol)	5 500[b)]	1.1	200 000

a) The total project cost is for 140 000 bbl/day GTL and 120 000 bbl/day natural gas liquids. The capital cost split to calculate the GTL contribution was estimated based on refining capital cost required for the natural gas liquids.
b) The capacity was not announced, but may be as high as 7000 bbl/day, reducing the cost to around US$160 000/bbl/day.

actual capital costs ~$100 000 per daily barrel are being reported now, less than 10 years later [11]. The economics of FT facilities, discussed in detail in Chapter 7, indicate that for a large-scale gas-to-liquids facility, the estimated capital cost is around $83 000 per daily barrel. Some of the more recently reported capital cost data for FT GTL facilities are summarized in Table 16.2.

c) *The investment risk is substantial, both financially and technically.* The main economic driver is the price differential between FT oil (from various carbon feed sources) and crude oil (Section 16.2.2). However, past price volatility shows that the financial risk is substantial and that the price differential of feed material relative to crude oil can quickly change from a large difference that makes FT facilities economically viable to a small one that renders them uneconomical (Figure 7.2). There is also a technical investment risk and even companies with long production experience in FT synthesis have had some serious technical difficulties associated with the commissioning of new and larger facilities [12].
d) *The large perceived environmental footprint.* Like many other energy enterprises, the FT facilities have not been known for their ecological friendliness, especially in terms of effluent disposal and usage of resources. The CO_2 and water footprints of FT facilities are discussed in Chapter 15.
e) *Conventional crude oil, sufficient for present needs, is still readily available.* Crude oil can easily be refined with mature technology at high efficiency to liquid products. Simply put: at present it is often seen as cheaper, less risky and more convenient to keep on using crude oil.

16.4
Opportunities to Improve Fischer–Tropsch Facilities

Despite the complexity of the FT liquefaction technology and despite volatility in the price of crude oil relative to coal and natural gas, FT facilities have operated successfully and profitably over many decades, thus the economic case in favor of FT facilities outweighs the case against it. While we do not dismiss the problems, we should use the opportunity for improvements to make FT facilities even better, as suggested in Sections 16.4.1–16.4.3:

a) The complexity of facilities should be reduced. This is a difficult task, because indirect liquefaction is inherently a multistep process. However, complexity arises not only from the number of steps but also from the connectivity between the steps. A process that is purely sequential, even if it contains many steps, is less complex than a process with fewer steps but high connectivity. The mathematical complexity of chemical processes can be calculated using digraph theory [11], and the opportunity should be taken to revisit the fundamental design of indirect liquefaction to reduce complexity.
b) We need to go back to fundamentals to find solutions to the challenges that impede progress. Even though FT synthesis has been practiced for many decades, some key problems are still unresolved (Chapters 8, 9 and 12). The same is true for many of the other technologies, both upstream and downstream from FT synthesis. At the heart of this aspect is that our understanding of the basic chemistry and surface science underlying the FT process is woefully inadequate, and many processes have been engineered without sufficient understanding. Indeed, the failure in the United States to develop a viable synthetic fuels technology has been ascribed to the lack of a long-term research commitment and to the premature attempts to commercialize technology from a weak base of understanding [13]. We need to take this opportunity to carry out the research that will bring the fundamental understanding to allow improvements of the overall process (see Sections 16.4.3 and 16.5).
c) Capital costs can be reduced by innovation. The drive leading to larger FT facilities is due to the economic gain derived from increasing size (the "economy of scale") (Section 7.4.1). A better way to reduce the capital cost significantly is to stop increasing the size of the "old" technology and instead to develop new technology based on better understanding of the process. While increasing the size may reduce the capital cost on a capacity basis, it still increases the total absolute capital cost of the facility. The absolute capital cost must be reduced in order to reduce the financial risk. There is an opportunity for innovation to provide new technology that will fundamentally reduce the capital cost. That will also enable smaller companies to participate.

The biggest detractor from investment in FT facilities is risk. The absolute cost (multibillion dollar investment) of a large-scale FT facility excludes all but governments and large companies. The risk of technically not achieving the objective and the associated financial loss are both considerable. In such an environment, there is little tolerance for innovation. In order to stimulate innovation and enable new technology, one must consider smaller scale FT facilities, because smaller facilities reduce the barrier for innovation (Section 16.4.1).

16.4.1
Opportunities Offered by Small-Scale FT Facilities

Smaller scale FT facilities (typically <100 000 t/a or <2000 bbl/day) represent a smaller risk than the large-scale industrial facilities discussed in Chapter 5. Reducing the size does not reduce the probability of failure, but it reduces the adverse consequences of failure, that is, the amount of money lost. Hence, smaller FT facilities present a smaller risk. By exploring innovative and new technology, the technical risk is increased, but the consequence of failure is not increased. Yet, the reward of success is considerably more, because it leads to a substantial advances; it is this reward of success that drives venture capital investments in the first place.

In addition to lowering the investment risk, there are also other advantages in stimulating interest in the development of small FT facilities:

a) In a smaller facility, there is less temptation to increase complexity by optimization of the utility network and process flow; thus, keeping the size small stimulates design that improves single pass efficiency and avoids recycle.
b) Due to the smaller size, the absolute capital cost decreases, i.e. the total amount of money required to build the facility becomes less. The capital cost per capacity may be higher than that for a large-scale facility, but this will drive innovation to reduce capital cost rather than relying on economy of scale.
c) Innovation is fostered by smaller facilities, because there is less money to lose if an idea does not work out. This is the same principle that is applied in piloting a new process. Piloting does not prevent failure due to technology scale-up, since the hydrodynamics and transport phenomena at small scale and large scale are often quite different in a complex way and cause uncertainty in scale-up calculations [14–16]. However, by keeping the facility small it reduces the technical risk associated with scale-up.
d) Since smaller is nimbler, the response to learning is faster. A small facility can be modified more easily and it is quicker to commission and decommission; heat-up and cool-down times are shorter and there is less material inventory in process equipment and lines. Smaller vessel and line sizes make it easier to transport, install, and remove equipment. Furthermore, the financial impact of downtime is less in absolute terms.
e) By reducing the absolute investment cost, it is possible to generate broader interest. Increased diversity stimulates innovation and creates a market for FT-specific technology. This reduces the overall development cost for all participants.
f) Really remote locations do not have the infrastructure that allows the transport of large vessels and do not provide non-monetary incentives to attract skilled workers. By reducing the size of the facility, it becomes possible to reach locations that would be very costly or impossible to access for the construction of a large-scale facility.
g) It is customary to construct an FT facility close to the feed source, because it is difficult to relocate the feed. However, if the feed is mobile, its value increases. The decision to construct an FT facility depends not only on the rate of feed that

can be provided but also on the size of the reserve, that is, the time before the natural resource runs out. Small-scale FT facilities therefore enable the exploitation of smaller sized feed deposits.

h) The cost of biomass waste feed is mainly related to its transportation cost, which indirectly determines its supply radius. Biomass waste has other drawbacks too, such as a low energy density, a widely distributed origin, and a seasonal nature. It is difficult and costly to provide biomass as feed to a large-scale FT facility [17]. Small-scale FT facilities could overcome this constraint.
i) The size of a small-scale FT facility can be increased using small units in parallel. Although this negates economy of scale, it increases operating robustness. For example, if 1 out of 10 small units in parallel fails, 10% production is lost; but if 1 out of 2 large units in parallel fails, half the production is lost.
j) Small-scale designs lend themselves to modularity, which cuts down on engineering cost and can lead to savings through "mass" production. It is this approach that is being favored by companies such as Velocys, which construct modular microchannel reactors [18]. The same principle can of course be applied to other reactor types too.
k) For a very small-scale design, it is possible to design the FT facility module in such a way that it is also mobile. This would enable exploitation of very small reserve natural deposits at a higher throughput (point (g)) or the conversion of biomass over a larger geographical area (point (h)). In addition to these, there are also potential military applications for example, by making supply logistics less onerous by converting local biomass into fuel.

16.4.2
Technical Opportunities in Syngas Generation and Cleaning

Synthesis gas generation and cleaning are widely applied and only a fraction of the installed technology is associated with FT facilities [2, 19]. Industrial units have been optimized over many decades, despite the lower level of experience with some industrial feed materials (e.g., biomass or waste), which are less used as gasification feed than coal or natural gas. Despite the widespread industrial application of syngas generation and cleaning, there are still opportunities for innovation, some of which are specifically related to the development of small-scale FT facilities:

a) *Efficient noncryogenic air separation.* When syngas is generated by a technology that involves direct partial oxidation of the feed (e.g., oxidative reforming and gasification), as is the case with most syngas produced in FT facilities, the use of pure O_2 is preferred. Cryogenic air separation is an efficient but costly process. For example, air separation accounts for 8% of the total capital cost in a gas-to-liquids FT facility (Figure 7.4). Since it is impractical to include a cryogenic air separation unit in the design of a small-scale FT facility, this gives an opportunity for the development of novel technologies, for example, membrane-based air separation.
b) *Small-scale indirectly heated syngas production technology.* The need for air separation is overcome when the oxidant is not mixed with the process feed, as

in steam reforming. However, indirectly heated syngas generation technology is bulky. The furnace configuration has served industry well, but there is an opportunity to develop a different and more compact technology that will be suitable for small-scale facilities. Such a design would have much wider benefit, since it could in principle displace fired heaters and furnaces in many applications.

c) *Inert gas removal from syngas.* Even when cryogenic air separation is employed, some inert gases (e.g., ~1% Ar and N_2) still find their way into the purified O_2. Whenever there is a recycle of unconverted syngas, the inert gases build up in the recycle and thereby determine the tail gas purge rate. The same is true of other "inert" gases produced during FT synthesis, such as the light paraffins. Cryogenic separation of methane and ethane, which cannot be recovered by pressure distillation at near ambient conditions, is expensive and often not included in the FT gas loop of even large-scale industrial FT facilities. The ability to remove Ar, N_2, CH_4, and C_2H_6 from syngas will lead to an improvement in the carbon efficiency of FT gas loop designs that include recycle streams.

d) *Cheap and environmentally benign /useful disposable absorbents for gas cleaning.* In large facilities, regenerable absorbents are employed for the removal of contaminants during gas cleaning and the absorbents are regenerated on-site. The contaminants are captured in a benign or useful form, such as elemental sulfur. For small-scale applications, the implementation of traditional acid gas removal and capture technology become impractical. In a small-scale facility, it is more likely that the sulfur removed from the syngas will be emitted as SO_x instead of being converted into elemental sulfur or sulfuric acid. In order to keep the environmental footprint of smaller facilities small, there is a need to develop contaminant capture technology on materials that are cheap, disposable, and will not cause harm. One example is to use a natural porous mineral that will bind the sulfur and produce a low-cost material for agricultural applications. The principle is also applied with spent solid phosphoric acid catalyst, thereby turning a solid waste into a useful product [20].

16.4.3
Technical Opportunities in Fischer–Tropsch Synthesis

There is considerable diversity in industrially applied FT technology (Chapters 3 and 5). This diversity is healthy, but it also indicates that there are many opportunities for improvement. The lack of a unified FT model, which accurately describes syngas consumption and compound distribution, makes conceptual studies difficult and also highlights our present lack of fundamental understanding. The same can be said of the description of straight-run FT syncrude. Straight-run FT syncrude displays nonideal phase behavior (potentially four phases containing both very polar and very apolar compounds) and a thermodynamic model that accurately describes such a complex mixture is still a challenge. There are other areas too where our knowledge of the fundamentals is still lacking.

In addition to the need for advancing our fundamental understanding, there are also opportunities for the direct improvement of FT technology. Some key opportunities are outlined:

a) Methane production is wasteful and especially so in GTL applications where methane is the feed. Both positive and negative deviations from the ASF carbon number distribution can be observed with respect to methane. Methane selectivity usually increases with time on stream and with a decrease in chain growth probability (α-value), but in some Fe-FT catalyst formulations, methane production is suppressed [11, 21]. For example, the nature of the pretreatment and the phases present in Fe-FT catalysts result in different methane selectivities, independent of the α-value [22]. Any improvements in FT synthesis that lead to lower methane selectivity will increase the overall efficiency of syngas conversion.
b) Although considerable variation in FT syncrude composition is possible, most industrial syncrudes are either HTFT or LTFT look-alikes. One of the main strengths of the FT process over crude oil, direct liquefaction, and methanol technologies is the ability to manipulate FT synthesis to tailor the product spectrum in terms of unsaturation, oxygenate content, branching, and carbon number distribution. This aspect of FT synthesis has not yet been exploited and holds tremendous opportunity, especially for the production of petrochemicals [11]. In other words, based on what is wanted, can produce a syncrude that simplifies refining to the desired products can be produced.

16.4.4 Technical Opportunities in FT Syncrude Recovery and Refining

Few technologies have been developed specifically for the refining of FT syncrude, even though syncrude requires a different combination of technologies and a different refining approach to conventional crude oil refining [11, 23]. There are opportunities in the selection, modification, and development of new technology for syncrude refining, as well as in the configuration and design of syncrude recovery and refining. Some key opportunities are listed in Sections 16.4.4.1–16.4.4.3.

16.4.4.1 Syncrude Recovery Design

A straightforward stepwise cooling is the approach that has thus far been taken with syncrude recovery after FT synthesis (Figures 4.4 and 4.5). The fractions are produced purely for recovery and not with downstream refining in mind. There is consequently an opportunity to improve selective recovery and separation and thereby improve refining efficiency and carbon efficiency, for example, in syncrude recovery and cooling by distillation, instead of cooling and phase separation. If one looks at a typical atmospheric distillation unit employed in conventional crude oil refining (Figure 16.1), one notices that it does not contain a reboiler, but employs a fired heater as feed preheater. After FT synthesis, the FT syncrude is already "preheated." By exploiting the "preheated" state and removing the heat through distillation at FT gas loop conditions, one not only achieves fractionation but also avoids the reboiler temperature constraints associated with more reactive syncrudes [11].

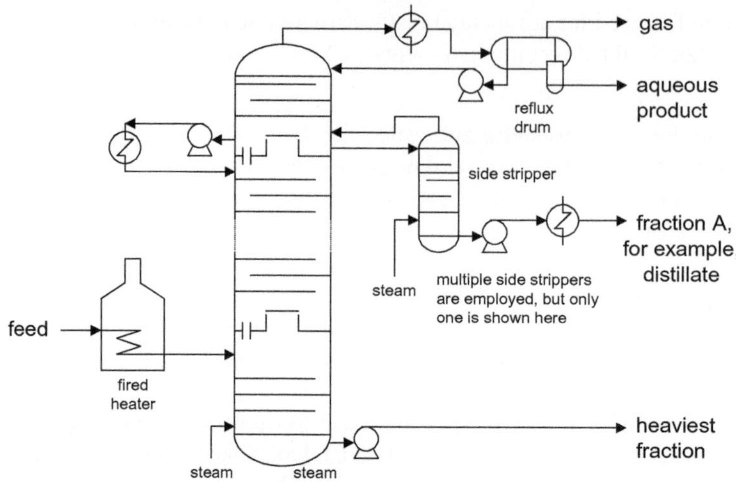

Figure 16.1 Typical atmospheric distillation unit as found in crude oil refineries.

There can be a substantial energy saving with respect to postrecovery distillation, as well as other refining benefits.

16.4.4.2 Tail Gas Recovery and Conversion

In small-scale FT facilities and in FT facilities at remote locations far from petrochemical markets, the light gases have less value as products. Yet, these hydrogen-rich products are valuable, especially in coal-to-liquids, biomass-to-liquids, and waste-to-liquids facilities where the feed has a low effective H : C ratio [9]. Direct conversion of the light hydrocarbons into liquid products (not via syngas) will improve both the amount of syncrude recovered from the tail gas and the carbon efficiency of the process. Complete recovery is unlikely. Some catalytic pathways that can be considered for such conversion are aromatization to produce aromatic naphtha, oligomerization to produce olefinic naphtha/distillate, and selective partial oxidation to produce light oxygenates for coprocessing with the FT aqueous product-derived nonacid oxygenates.

16.4.4.3 Aqueous Product Refining

The dilute solution of FT oxygenates found in the aqueous product poses a challenge to refining, especially for recovery of the carboxylic acids, since it is impractical to remove the water from the carboxylic acids or to efficiently remove the acids from the water. Selective adsorption and extraction are the most obvious routes, but finding an efficient adsorption or extraction cycle has been surprisingly elusive. Similarly, conversion of the carboxylic acids into carboxylates or esters to facilitate precipitation or to use volatility or polarity changes (for liquid–vapor separation or liquid–liquid separation) has not proved effective. Unfortunately, the carboxylates of short-chain carboxylic acids are water-soluble and the aqueous matrix inhibits esterification with alcohols, or the application of ammonium carboxylate to amide conversion chemistry.

Propanoates and butanoates are amphiphilic, but it may not be possible to extend the use of techniques such as foam fractionation to ethanoates.

16.5
Fundamental Studies: Keys to Improved FT Processes

While good engineering and good management are crucial to a successful FT project, we must not forget the chemistry and surface science that underpin the FT process and it is here that the basic research and new thinking can bring about much needed improvements and greater understanding. We now know a great deal about surface reactions largely from the advances in instrumentation that have taken place over the past decades and due to the work of the researchers that have employed them. In this context, we should cite the name of Gerhart Ertl [31] who developed good pictures of simple atomic processes on surfaces that can explain the Haber–Bosch ammonia synthesis over an iron catalyst. Despite those advances, there are still formidable lacunae in our understanding of heterogeneously catalyzed reactions such as the FT synthesis dealt with in this book. The techniques to study surface reactions are becoming available routinely and their application should now allow us to answer some of the long-standing fundamental questions, such as how apparently small variations in the nature of the catalyst surface can change the whole reaction, from methanation to hydrocarbon synthesis to methanol synthesis. The variations must ultimately be under electronic control, but we know very little about the whys and wherefores and how to manipulate the reactions more effectively. This is where we need new knowledge, as is discussed in Sections 16.5.1–16.5.6.

Thus, while it is generally agreed that many FT reactions take place at interfaces, in particular those between the active metal and the oxide support, we do not know how and when this happens and just what the factors are that control it. To what extent do the characteristics of various surfaces such as the composition, polarities, and shapes affect the surface reactions and their rates. Oxide and metal surfaces bear charges and these must have significant influences on the interactions with (charged) surface intermediate species. Another question is whether all the heterogeneously catalyzed reactions proceed by interactions between chemisorbed species or do some involve direct interactions between chemisorbed species and gas- or liquid-phase atoms or molecules. Such pictures have their counterpart in the interactions between solute and solvent in solution chemistry. Testing these and related ideas will pay good dividends.

16.5.1
New Instrumentation

As in other areas of science and engineering, new techniques and new types of instrumentation drive innovation. Thus, we are now able to analyze surface reactions under real catalytic conditions rather than under ultrahigh vacuum, as was the situation previously.

16.5.2 New Catalysts and Supports

Classical FT catalysts were made of bulk metals or by depositing metals from solution onto naturally occurring support materials such as kieselguhr. Although newer methods such as the gas-phase metal nanoparticle deposition or the use of "egg shell" catalysts have been developed, viable large-scale methods of synthesis and their long-term stabilities have not yet been evaluated. Since catalysts are subjected to quite extreme conditions of temperature and pressure, they do deactivate and degrade. To extend their useful working lives, new support materials such as synthetic zeolites, carbon nanofibers, and carbon nanotubes are now being studied, as are various metal and promoter combinations.

16.5.3 Isotopic Labeling

Although isotopic labeling has been widely used to investigate FT reactions, it has so far been largely limited to the use of ^{13}C- and ^{14}C-labeled carbon monoxide. Expansion to ^{18}O-labeled CO would give valuable information, especially in understanding oxygenate formation in FT synthesis. The use of D (^{2}H) labels in the analysis of FT mechanisms could bring huge advances in our understanding. However, the conditions under which the reactions are run must be carefully and critically monitored to avoid the random H/D exchange processes that can occur easily over many catalysts at working temperatures. Further extensions of the SSITKA techniques (Section 12.3.4), where the products obtained when inlet gas streams of different isotopic composition are alternated, could be very valuable.

16.5.4 Surface Microscopy

The study of catalyst surfaces at the atomic level by various forms of microscopy is only in its infancy, but could bring great benefits, for example, in pinpointing and characterizing the most active surface sites at oxide–metal interfaces. This could lead to answers to questions such as what features of the surface determine whether $C + C$ coupling (leading to longer chains) or hydrogen transfer (chain termination) is preferred in a FT reaction. It will also be valuable to probe how and why some metal oxide supports exhibit SMSI (strong metal–surface interactions) (Section 11.3) that can significantly modify hydrogenation properties of metals.

16.5.5 Analytical Methods

The success of an FT process also depends on getting comprehensive, accurate, and fast analytical data on the products formed. Although modern techniques, especially various forms of GC combined with mass spectrometry, are very good, the quantitative determinations, for example, of carboxylic acids and other oxygenates need to be improved.

16.5.6
Greener Procedures

While much has been done to make FT reactions more environment-friendly, efforts to reduce waste, improve selectivity, and manage heat flows in a facility can still be improved with the help of new ideas based on basic research.

16.6
Challenges for the Future

Ideally all challenges should be seen as opportunities, but there are some issues that may indeed complicate the future development of FT-based processes [13, 24–26]. However, the fundamentals that stimulated interest in synthetic hydrocarbon liquids have not changed. We here analyze some of the important factors that future work in FT will need to consider.

16.6.1
Hiatus Effect

An important challenge arises from the need for continuity, of good lead times, and good subsequent performance of the industry. While it is agreed that there is a need for more FT facilities in future, insufficient effective and sustained basic research has hindered advances in FT synthesis everywhere. Historically, bursts of local activity in research and development preceded rapid commercial exploitation, which in turn subsided as the favorable economics that initially gave the impetus for commercialization changed and lost their validity. As a consequence, significant parts of the learning and expertise that had been developed in the "good" times were lost with each change in the economic situation. Furthermore, the premature commercialization created a bad impression with the public and investors alike. The concern with respect to synthetic liquid development has been summarized by the concluding remarks and the plea for sustained R&D in the book by Crow *et al.* [13]: "The ultimate need for synthetic fuels seems a certainty. . . . [Yet,] it is likely that there won't be twenty to twenty-five years of lead time to any long-term oil supply interruption, [and] it is fair to conclude that highly inefficient [liquefaction] options will be our only means of meeting the need. . . . Thus, the choice is to launch a realistic, long-term R&D program twenty to twenty-five years before the need for low-cost synthetic fuels arises or proceed with [outdated] technology . . . "

In a study to evaluate the technical failure of coal liquefaction technology in the United States, it was found that the most critical factor in the successful development of process technology is the continuous support of a R&D program over an extended period of time. Any discontinuity in research and development seriously undermines the probability of success; this was called the "Hiatus Effect." Person-to-person knowledge transfer was found to be a key factor for success and once a team was dismantled (even if the personnel were retained), the rate of loss of knowledge was rapid. Even with proper archiving of the information, considerable research duplication and delays in progress have resulted from the hiatus effect [13]. Put another way, knowledge can be kept in books, but understanding is seated in people. Unless sustained interest is maintained, the knowledge remains, but the understanding is lost. In order to regain that understanding, some duplication of effort is inevitable.

It was further found that significant cost and technical risk reduction in a process required around 20 years of sustained and uninterrupted research and development [13]. The commercialization of a chemical process before completing this period runs a risk of being too early and too soon, and may actually appear to demonstrate that a potentially good and useful technology is not technically and/or economically viable. Thus, the challenge of how to ensure timely and sustained investment in basic research to support the future need for a synthetic liquids (XTL) industry is still real.

16.6.2 Practical Constraints

There is a challenge in the implementation of a large-scale FT-based industry. Global synthetic liquids production capacity is at present less than 1% of the total crude oil-derived transportation fuel and petrochemical production capacity. Expanding synthetic capacity will take place in competition with other industries for resources and skilled labor. The main potential practical constraints to an increase in construction of FT-based facilities can be evaluated (Sections 16.7.1–16.7.6) [27].

16.6.2.1 Critical Materials Availability

Of the critical construction materials identified for meeting the 1992 requirements mandated by the US Energy Securities Act, the availability of nickel was the most significant. It is therefore interesting to note that it was the price of nickel that caused much of the construction cost increase in 2006–2007, when the price of nickel reached US$50/kg and the cost of projects such as Pearl GTL and Escravos GTL rapidly exceeded their original capital cost estimates.

16.6.2.2 Equipment Availability

The global manufacturing capacity is limited and the procurement of some specialized types of equipment may become constraining in construction projects.

16.6.2.3 **Trained Manpower**

Skilled labor for engineering, construction, and operation must be trained in advance of any ambitious FT expansion programs. For one FT facility of 60 000 bbl/day, the peak workforce requirement during construction was 2000 people [28]. There is a finite labor pool and the impact of drawing experienced labor away from other industries can have dire consequences. It has been estimated that in order to establish a synthetic fuels industry that is equivalent to 10% of the current global crude oil refining capacity, a workforce of around 250 000 people is required in constant employment, with almost the same number of construction workers during peak construction [27], Loss of key personnel can often lead to later production problems [29].

16.6.2.4 **Water Availability**

This is a location-dependent constraint and has been discussed in Chapter 15.

16.6.2.5 **Environmental Requirements, Permits, and Licensing**

Environmental impacts specific to FT-based facilities are less well studied than those of crude oil refineries and power plants, but in general it may not be possible to provide sufficient details to satisfy permission requirements in countries with highly regulated systems.

16.6.2.6 **Socioeconomic Impacts**

The rapid development of towns associated with large construction projects in otherwise sparsely populated regions can create socioeconomic problems. This is often associated with insufficient secondary services such as housing, recreational facilities, and social support systems [27, 28].

16.6.3
Politics, Profit, and Perspectives

There is a challenge related to the potential conflict between national interest and politics, the profit motive, and environmental perspectives. All three of these topics (strategic, economic, and environmental) have been cited as justifications for developing FT facilities (Section 16.2), but in practice they are not necessarily aligned.

Even though we know that at some point in the future there will be a serious imbalance in crude oil supply and demand, the ramifications of which may be very serious, as long as oil continues to flow relatively easily, it will be difficult to persuade people that a crisis is likely. This situation has all the hallmarks of a "predictable surprise" in the making (Box 16.1) [30].

It should be recognized that as resources, people, and finances are limited, there are projects other than FT that appear to yield better short-term returns. Since free market incentives alone cannot adequately address the problem [25], government involvement is required to promote timely, sustantial, and sustained investment in

Box 16.1 "Predictable" and "Surprise" Events

A *predictable event* is an event that could be anticipated in advance based on available information, even though the exact timing and extent of the event may be uncertain. A *surprise* is an event that is unanticipated. Although a *predictable surprise* seems like an oxymoron, it describes an event that is predictable, but its immediacy is underestimated. The following are some of the traits of predictable surprises that can occur and lead to disaster [30]:

a. Leaders know that the problem exists, but fail to respond in time.
b. Society knows that the problem exists, but fails to take action.
c. Fixing the problem costs money in the present, while benefits will accrue only in the future.
d. The cost of fixing the problem is real, but the reward (if any) is uncertain.
e. Maintaining the status quo is often the path of least resistance.
f. Vocal minorities that benefit from inaction often selfishly subvert the actions of leaders to address the problem.

basic research and development to improve synthetic liquids technology before a crisis is reached.

The challenges raised in Sections 16.6.1 and 16.6.2 need timely action, despite a delay in demonstrable benefit.

16.7
Conclusions

Over the past 150 years, there has been a significant change in the nature and importance of manufactured goods and transportation, which has been accompanied by a gradual shift toward crude oil as the raw material for the production of most transportation fuels and petrochemicals. This dominance was partly due to the low cost and availability of crude oil, but the ease and efficiency of crude oil refining also contributed. Although it is unlikely that crude oil will soon be replaced as the chief raw material for many transportation fuels and petrochemicals, it is unreasonable to expect that crude oil will be able to meet all demands for much longer at the current rate of global consumption. There is consequently a need for diversification of the raw materials that are employed to produce transportation fuels and petrochemicals.

One of the most attractive features of indirect liquefaction technology that use synthesis gas as an intermediate product, is that it decouples the synthetic process from the nature of the raw material feed. Indirect liquefaction by FT synthesis will play an increasingly important role in the future, since it is capable of producing transportation fuels and petrochemicals that are compatible with the global transportation and manufacturing infrastructure.

Although there are some challenges to overcome, there are many opportunities to improve the technologies used in the production of synthetic liquids by FT synthesis. As with any conversion process, the energy loss arising as a consequence of the second law of thermodynamics must be taken into account. An FT based facility has a conversion efficiency on a carbon basis close to that of electricity generation. While the suggestion that efforts should be exclusively focused on the development of "clean" carbon-free technologies to replace crude oil-based processes may be environmentally laudable, it is unlikely that the energy infrastructure will change overnight. Petrochemicals can never be carbon free and it is unrealistic to imagine a swift and complete transition from carbon-based energy carriers. Fischer–Tropsch synthesis and synthetic liquids (XTL) processes have a vital future role to play.

References

1 MacRae, K.M. (1991) *New Coal Technology and Electric Power Development*, Canadian Energy Research Institute, Calgary, p. 200.
2 Higman, C. and van der Burgt, M. (2008) *Gasification*, 2nd edn, Elsevier, Amsterdam.
3 Hogue, C. (2007) *Chem. Eng. News*, **85** (41), 11.
4 Hanson, D.J. (2011) *Chem. Eng. News*, **89** (43), 28–31.
5 Hanson, D.J. (2011) *Chem. Eng. News*, **89** (49), 33–34.
6 Anderson, L.L. and Tillman, D.A. (1979) *Synthetic Fuels from Coal: Overview and Assessment*, John Wiley & Sons, Inc., New York.
7 Gary, J.H., Handwerk, G.E., and Kaiser, M.J. (2007) *Petroleum Refining: Technology and Economics*, 5th edn, CRC Press, Boca Raton, FL.
8 Dry, M.E. and Steynberg, A.P. (2004) *Stud. Surf. Sci. Catal.*, **152**, 406–481.
9 De Klerk, A. (2011) *ACS Symp. Ser.*, **1084**, 215–235.
10 Nicholls, T. (ed.) (2003) *Fundamentals of Gas to Liquids*, 1st edn, Petroleum Economist, London.
11 De Klerk, A. (2011) *Fischer–Tropsch Refining*, Wiley-VCH Verlag GmbH, Weinheim.
12 (2008) *Petrol. Econ.*, **75** (6), 36–38.
13 Crow, M., Bozeman, B., Meyer, W., and Shangraw, R., Jr. (1988) *Synthetic Fuel Technology Development in the United States: A Retrospective Assessment*, Praeger, New York.
14 Sie, S.T. (1991) *Rev.I. Fr. Petrol.*, **46** (4), 501–515.
15 Donati, G. and Paludetto, R. (1997) *Catal. Today*, **34**, 483–533.
16 Krishna, R. (2000) *Rev.I. Fr. Petrol.*, **55** (4), 359–393.
17 Zwart, R.W.R., Boerrigter, H., and Van der Drift, A. (2006) *Energy Fuels*, **20**, 2192–2197.
18 Lerou, J.J., Tonkovich, A.L., Silva, L., Perry, S., and McDaniel, J. (2010) *Chem. Eng. Sci.*, **65**, 380–385.
19 Liu, K., Song, C., and Subramani, V. (eds) (2010) *Hydrogen and Syngas Production and Purification Technologies*, John Wiley & Sons, Inc., New York.
20 van der Merwe, W. (2010) *Environ. Sci. Technol.*, **44**, 1806–1812.
21 Torres Galvis, H.M., Bitter, J.H., and De Jong, K.P. (2012) Patent Appl. WO 2011/049456.
22 Storch, H.H. (1954) *The Chemistry of Petroleum Hydrocarbons*, vol. 1 (eds B.T. Brooks, C.E. Boord, S.S. Kurtz, and L. Schmerling), Reinhold, New York, pp. 631–646.
23 De Klerk, A. (2008) *Green Chem.*, **10**, 1249–1279.
24 Thumann, A. (ed.) (1981) *The Emerging Synthetic Fuel Industry*, Fairmont Press, Atlanta.
25 Harlan, J.K. (1982) *Starting with Synfuels: Benefits, Costs, and Program Design Assessments*, Ballinger Publishing Co., Cambridge, MA.

26 Hoffman, E.J. (1982) *Synfuels: The Problems and the Promise*, Energon Co., Laramie.

27 Hunt, V.D. (1983) *Synfuels Handbook*, Industrial Press, New York.

28 Mako, P.F. and Samuel, W.A. (1984) *Handbook of Synfuels Technology* (ed. R.A. Meyers), McGraw-Hill, New York, pp. 21–243.

29 Wessels, P. (1990) *Crescendo tot sukses. Sasol 1975–1987* (English trans. *Crescendo to Success: Sasol* 1975–1987), Human & Rousseau, Cape Town.

30 Bazerman, M.H. and Watkins, M.D. (2004) *Predictable Surprises: The Disasters You Should Have Seen Coming and How to Prevent Them*, Harvard Business School Press, Boston.

31 Ertl, G. (2008) *From atoms to complexity.* Nobel Prize Lecture, December 8, 2007.

Glossary

\$10Bn	ten billion dollars (US) $\$10^9$
\$Bn	billion dollars
Acetic	ethanoic (acid, CH_3COOH)
ADU	atmospheric distillation unit
AES	Auger electron spectroscopy
Alkylate	mixed branched chain paraffins (mostly *iso*heptane and *iso*octane) with a high octane rating
ASU	air separation unit
ATR	autothermal reforming
atm	1 atmosphere pressure = 0.101 MPa = 14.6 psi = 101325 N/m^2
barrel	oil industry volumetric unit, bbl (1 bbl = 0.158987 m^3)
BATANEC	best available technology at nonexcessive cost
bpd	barrels (of oil) per day; (50 bpd $\sim$ 1 t/a)
Bcm	billion (10^9) cubic meters
BFB	bubbling fluidized bed
BTL	biomass to liquid (technology)
Btu	British thermal unit (about 1055 J)
"C"	total carbon present in whatever molecular form
CFB	circulating fluidized bed
COD	chemical oxygen demand in water treatment
CPO	catalytic partial oxidation
CPR	compact reforming
CSTR	continuously stirred tank reactor
CTL	coal to liquid (technology)
DME	dimethyl ether, $(CH_3)_2O$
Doubly promoted catalyst	iron FT catalyst containing alkali (K^+) + silicate
EELS	electron energy loss spectroscopies
E-factor	kg waste per kg of desired product
EPC	engineering, procurement, and construction costs
ESCA	electron spectroscopy for chemical analysis

Greener Fischer-Tropsch Processes for Fuels and Feedstocks, First Edition. Edited by Peter M. Maitlis and Arno de Klerk.

ET	entrained flow (gasifier)
Ethylene	ethene, CH_2CH_2
EXAFS	extended X-ray absorption fine structure spectroscopy
FB	fixed bed
fcc	fluid catalytic cracking
FT	Fischer–Tropsch
FT-P	Fischer–Tropsch process
FT-S	Fischer–Tropsch synthesis
FT-HS	Fischer–Tropsch hydrocarbon synthesis
Fracking or *hydrofracking*	horizontal drilling and hydraulic fracturing
GTL	gas to liquid (technology)
"H"	total hydrogen present in whatever molecular form
HER	heat exchange reforming
HREELS	high-resolution electron energy loss spectroscopy
HREM	high-resolution electron microscopy
HTFT	high-temperature Fischer–Tropsch
HTSFTP	high-temperature slurry Fischer–Tropsch process
EIA	United States Energy Information Agency
IEA	International Energy Agency
Intensity factor	material or energy use, or effluent per amount of value addition
kb/d	thousand barrels per day, also kbbl/d
LNG	liquid natural gas
LTFT	low-temperature Fischer–Tropsch
MON	motor octane number
Mt/a	million tons per annum
Ncm or Nm^3	normal cubic meters: gas volume at standard conditions, 0 °C and 1 atm
PES	photoelectron spectroscopy
pfr values	product formation rate, often in $\mu mol/(g_{catalyst}\ h)$
POX	noncatalytic partial oxidation
QUAD	quadrillion (10^{15}) BTUs = 1.055×10^{18} J
RON	research octane number
Scf	standard cubic foot (US standard gas volume at 60 °F (15.5 °C), 1 atm)
Scm	standard cubic meter (Sm^3)
SFM	scanning force microscopy
SMR	steam methane reforming
SNG	synthetic natural gas
SPA	solid phosphoric acid
STM	scanning tunneling microscopy
syngas	mixture of hydrogen and carbon monoxide (also water gas)

t/a	tons per annum
TEM	transmission electron microscopy
THF	tetrahydrofuran (C_4H_8O)
ton	7.33 barrels (approximately, depending on oil density)
TPD	temperature programmed desorption
Tscm	trillion (10^{12}) standard cubic meters
Turndown ratio	ratio of actual feed rate to design feed rate
VLE	vapor–liquid equilibrium
WGSR	water-gas shift reaction
WHSV	weight hourly space velocity
WTI	World Trade Institute, Berne
WTL	waste (organic, plastics) to liquid (technology)
XPS	X-ray photoelectron spectroscopy
ZLED	zero liquid effluent discharge

Index

Greener Fischer-Tropsch Processes for Fuels and Feedstocks, First Edition. Edited by Peter M. Maitlis and Arno de Klerk.